AF686978

NET ZERO ENERGY BUILDINGS

KARSTEN VOSS
EIKE MUSALL

NET ZERO ENERGY BUILDINGS

INTERNATIONAL PROJECTS OF CARBON NEUTRALITY IN BUILDINGS

EnOB
Forschung für
Energieoptimiertes Bauen

DETAIL Green Books

IMPRINT

AUTHORS AND EDITORS
Karsten Voss, Prof. Dr.-Ing.; Eike Musall, M.Sc.arch.

CO-AUTHORS
Andreas K. Athienitis, Prof. Ph.D.; Soara Bernard, M.Sc.arch., Dipl.-Ing.; Armin Binz, Prof.; Dylan Brady; Marko Brandes, Dipl.-Ing.; José A. Candanedo, Ph.D.; Arnulf Dinkel, Dipl.-Ing. Architekt; Sabine Erber, DI; Shane Esmore, Director & Principal Sustainability Consultant; Norbert Fisch, Univ.-Prof. Dr.-Ing.; François Garde, Prof.; Sonja Geier, DI; Michael Gies, Dipl.-Ing.; Monika Hall, Dr.-Ing.; Ferdinand Hammerer; Sebastian Herkel, Dipl.-Ing. Maschinenbau; Patrick Jung, Dipl.-Ing., Visiting Prof.; Florian Kagerer, Dipl.-Ing.; Beat Kämpfen, Dipl. Architekt eth/sia, M.A.; Jens Krause, Dipl.-Ing.; Jörg Lange, Dr.; Henrik Langehein, Dipl.-Ing.; Beatrix Lehnert, Design Manager; Aurélie Lenoir; Ingo Lütkemeyer, Prof. Dipl.-Ing. Architekt; Masa Noguchi, Ph.D., Senior Lecturer; Stefan Plesser, Dipl.-Ing. Architekt; Dietmar Riecks, Dipl.-Ing. Architekt; Audrius Ringaila, Invited Designer; Lars Rössing, Dipl.-Ing. Architekt; Tanja Siems, Prof. Dr.-Ing.; Katharina Simon, M.Sc.arch.; Armin Themessl, Ing.; Bert Tilicke, Dipl.-Ing. Architekt/Landschaftsarchitekt; Martin Ufheil, Dipl.-Ing.; David Waldren, Development Manager; Gerhard Zweier, Dipl.Ing. Architekt

ASSISTANTS
Melina Schulz, B.Sc.arch., M.Sc.arch.cand.; Masato Takagaki, Bachelor of Law, B.Sc.arch.cand.; Dennis Hagen, M.Sc.arch.; Jana Eggermann, B.Sc.arch.cand.

FUNDING
This publication was produced in the context of collaboration between experts of the International Energy Agency (IEA) in the framework of the programmes Solar Heating & Cooling (SHC, Task 40) and Energy Conservation in Buildings & Community Systems (ECBCS, Annex 52) under the title "Towards Net Zero Energy Solar Buildings". The official publication is the English edition.

SOLAR HEATING & COOLING PROGRAMME
INTERNATIONAL ENERGY AGENCY

International Energy Agency
Energy Conservation in Buildings and Community Systems Programme

This book and the German collaboration in the framework of the IEA were aided as part of the research initiative EnOB – Reseach for Energy-optimised Construction, in accordance with a decision of the Bundestag, by the German Federal Ministry of Economics and Technology. The Austrian collaboration in the framework of the IEA and in the production of this publication was aided by the Austrian Federal Ministry of Transport, Innovation and Technology, the Swiss contribution by the Swiss Federal Office of Energy.

Partial funding for the translation into English was provided by:
- Government of Canada's Program of Energy Research and Development (PERD) under the project management of CanmetENERGY research centre. PERD Is a federal interdepartmental program operated the Department of Natural Resources Canada.
- Grocon group of companies, Australia.

EDITORIAL SERVICES
Editors:
Cornelia Hellstern, Dipl.-Ing. (project head); Kim Ahrend, Dipl.-Ing. Architektin; Jana Rackwitz, Dipl.-Ing.

Editorial assistants:
Katinka Johanning, Dr.; Michaela Linder, State Certified Translator; Jakob Schoof, Dipl.-Ing.

Translation into English:
J. Roderick O'Donovan, B. Arch., Vienna (A)
Premier Focus Inc., Waterloo, Ontario (CDN)
Sevag Pogharian Design, Montreal, Quebec (CDN)

Proofreading:
Mark Kammerbauer, Dipl.Ing., M. Sc., Munich

Drawings:
Ralph Donhauser, Dipl.-Ing.

Graphic design:
Michael Vitzthum, Dipl.-Des. (Cover)
Cornelia Hellstern, Dipl.-Ing. (Layout)

DTP & Production:
Roswitha Siegler, Simone Soesters

Reproduction:
Martin Härtl OHG, Munich

Printing and binding:
Kösel GmbH & Co. KG, Altusried-Krugzell

This work is subject to copyright. All rights reserved, whether the whole or part of the material is concerned, specifically the rights of translation, reprinting, recitation, reuse of illustrations and tables, broadcasting, reproduction on microfilm or in other ways and storage in data processing systems. Reproduction of any part of this work in individual cases too, is only permitted within the limits of the provisions of the valid edition of the copyright law. A charge will be levied. Infringements will be subject to the penalty clauses of the copyright law.

Bibliographic information published by the German National Library: The German National Library lists this publication in the Deutsche Nationalbibliografie; detailed bibliographic information is available on the Internet http://dnb.d-nb.de.

Institut für internationale Architektur-Dokumentation GmbH & Co. KG
Hackerbrücke 6, D-80335 Munich
www.detail.de

© 2011, first edition
ISBN: 978-3-0346-0780-3

MIX
Paper from responsible sources
Papier aus verantwortungsvollen Quellen
FSC® C017373

CONTENTS

PREFACE

Discussions on the appropriate energy policy for the future and the growing concerns about climate change regularly focus on the built environment in particular. On the one hand, the construction, maintenance, and operation of buildings throughout their life cycle consumes large amounts of energy and causes emissions. On the other hand, we are already aware of and have tested measures for all kinds of buildings that can dramatically reduce the level of consumption and emissions.

However, the net zero energy and plus energy buildings dealt with in this book go further than these concepts. They indicate how an equalised annual energy balance can be achieved by bringing together architectural design, energy efficiency and the local use of renewables. They stand for independence from finite resources and immunity to fluctuating energy prices. A zero-carbon building does not contribute to climate change.

INTERNATIONAL ENVIRONMENT In the new version of the building guidelines published in 2010 the European Union calls upon member states to introduce the energy standard "Nearly Zero Energy Building" for all new buildings by no later than the end of 2020. The building technology programme of the United States of America formulates the goals of arriving at marketable zero energy residential buildings by 2020 and non-residential buildings by 2025. Nevertheless, standards that precisely define the goals in relation to the respective national building practice standards do not yet exist. The Swiss MINERGIE-A Certificate, which was released in March 2010, has become a pioneer in this important area of establishing definitions.

A proposal towards a calculation process in the context of German standardisation has been formulated and accompanied by a relevant calculation tool. The basic material needed to acquire a general understanding of the theme is conveyed in Section A of this book, "BACKGROUND INFORMATION".

CLIMATE NEUTRAL BUILDINGS Under extreme conditions, autonomous buildings point the way. Far removed from any kind of energy infrastructure and without a connection to an energy grid, they are generally entirely self-supplying by means of renewables. But for the broad mass of buildings connected to the grid, this can't represent the model of the future.
The long-term storage of energy, in particular of electricity, is a significant technological bottleneck. Equipping buildings to produce their own electricity is not only technologically demanding, the maintenance of such systems is complex and expensive, which means that connecting the building to an electricity grid offers a significant advantage. However, a building can only be described as climate neutral if the electricity grids are based to 100 % on renewables. But today and in the long term there is too little of this certified green electricity to waste it through inadequate building efficiency. The planned introduction of electro-mobility will increase the demand for clean electricity, allowing even less room for wastage.

PROJECTS AND LESSONS LEARNED Buildings have sufficient surface area, space, and infrastructure to operate their own plants to generate energy and feed it into grids. Here, photovoltaic systems and (above all, in larger and more energy intensive non-residential buildings) combined heat and power plants integrated in the building and run on biomass are suitable. Many of the projects presented in this book are so-called all-electric houses. This applies in particular to the residential buildings. Their energy systems are restricted to photovoltaic systems and heat pumps, so that electricity is their only energy source.

Because net zero energy buildings manage without long-term storage of electrical energy, the national power grid takes on this function and balances seasonal fluctuations in energy generation in relation to varying energy demands. The buildings presented in Section B of this book, "PROJECTS AND LESSONS LEARNED", differ clearly with regard to the extent to which this “service” for grid based balancing is used and how flexibly the building energy system can react to the demands of the grids. In the future, to ensure an optimally functioning grid infrastructure, also with a substantially higher quota of electricity from renewables (smart grids), buildings will have to be more intensively integrated in generation and load management than has been the case thus far.

The case studies show that very high energy efficiency is imperative for a realistic chance of achieving an equalised annual energy balance. Through the interaction of architecture, building construction, and energy technology the studies presented utilise diverse possibilities: from the geometry to the U-values of the parts of the building envelope to the performance of combined heat and power units or photovoltaic arrays. The 23 projects selected present buildings of different sizes, typologies, locations, and construction methods, ranging from residential and non-residential buildings to housing developments and even an entire city. Many projects use the passive house standard as their starting point, while the first renovation projects point the way towards zero energy for an existing building fabric. Designing and building a net zero energy building means that from the very start energy demand and energy generation must be consistently kept in balance: If the demand in the annual sum exceeds the possibilities for energy generation, further savings must be implemented. Here, an integrated planning team of architects, structural designers, and energy engineers is the decisive and essential requirement. This book documents in detail the results of energy monitoring and the experience gained from the planning and use phases, as well as the individual steps on the way to an equalised energy balance. The fact that a number of buildings don't achieve this balance in practice emphasises the difficulty of this task and the gap between planning and reality. However, this does not reduce their value. Cross-sectional analyses of more than 50 further projects carried out all around the world supplement the overview and broaden knowledge of possible strategies. Research has revealed the dynamics with which the field of zero energy is currently being developed.

NETWORK OF RESEARCHERS The collaboration in the international energy agency IEA titled “Towards Net Zero Energy Solar Buildings” involves representatives of 19 nations who participate in an intensive dialogue on suitable definitions and assessment procedures, discussing the experience they have gained from national demonstration projects, and publishing their findings. These activities also underline the international dimensions of the theme and its growing importance.

ACKNOWLEDGEMENTS We thank the numerous authors whose contributions are of such significant importance for the success of this book. With their buildings, committed clients and designers have created the conditions under which net zero energy buildings can become reality.

The work on this book was assisted by the German Federal Ministry of Economics and Technology, the Swiss Federal Office of Energy, and the Austrian Federal Ministry of Transport, Innovation and Technology. The English language edition is a slightly modified follow-up of the German language book published in June 2011. The translation was supported by the Government of Canada's Program of Energy Research and Development as well as the Grocon group of companies in Australia.

Wuppertal October 2011
Karsten Voss, Eike Musall
The editors

The IEA joint “Solar Heating and Cooling (SHC) Programme Task 40/Energy Conservation in Buildings and Community Systems (ECBCS) Programme Annex 52: Towards Net-Zero Energy Solar Buildings” (NZEBs) is a 5-year international collaboration between approximately 75 national experts from 19 nations in Europe, North America, Oceania, and Southeast Asia. It seeks to study current net-zero, near net-zero and very low energy buildings and to develop a common understanding of a harmonised international definitions framework, tools, innovative solutions and industry guidelines to support the conversion of the NZEB concept from an idea into practical reality in the marketplace. I am pleased to present the English edition of “Net-Zero Energy Buildings”, a major accomplishment in this field, and which encapsulates the many and varied concepts and views of defining net-zero energy buildings by government research organizations, international and regional research centres, academia, and industry that have been discussed in this Task/Annex since its inauguration in the fall of 2008. I am confident this book will find many interested readers.

Varennes, Canada October 2011
Josef Ayoub
Operating Agent, IEA SHC Task 40/ECBCS Annex 52,
CanmetENERGY/Natural Resources Canada

8

BACKGROUND INFORMATION

TOWARDS CLIMATE NEUTRAL BUILDINGS

A DESCRIPTIVE EXAMPLE In Germany, buildings and their use account for approximately one-third of total energy consumption and emissions. The majority of this demand is generated by living in residential buildings and the remainder by so-called non-residential buildings, i.e. for commercial uses, trade, and services. Only half of emissions is generated on site by fossil fuel combustion. The other half is produced by power plants and heating stations that supply buildings with power and district heating [1]. Residential buildings are the clear leader due to their quantity: they constitute the majority of existing buildings (Fig. A 1.01).

In existing residential buildings, heating comprises the highest share of energy consumption. In non-residential buildings, the impact of electrical power consumption is growing. Significant energy loads are caused by lighting, ventilation, and cooling, in addition to use-specific appliances, such as computers and production machinery (Fig. A 1.02). This trend is emphasized in new construction, since heating consumption is significantly lower here than in existing buildings. As result, electrical power consumption, particularly of household appliances ("white goods"), is becoming the dominant factor in residential buildings as well. Heating has a typically high impact in central Europe's climate, leading to a pronounced seasonal character of energy consumption profiles of buildings: annual electrical power consumption is contrasted by high heating consumption in winter, (still) predominantly covered by fossil fuel combustion. This results in a very distinct seasonal imbalance between consumption and on-site solar energy provision. Addressing this imbalance is one of the major challenges on the path towards climate neutral architecture.

BUILDINGS ENDURE Approximately 70 % of existing residential buildings in Germany are older than 30 years of age. They were built at a time when no significant energy efficiency or energy savings requirements were in place [2, 3]. Current annual renovation rates hardly reach one percent. Without additional efforts, an energy upgrade of all existing buildings will take more than 100 years to complete. From today's point of view, this is much too long to deal with finite resources or to turn around climate change. This is why incentives for increasing renovation rates, such as financial benefits or tax credits, are of particular interest. Such incentives gain importance due to rising energy costs, since these lead to renovation measures becoming economically feasible.

Renovations of high architectural quality can further drive a related future market. An example for this is the renovated "Neue Burse" dormitory at the University of Wuppertal. More than 600 students currently live there. In 2003 the deteriorating, 1970s-era

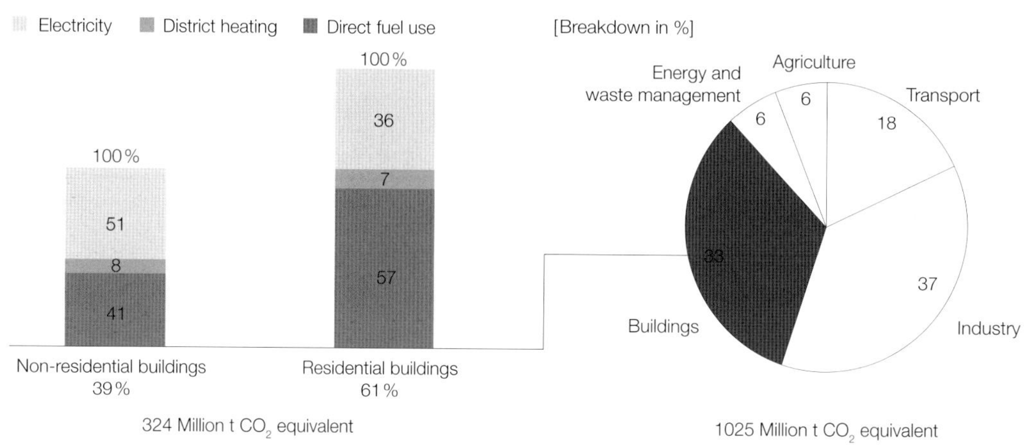

A 1.01

building was renovated in a two-stage approach as part of a demonstration project. The new building envelope meets the passive house standard. Heat recovery ventilation contributed to a reduction of energy consumption for technical infrastructure, emissions, and energy costs by more than 50% [4]. The renovated building is currently one of the most economically efficient dormitory buildings. At this standard, occupants' energy consumption for hot water and electricity comprises the major share of the energy balance, since the typical rent for student housing includes a utility flat rate (Figs. A 1.04 and A 1.05, p. 12).

DIFFERENT CLIMATE, DIFFERENT TASKS The presented data exemplarily describe the situation in Germany. This is comparable to the situation in most central European countries. In regions with higher annual temperatures and high humidity, air conditioning comprises the major share of energy consumption. This results in different consumption profiles and energy provider structures.

In large parts of Asia and the Arab world, greater economic dynamics cause extensive new construction activity at scales unimaginable in the current European context. Day after day, entire cities and urban quarters are created out of thin air, and glazed office towers ascend ever higher into the sky. Growing prosperity causes an enormous demand for small air conditioning units and an increase in power consumption in existing buildings. Low or even no requirements at all on energy consumption in new construction constitute a wasted opportunity that will lead to disproportionately high levels of energy consumption in upcoming decades. In Asia, blindly copying European or North American architecture and construction techniques leads to inadequately high consumption rates. This sets the wrong tone in the pioneering spirit of these emerging markets. Typical office buildings there consume approximately five times more energy than their average central European counterparts.

However, the signs of the time are recognized here as well, as early examples demonstrate: an ecological model city for 50,000 inhabitants is under construction in the United Arab Emirates as a large scale model project for the post-oil era (see Masdar Urban Development Project, p. 108ff.). The first buildings are already complete (Fig. A 1.03). 2022 World Cup officials plan to build CO_2 neutral soccer stadiums. A pilot stadium for 500 spectators has already been designed and completed in 2010. Solar cells and parabolic collectors placed along the stadium perimeter provide solar energy for cooling. During non-service hours, the photovoltaic system feeds energy into the power grid. During service, the grid covers stadium peak demands (Fig. A 1.06, p.13).

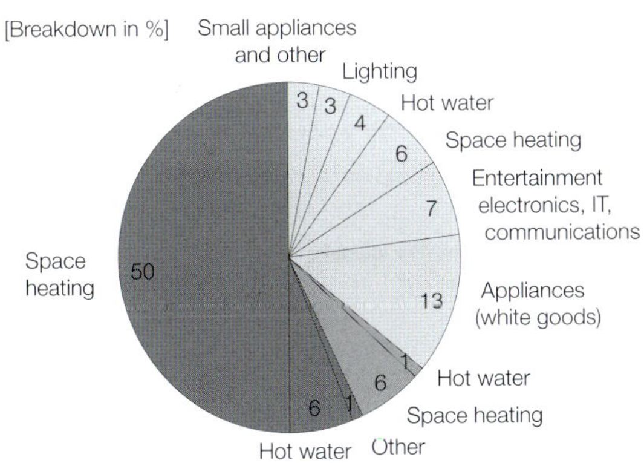

a 208 Million t CO_2 equivalent

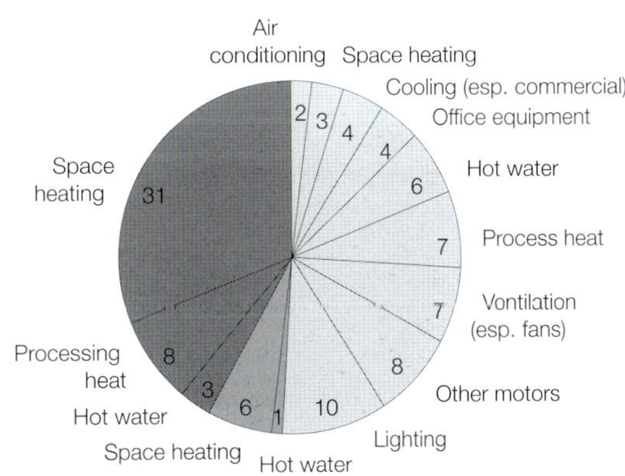

b 124 Million t CO_2 equivalent

A 1.02
A 1.03

A 1.01 Percentage and breakdown of greenhouse gas emission as CO_2 equivalent in Germany. Approximately one third of total annual emission of about one million tonnes is attributable directly or indirectly to the building sector.

A 1.02 Structure of climate gas emission as CO_2-equivalent forexisting residential and non-residential buildings in Germany (based on German energy survey 2004 consumption data of the "household" and "commercial, trade, and services" sectors).
a Residential buildings
b Non-residential buildings

A 1.03 Student dormitory in the ecological model city of Masdar City, United Arab Emirates (UAE) 2007, Foster + Partners.

A 1.04
A 1.05

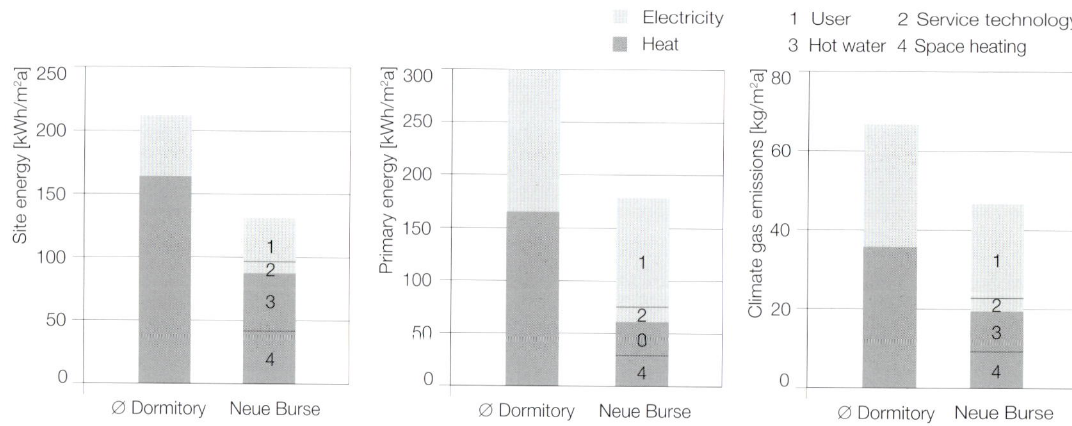

THE BUILDING SECTOR IN THE SPOTLIGHT OF ENERGY AND CLIMATE POLICY Many architectural, structural, and technical equipment designs for constructing new, energy-efficient buildings and for renovating existing buildings are already tried and tested. Therefore, it is easy to understand why political decisionmakers today particularly emphasize the relevance of the building sector for energy savings and climate protection regulations. It seems that goals are easier to achieve here than in other sectors. In the case of mobility, technological aspects still dominate discussions on post-oil era sustainable individual transport. Early electric vehicles display difficulties, for instance during driving tests, since currently available battery technology doesn't yet permit broad use. Questions on supplying emission-free electricity at necessary scales are still unanswered.

The situation is different in the building sector. At increasing scales, zero energy buildings, energy plus buildings, and entire related housing projects are under construction and in operation in Europe and worldwide. The project section of this book will illustrate the various reasons for these developments. In Europe, zero energy buildings are considered the logical continuation of a long chain of developments from low-energy houses towards passive houses. The claim is made that zero energy buildings completely balance their annual energy consumption and related CO_2 emissions. Compared to cars, the technological advantage of buildings is their connection to the electrical power grid.

AMBITIOUS POLICY GOALS Against the background of climate change, global population increase, and finite resources, it comes as no surprise that energy policy goals are so ambitious.

At present, terminology related to net zero energy buildings is important for strategy papers on energy policy in many countries. This is partly due to positive connotations of the term "zero energy". In the context of finite resources and increasing energy costs, it suggests independence, no costs, or an orientation towards the future. Also, "zero" leaves no room for discussion on quantification of suitable parameters. The interpretation of energy parameters as quantitative target definitions remains the domain of experts and offers no real basis for communication with the general public. However, at first glance, "zero" seems to doubtlessly demand highest possible standards regardless of building type or climate, only to be superseded by the term "plus".

EUROPEAN UNION The amendment to the EU Energy Performance in Buildings Directive (EPBD) prominently addresses the related subject matter. The first directive in 2002 initiated the introduction of integrated energy balancing methods throughout Europe. These were required to take air conditioning, lighting, and use of renewable energy into account [5]. In the 2010 recast the European Commission introduced the term "near zero energy building" and specified timeframes for the implementation of related construction standards within its member states:

- "Article 2(2): Definitions:
  Nearly zero-energy building means a building that has a very high energy performance .... The nearly zero or very low amount of energy required should be covered to a very significant extent by energy from renewable sources, including energy from renewable sources produced on-site or nearby.
- Article 9: Nearly Zero Energy Buildings:
  Member States shall ensure that:
  a) by 31 December 2020, all new buildings are nearly zero-energy buildings; and
  b) after 31 December 2018, new buildings occupied and owned by public authorities are nearly zero-energy buildings.

  Member States shall draw up national plans for increasing the number of nearly zero-energy buildings. These national plans may include targets differentiated according to the category of building." [6]

A 1.04 The "Neue Burse" student dormitory, an award-winning application of the passive house strategy in the renovation of a 1970s-era building. University of Wuppertal, Germany, 2003, Architect: Contor Müller Schlüter.

A 1.05 Comparison of consumption characteristics of 24 student dormitories, 2005 to 2008 mean values in Germany (total values) with metered values of the "Neue Burse" dormitory from 2007 (partial target values). Typical for residential use, thermal energy consumption for hot water is high, and user demands dominate electricity consumption. Primary energy factors and emission factors according to A 2.07, p. 31.

A 1.06 CO_2-neutral 500-seat model stadium for the 2022 soccer World Cup in Qatar, 2010, Arup Associates. Solar cells and collectors adjacent to the stadium supply energy for solar cooling.

A 1.06

As is typical in such strategy papers, interpreting the implementation of measures and methods of calculation are left to member states. How close "nearly zero" to "zero" should be remains unanswered. The draft of the directive still contained the phrase "zero energy buildings", but this was apparently considered too ambitious as a goal. A provision on the passive house standard already effectively implemented in European countries such as Austria, Germany, and Switzerland (there comparable with the "MINERGIE-P" standard), was discussed, yet rejected. While the passive house standard is defined by a method of calculation [7], the concept of a "near zero energy building" lacks authority, while further procedures are open to interpretation by each respective nation. Despite numerous efforts to streamline construction standards internationally, the respective national planning and policy context remains decisive for related definitions. In 2011 this process is only in its beginnings. The chapter "Energy Balancing: Practice, Standardization, and Legislation" (pp. 40ff.) reflects on this subject in detail.

GERMANY The currently ongoing energy research program [8] and the German federal government's energy concept, agreed upon in September 2010, particularly address the building sector. It is remarkable that, in both cases, the main focus is climate protection: not "zero energy buildings," but instead "zero emission" or "climate neutrality." However, this can't hide the fact that the focus is on energy savings, at least until an affordable, climate-neutral energy source available on a broad scale is in sight. The 2010 energy concept of the German federal government states:

- "The central goal is a long term reduction of heating demands of existing buildings, in order to attain a nearly climate-neutral stock of existing buildings by 2050. Climate neutrality entails that buildings have very low energy demands and that remaining energy demands are predominantly covered by renewable energy sources [...]
- The amendment of the Energy Savings Directive of 2012 introduces the concept of "climate-neutral building" as standard for all new construction until 2020 on the basis of primary energy parameters. The related renovation roadmap for existing buildings set up in this amendment begins in 2020 and leads to a step-by-step target of reducing primary energy demands by 80 % until 2050. Maintaining cost-efficiency is required [...]
- The federal government will assume a pioneering role in reducing energy consumption for its existing and future new construction stock." [10]

INCENTIVE CONCEPT FOR ENERGY-OPTIMIZED CONSTRUCTION Within the Research Focus Energy Optimized Construction, or "Energieoptimiertes Bauen" (EnOB), the German Federal Ministry of Economics and Technology (BMWi) has continually addressed energy research since 1995, among others via multi-faceted demonstration projects [11]. Initial zero energy and zero emission buildings can be found among numerous administration buildings, schools, production facilities, and residential buildings. This activity also includes the successful participation of student teams in the Solar Decathlon program in America and Europe (see Solar Decathlon Europe, p.168ff.). For 2011, climate neutrality and energy efficiency are keywords for the "Architecture with Energy" prize of the BMWi, awarded for the second time since 2009 [12].

SWITZERLAND Since 2009 the "Energieleitbild Bau" (guiding energy paradigm for construction) of the Swiss Association of Engineers and Architects (Schweizerischer Ingenieur und Architektenverein, or SIA) demands a sustainable basis for Switzerland's building stock as well as intelligent handling of energy resources [13]. Its implementation is based on SIA's "Effizienzpfad Energie" (efficiency path for energy) [14, 15] that, due to clearly specified limits and targets, supports quantitative definition and implementation of energy-related goals for the "2000-watt society" in the building categories of housing, office, and education [16, 17].

A 1.07 Comparison of continuous power consumption per person.
A 1.08 2000-watt society as strategic goal.
A 1.09 The canopy structure of an open-air ice skating rink next to the UPC Arena in Graz, Austria, features 1400 m² of solar collectors that have supplied over 500 MWh of climate neutral heat into the district heating grid since 2002.
A 1.10 Various neologisms represent comparable goals in a national context.
A 1.11 Mobility and buildings

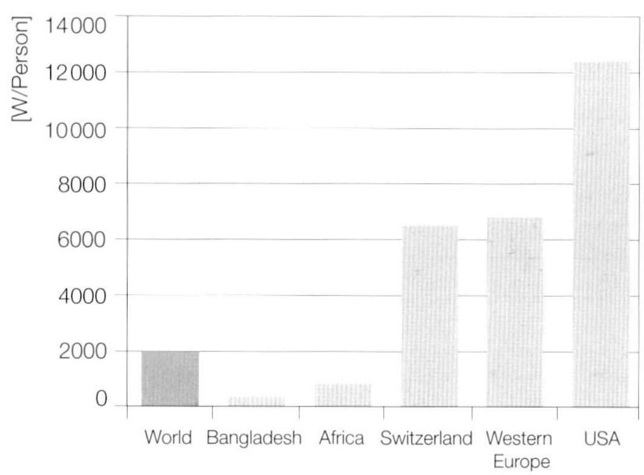

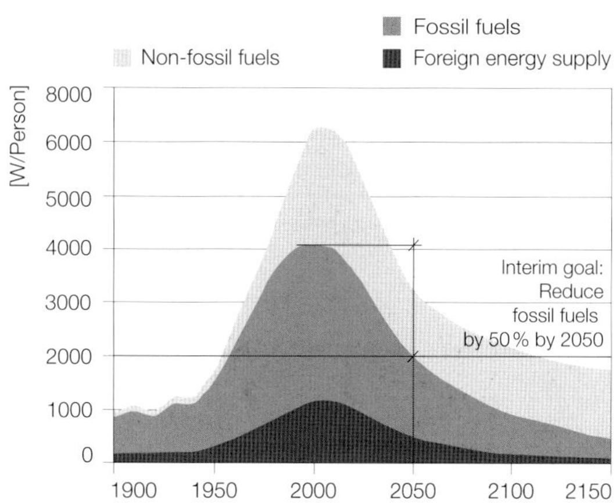

A 1.07
A 1.08

2000-WATT SOCIETY The 2000-watt society is a strategy developed in the late 1990s at the Swiss Federal Technical University in Zurich (ETH). Since then it has become the official planning paradigm not only of various Swiss cities and communities, but also the federal government and several municipalities [18]. It aims at sustainability as well as just and balanced consumption of global resources and addresses all aspects of life: consumer behaviour, mobility (Fig. A 1.11), recreation, and of course, construction and housing as well. In the 2000-watt society, 2000 W (primary energy) of continuous power will be available per person, which corresponds to current global average figures (Fig. A 1.07). Of this, a maximum of 500 W may be provided by fossil fuel sources (Fig. A 1.08). This policy entails that greenhouse-gas emissions equivalent to one ton of CO_2 [19] annually are permissible per person. This represents an implementation of the Intergovernmental Panel on Climate Change (IPCC) recommendation for limiting global warming at 2 °C [20]. Swiss rates will lead to a reduction of the total energy consumption per person from approximately 6500 W today to 2000 W, and of CO_2 emissions from 8.5 tons to 1 ton by 2150. For this purpose, three-quarters of the 2150 demand are supposed to be covered by renewable energy sources. To reach these goals, total energy consumption must be reduced by half and greenhouse-gas emissions must be reduced by a factor of four by 2050. The comparatively long time frame reflects the large scale of required changes.

The 2000-watt society has found broad acceptance in Switzerland as a concept within political programs and within institutional documents on target specifications. The SIA has made the 2000-watt society the basis of its "Efficiency Path for Energy."

A PATH TOWARDS ENERGY EFFICIENCY The Efficiency Path for Energy ("Effizienzpfad Energie") sets target specifications for primary energy and for greenhouse gas emissions [21]. The targets are set to meet preconditions for reaching the 2050 interim goal of the 2000-watt society. They comprise target values for three usage types: production (embodied energy for production/replacement and disposal of building materials), operation (air-conditioning, hot water, lighting and equipment) and building-induced mobility (everyday mobility). Here, production, mobility, and energy for service and operations as classic use type receive equal emphasis. The Efficiency Path for Energy permits calculation of total energy consumption and corresponding greenhouse gas emissions for these three use types already at early planning phases.

AUSTRIA In Austria, the term "energy plus building" is much more prevalent than the term "zero-energy building," which has received attention only since the "Energy Strategy" was presented in March 2010 [22]. One of the three pillars of this strategy aims at the "nearly zero energy building" standard. This standard originates in the "Building of Tomorrow" technology program and was used as basis for the further development of the energy plus building concept [23].

STRATEGY PROCESS ENERGY 2050 The "Strategieprozess Energie 2050" (Strategy Process Energy 2050) project initiated by the Austrian Federal Ministry for Transport, Innovation and Technology (BMVIT) aims at identifying possible future developments within a long-term perspective as basis for targeted research and development projects. The focus is on energy efficiency, renewable energy sources, and intelligent energy systems. One of seven focal subjects involves "Energy in Buildings"; it envisions an integration of buildings into both electrical and heating grids [24].

District heating in Austria dates back 60 years. However, by 1982, supply only reached 2 %. The completion of the 1982 Act Promoting District Heating triggered growth in the development of heating grids. A number of large projects and approximately 7400 smaller systems contributed to district heating or local heating grids providing 19 %of heating in Austria in 2008. District heating boomed particularly in urban areas: as the "district-heating city" of Austria, Linz boasts a supply rate of 60 %, while Vienna achieved "only" 34 %. In rural areas, grid energy sources are primarily based on regionally available biomass [25]. Many of these regional and local heating grids are also powered by large scale solar systems. The largest feed into a heating grid in Austria takes place in Graz [26]. More than 10,000 m² of collectors feed approximately 4 GWh annually into the district heating grid (Fig. A 1.09).

BUILDING OF TOMORROW For more than ten years, the "Building of Tomorrow" research and technology program of BMVIT in Austria has been a driving force for innovation in the construction sector [27]. This was made possible by a combination of research grants and economic subventions. The first program phase between 1999 and 2007 focused on the passive house standard and low-energy solar houses. The pioneering effect of these demonstration projects significantly contributed to Austria, in comparison to other states of the EU, currently having the highest density of completed passive houses, equalling a total usable area of 3.2 million square meters. The second phase, Building of Tomorrow Plus, started in 2008. It now concentrates more on technologies and the implementation of energy plus buildings and residential areas.

USA In the United States, the focus is on significantly reducing the energy consumption of buildings to the point that renewable energy can meet remaining loads. These efforts are implemented by the Department of Energy (DOE [28]). Other organizations actively pursue this goal as well, including the American Society of Heating, Refrigeration & Air-Conditioning Engineers (ASHRAE), the American Institute of Architects (AIA), and the U.S. Green Building Council (USGBC). These energy policies are related to the goal of increasing independence from energy imports (Fig. A 1.07). In 2007, Congress passed the Energy Independence and Security Act.

DOE's current building technologies program states [29]: "The long-term strategic goal is to create technologies and design approaches that lead to marketable zero-energy buildings by 2020 and to zero-energy commercial buildings by 2025." Definitions for Net Zero Energy have been established to include all building laods and a hierarchy of energy supply after energy efficiency has been maximized [30, 31].

Four critical metrics have been established as part of the definition:

- Net Zero Site Energy
  The control boundary is established around the site and includes purchased energy crossing the boundary at the site. It includes all building loads.
- Net Zero Source Energy
  Even though the measurement point is at the site, these values are scaled to account for inefficiencies in the supply system and reflect energy consumption at the primary or source energy. Currently these source-to-site weighing factors are fixed annually and don't account for seasonal or time-of-day dependencies [32].
- Net Zero Emission
  The balance is based on emission factors and not primary energy.
- Net Zero Site Energy Cost
  Buildings with no energy costs: consumption costs don't exceed income from feeding electricity into the grid.

The new building of the Research Support Facility completed in 2011 for the National Renewable Energy Lab (NREL) in Golden, Colorado, is currently the largest structure claiming to be a net-zero energy building (Fig. A 1.13, p. 16).

CANADA Canada lacks an official roadmap for the building sector addressed towards Net ZEB. In 2006, the Canadian government, through its Canada Mortgage and Housing Corporation (CMHC), announced the launch of the EQuilibrium Housing Initiative (formerly known as the Net Zero Energy Healthy Housing initiative). In the EQuilibrium housing initiative, the private and public sectors are supposed to jointly develop homes that address occupant health and comfort, energy efficiency, renewable energy production, resource conservation, reduced environmental impact, and affordability. EQuilibrium housing integrates a wide range of technologies, strategies, products, and techniques to reduce the environmental impact of homes to an absolute minimum. At the same time, EQuilibrium housing also features commercially available, on-site renewable energy systems to provide clean energy

D	Nullenergiehaus, Plusenergiehaus, Nullemissionshaus
E	net zero energy building, equilibrium building, nearly zero energy building, carbon neutral building
F	bâtiment zéro énergie, les maisons equilibrium, bâtiment à énergie positive, bâtiment zéro émission

A 1.09
A 1.10
A 1.11

INFOBOX **MOBILITY AND BUILDINGS** Building induced mobility is the sum of all distances travelled on foot, by vehicle or aircraft and induced by a building in use. In addition to energy used for driving vehicles, embodied energy that is expended for the manufacture and upkeep of vehicles and traffic infrastructure in use is taken into account. Differentiation takes place between daily mobility – all distances in connection to daily activities with maximum transit times of three hours within typical surroundings – and non-daily mobility – all distances in connection to day trips (transit time of more than three hours and beyond typical surroundings) as well as trips that include overnight accommodation. The definition is based on Swiss average values and takes into account the influences of different location-dependent aspects, or how the availability of cars or public transport fees influence basic mobility, with correction factors. The reference values featured in the Efficiency Path for Energy apply to daily mobility [33].

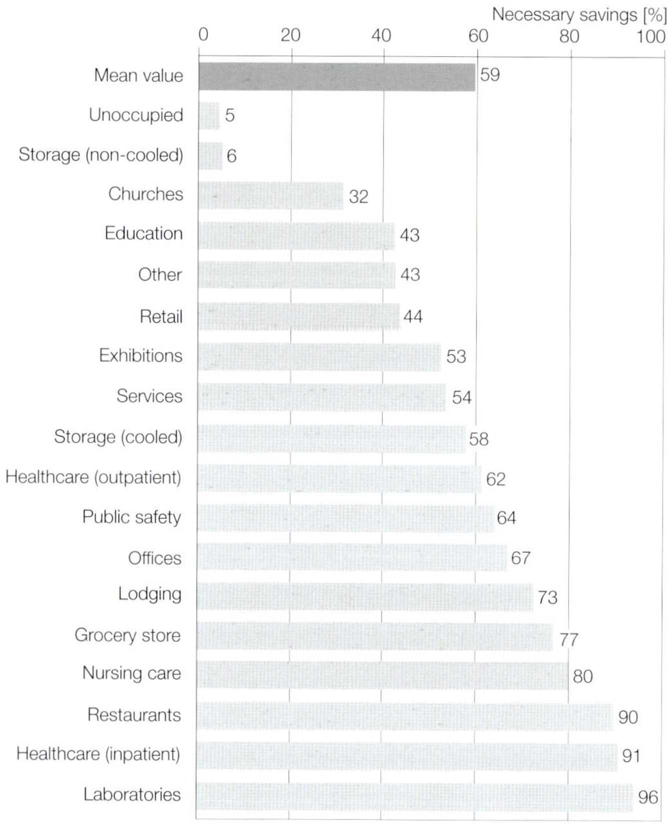

A 1.12
A 1.13

and to help reduce annual consumption and costs. The ultimate goal is a highly energy efficient building with low environmental impact that provides healthy indoor living for its occupants and produces as much energy as it consumes annually [34]. The initiative is intended to significantly increase consumer interest in and awareness of the important role that solar and other renewable energy technologies can play in meeting Canada's commitment to a clean energy future and healthy communities. By the beginning of 2011, there were fifteen approved EQuilibrium projects in Canada (see Ecoterra Home, p. 60 ff.).

ZERO ENERGY BUILDINGS ARE ENERGY-EFFICIENT Regardless of the sources, the energy efficiency of buildings is covered – the reduction of energy demands – is a central element of any sustainability strategy. Even if energy demands were covered in a climate-neutral way by renewable energy,

- the basically finite availability of many renewable energy sources
- often variable temporal availability (sun, wind, water)
- manufacturing expenditure of energy converters, and
- economic considerations

lead to a clear recognition: net zero energy buildings are primarily energy-efficient buildings!

One American study shows this in an impressive way [35]. If the total annual electricity consumption of existing buildings were to be covered solely by solar electricity generated by available rooftop areas, an average reduction in electricity consumption of almost 60 % would be necessary (Fig. A 1.12). Predominantly single-story construction prevalent in the United States is suited for this comparison, since the ratio of roof area per usable floor area is high. Heating consumption isn't even taken into account here. Conclusion: without a consistent efficiency strategy, a path towards net zero energy buildings isn't available!

THE PASSIVE HOUSE In terms of energy efficiency, the passive house (Passivhaus) concept has assumed a leading position in Europe, especially in the sector of residential construction [36]. In late 2010, this type of construction comprised buildings with over seven million square meters of usable area in all of Europe. In Switzerland, the passive house has become known under the "MINERGIE-P" label [37]. With the development of high-performance thermal insulation components for building envelopes, as well as energy-efficient heating, ventilation, and air-conditioning units, important incentives have been set for new construction and renovation in the past 20 years. Updating of legal requirements will lead to components and systems that are currently still considered special features of passive house construction becoming future standards for new construction and energy-related building renovations.

Following an initial focus on a target figure for heating power and space heating demands [38], the passive house concept for central European climates aimed at a uniform primary energy parameter of 120 kWh/m² [39] early on. This reflects the fact that, if heating demands are very low, energy demands are dominated by hot water supply and particularly by the electricity consumption of household appliances. According to the passive house plan, more than 40 % of primary energy is consumed by household appliances (recommended PHPP value: < 50 kWh/m²a) [40]. Since, in most cases, more than only electrical power is used in buildings, primary energy is more suitable as an indicator than the sum of electricity and other energy sources such as natural gas, oil, etc. The advantage of conversion into primary energy is that varying energy expenditure for providing electricity and heat is taken into account. However, conversion factors aren't physical constants, but differ according to the structure of national electricity-generating systems (see Fig. A 2.08, p. 31).

THE SEARCH FOR SIMPLE RULES DESPITE COMPLEX REQUIREMENTS In practice, the transition to calculating energy demands based on primary nergy is already increasing complexity for practical applications in residential buildings. Prior to this, specifications for maximum permissible heat transfer coefficient values of components or of area-related heating demands were easily comprehensible. In contrast, interpreting primary energy limit values to produce clear recommendations is only possible in the context of on-site energy supply systems and based on experience. The data in the project examples in this book provide actual and practical solutions to abstract requirements (see Projects and Experiences, p. 56ff.).

Target definitions and energy demand calculations for non-residential buildings are even more complex. According to the first building directive of the EU in 2002 [41], all member states are obligated to adapt their national standardisation and legislation so that electrical energy demands of lighting and air conditioning are also taken into account within mandatory total energy parameters. For the first time, clients and system manufacturers have an incentive to consider energy efficiency as a decisive developmental aspect in these sectors. Such a parameter system has been in place only in Switzerland since the beginning of the 1990s: the SIA 380-4 – Electrical Energy in Building Construction directive [42]. Due to the diversity of uses, the total energy parameters for non-residential buildings are differentiated to varying degrees according to type. For mixed use, total energy parameters are determined from data for individual building parts, emphasized according to area.

Against this background, demonstration projects play an important pioneering role. Beyond calculation results, they offer a basis for defining total energy target values for national standards via operative performance. One excellent example is the Forum Chriesbach research and administrative building, completed in late 2006 in the Swiss municipality of Dübendorf and awarded numerous prizes (Figs. A 1.14 and A 1.15, p. 17) [43, 44]. The waste heat recovery from commercial chillers and the output of a 50 m² tube collector system cover approximately 40 % of heating demands. The roughly 12,000 m² building draws just under 6 kWh/m²a heat from the district-heating grid. Through an earth-air

A 1.12 Necessary reduction in electricity consumption of existing buildings in the United States if corresponding roof-mount photovoltaic systems are to annually balance consumption.

A 1.13 The 1.6-MW$_p$ photovoltaic systems of the 20,500 m² "Research Support Center" of the National Renewable Energy Laboratory (NREL) in Golden, Colorado, USA 2011, RNL Design, aim at a zero-energy balance.

A 1.14 Energy flow diagram for the heat supply of Forum Chriesbach based on metered values from 2007 (values refer to a heated floor area of 11,170 m²)

A 1.15 The Aquatic Research Institute of ETH Zurich (Eawag) is located in the Forum Chriesbach office and research building, conceived as a passive house building (MINERGIE-P). Dübendorf, Switzerland 2006, Architect: Bob Gysin + Partner, Energy planner: 3-Plan Haustechnik AG.

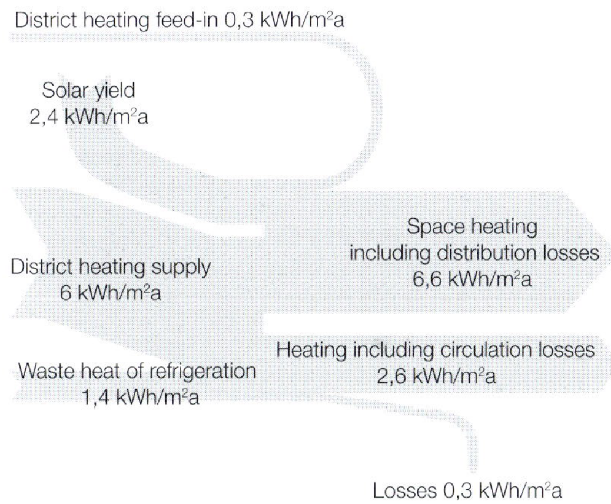

A 1.14
A 1.15

A 1.16

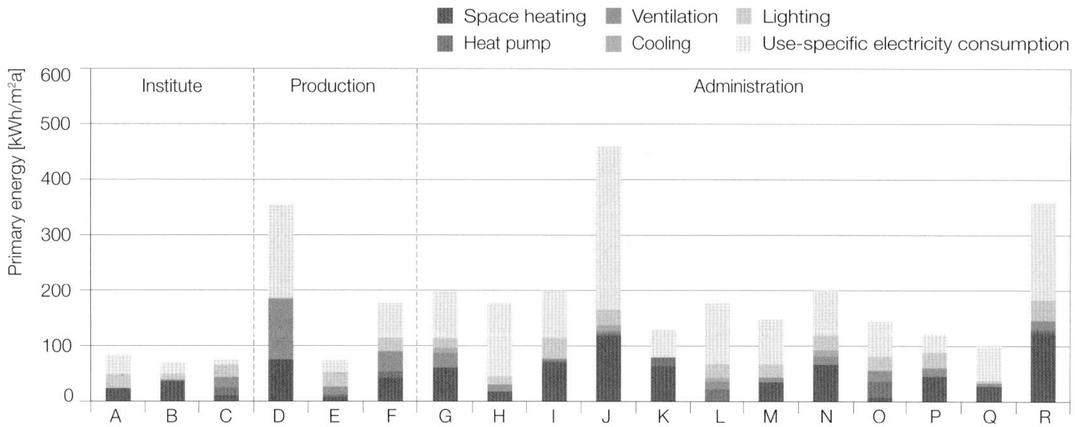

A 1.16 Measured primary energy parameters for the demonstration building of the EnOB development programme (data are standardized according to heated net floor area). For the administrative building, mean primary energy target values are 98 kWh/m² without and 188 kWh/m² with use-specific consumption.

A 1.17 Adobe construction office building as passive house, Tattendorf, Austria, 2006, Architekturbüro Reinberg

A 1.18 Cooling curve for the Tattendorf passive house. In late 2007 and early 2008 the building wasn't occupied.

heat exchanger and passive cooling, the building interiors require almost no active cooling. The power demand without central servers is just below 17 kWh/m²a. In comparison, the output of the 77 kW_p photovoltaic system is at 6.4 kWh/m²a annually.

For some building types, comparative studies already provide sufficient results. Primary energy parameters of energy efficient office and administrative buildings in central European climates today are of the order of 80 to 100 kWh/m²a without appliances and equipment. Including equipment, values of 120 to 190 kWh/m²a are attained during service (Fig. A 1.16) [45, 46], as in approximately half of typical new construction. Individual projects can be even significantly more efficient. Economic comparisons of demonstration buildings show that their investment costs are no more than 5 % above those of comparable buildings, while energy costs comprise only half the average value [47].

ZERO ENERGY EXTREME: SELF-SUFFICIENT BUILDINGS At first glance, and considering these degrees of energy efficiency, a building that can supply its own energy seems to be just a step away. A self-sufficient building, i.e. a building that isn't connected to energy infrastructure, guarantees continuous energy supply based on the size of the building's own, typically solar energy system, and particularly of energy storage devices, without reverting to other, external resources. Particularly in European climates, the seasonal balance between energy supply and demand requires significant technical investment in energy storage. In climates without pronounced seasonal differences, significantly simpler concepts are possible. Designs at scales of complex solutions for multiple buildings and for entire residential areas are easier as well. The aim isn't the sum total of self-sufficient buildings, but instead the self-sufficiency of the entire complex. The temporal and structural differences in energy demands of individual buildings permit time-related compensation, and thus, reduction of storage volume.

SELF-SUFFICIENT HEATING SUPPLY – POSSIBLE, BUT BETTER NOT ENTIRELY The interior temperature of a solid-construction passive house typically doesn't fall below 10 °C in central European climate, even without any heating (Figs. A 1.17 and A 1.18) [48]. This temperature is the result of balancing heat gains from the use of solar energy and waste heat generated by individuals and devices with heat losses from transmission and ventilation. Heat storage capacity of building construction ensures, even during longer phases of absence of use and at lower heat gains, that interior temperatures don't decline drastically or rise to extremely high interior temperatures when solar gains are high. In practice, interior temperatures have upper limits based on user behaviour (window ventilation, sun protection), so that during times of high solar intake unlimited storage charging isn't possible. Nevertheless, would it be possible to increase heat storage capacity of necessary structural elements of a building to such a degree that interior temperatures no longer fall below 20 °C solely by passive use of solar energy?

PHASE CHANGE MATERIALS Phase change materials (PCM) in the form of micro-encapsulated paraffins can be added to mineral materials of walls and ceilings. Such paraffins (waxes) absorb heat within a defined temperature range, such as from 21 to 22 °C during the melting process, and discharge this heat again during solidification. Special salt hydrates have even higher heat storage capacities and are being processed as components e.g. in suspended ceilings. Calculations show, however, that even if all available surfaces in a building were improved in this way or an extremely solid construction type was selected, the resulting storage capacity would indeed contribute to a further reduction of heating demands, but would not make heating systems redundant. In summer, in connection with sun screens and nighttime ventilation, more significant effects and even entirely avoiding active cooling systems are possible (see company headquarters in Kemptthal, p. 120ff.). This is primarily because heat is stored

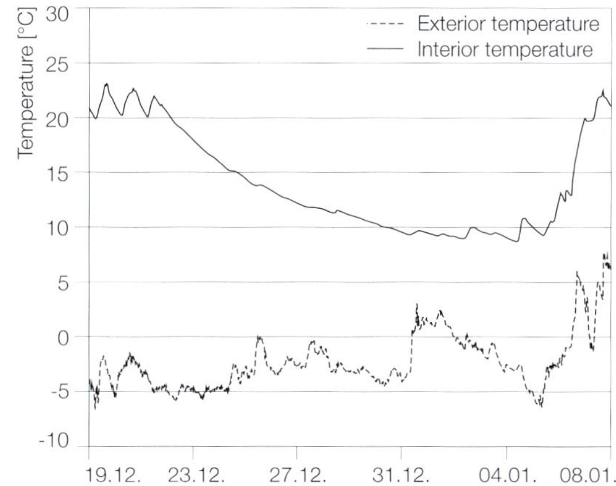

A 1.17
A 1.18

and discharged only in daily cycles. These systems don't permit buffering of longer heating phases.

SOLAR COLLECTORS AND HOT-WATER STORAGE DEVICES In active solar energy use via collectors, a water tank serves as heat storage. As result, the maximum acceptable interior temperature isn't applicable as the upper limit for storage temperature, while the heat insulation of water tanks permits longer storage phases.

A simple example, the full heating supply of a 150 m² house for four occupants, indicates the dimensions required [49]. This system is based on optimally oriented flat tube collectors and a large-volume water storage tank within the building (Figs. A 1.19 and A 1.20, p. 20). If both hot water heating and the entire space heating demands (25 kWh/m²a) are to be covered by solar energy, then the required collector area needs to be at least 30 m². Even more importantly, the storage volume needs to be at least 50 m³. Together with the necessary heat insulation of the storage device, required space equals 5–10% of the building volume. Geometric optimization is, in practice, a very significant factor in determining how much usable space is actually needed.
In Europe, the pioneer of self-sufficient system concepts was the Jenni house in 1991 in the Swiss municipality of Oberburg. The 200 m³ house has a collector surface area of 84 m² and a thermal energy storage device holding 118 m³ [50]. The storage device proved much too big. The numerous follow-up projects mainly in Switzerland and Austria operate successfully with significantly smaller systems supplemented by wood furnaces. Advantageously, smaller storage devices also require lower embodied energy. Considering the limited number of charge/discharge cycles, this comprises a significant parameter in the life cycle balancing of large metal storage devices. Large water storage devices made of plastic, however, haven't proven effective to date in terms of long-term stability, due to occasionally high summer storage temperatures.

One example for solar heating with high solar fraction is the residential system for the Samer Mösl passive house residential complex in the Austrian city of Salzburg (Figs. A 1.22 and M 1.23, p. 21) [51]. The 60 residential units were completed in 2006 by Heimat Österreich (public housing and residential development corporation) as public housing. The climate-neutral heating supply is based on a solar system with 200 m² flat tube collectors mount on the building's flat roof and a 21 m³ buffer storage device connected to a 100 kW pellet boiler. The 11 m tall storage tank with a diameter of 1.6 m (measured without insulation) is integrated vertically in the staircase and extends across all three stories plus the basement. The solar energy system covers approximately half of heating demands.

When a system is designed to cover demands to 100% according to a location's average climate data, experience has shown that, depending on actual weather patterns, energy is sufficient in one year (no additional heating demand). However, in the following year there can be a shortfall [52]. This is primarily due to the sequence of sunny and cloudy days. Long-lasting periods of bad weather or fog are critical. Many successful projects are located in mountain regions that experience more sunny winter days. To omit any service gaps, the required storage volume increases significantly. This effect impacts hot water use first, since higher temperatures are required here than for space heating. In general, heat storage devices for large buildings or residential areas show better results, because the ratio of area loss to storage volume is significantly improved.

STORAGE WITH HIGHER HEAT CAPACITY In theory, alternatives to large water storage tanks exist with higher storage densities and lower heat losses. If one cubic meter of water is heated or cooled by 30°C, this corresponds to an energy content of 30 kWh. Such a heating process increases the heat energy in a 50 m³ storage tank by only 1500 kWh. In contrast, a suitable phase-change material, such as sodium sulphate, stores approximately 100 kWh/m³ when melting at a temperature of around 30 °C and releases this heat again during solidification. Storage devices with thermo-chemical substances may be

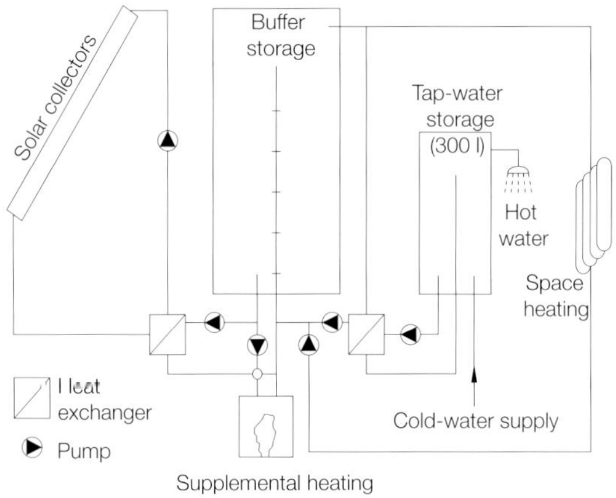

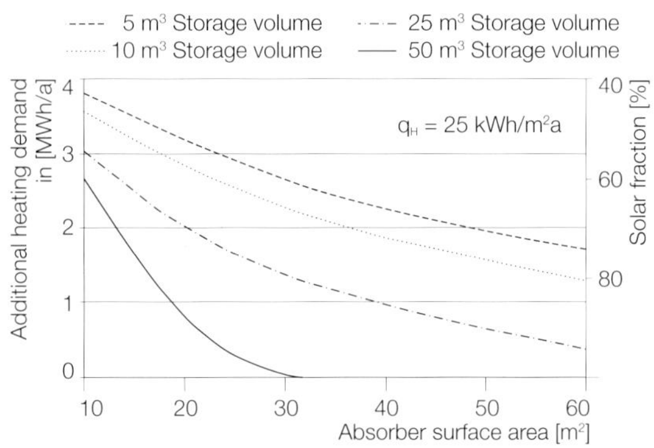

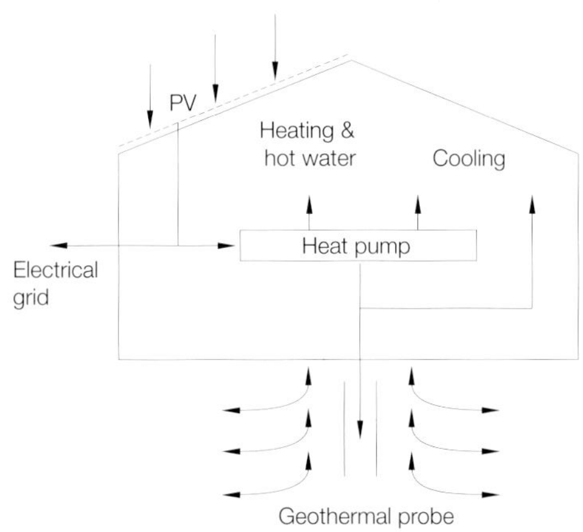

A 1.19
A 1.20
A 1.21

even more effective. They are based on heat absorption, i.e. when substances are separated. Heat discharge occurs when substances are reconnected. Their advantage is (theoretically) lossless storage across long time periods through consistent material separation. Storage with substances called zeolites that discharge water vapour (desorption) when heat is introduced and re-absorb this water vapour (adsorption) when heat is removed function according to a similar principle. Theoretically, storage densities around 500 kWh/m^3 are possible. In practice, however, these values haven't yet been achieved, and required additional technical expenditure may be significant (Fig. A 1.24, p. 22).

STORAGE IN COMBINATION WITH HEAT PUMPS
Another option is improved use of storage capacity in combination with a heat pump [53]. This way, low storage temperatures can also be used for supplying hot water and for space heating by being "pumped up" to usable temperature levels. This increases collector efficiency, since overall storage temperature levels are lower. However, heat pumps consume a considerable amount of electricity.

In connection with a heat pump, it is possible to use the ground as a seasonal low-temperature thermal energy storage device, instead of a large hot-water storage tank within a building, and to charge and discharge heat through geothermal heat exchangers. Different than with water storage tanks, heat is then drawn exclusively via heat pump, since ground temperatures at approximately 10°C are insufficient for heating buildings directly. Many "all-electric buildings" presented in this book are based on such a concept, in connection with grid-coupled photovoltaic systems to compensate for the power consumption of heat pumps. The great advantage of such concepts is the possibility of avoiding large-volume storage in residential interiors. However, they aren't self-sufficient, since they depend on electricity during winter (Fig. A 1.21). A geothermal heat exchanger may also be used for summer cooling.

Following the above featured examples, the conclusion is made that: very low space heating demands are currently achievable. Reducing them to zero is economically, ecologically, and technologically possible, but mostly not reasonable. In contrast, an additional heat source based on biomass is an interesting alternative to climate-neutral heating both in economic and ecological terms. The use of heat pumps leads to climate neutrality only in the case of CO_2-neutral electric power supply (see Projects and Experiences, p. 56ff.).

SELF-SUFFICIENT POWER SUPPLY – ONLY IF NECESSARY Houses without grid connections are rare in industrialized countries. There are reasons for this. In contrast to heating supply, the supply of electricity has a sharply defined limit: without it, systems and appliances don't work. In the case of insufficient heating supply, interior temperatures will merely be cooler during winter.

POWER SUPPLY AS AN INSULAR APPLICATION
Examples of so-called insular houses for the most part comprise remote mountain inns or lodges, hermitages, vacation houses, or research stations. In these cases, an insular solution is the most cost-effective alternative, compared to a kilometres-long power line with a single building as receiver at its end. Without its own electric grid connection, occupants quickly recognize the significance of a central power provider with a functioning grid, because insular systems are technically complex and require regular maintenance and inspection. Self-sufficient power-supply systems achieve high reliability only when multiple energy sources are in use, for example when a wind turbine supplements a photovoltaic system. (Bio) diesel generators are often used as emergency power supplies.
One example for a contemporary insular application is the Neue Monte Rosa Hütte (New Monte Rosa Lodge) (Fig. A 1.26, p. 22) [54]. Featuring 120 beds, it was completed in late 2009 at an elevation of 2883 m above Zermatt, Switzerland, with a view of the Matterhorn. It is based on a nearly 900 m^2 MINERGIE-P-certified passive house. Apart from cooking gas delivered by helicopter in canisters, 90 % of future energy demands are supposed to be covered by solar energy. A 16 kW$_p$ solar generator integrated in the roof structure and 60 m^2 of solar collectors adjacent to the building are in use for this

purpose. Excess energy is stored in a 200 m³ water tank and in batteries (capacity 288 kWh). A combined heat and power unit (8.5 kW_{el}, 19 kW_{th}) fuelled by rapeseed oil serves as a supplemental power and heating source. The dimensioning of units benefits from the fact that the lodge is closed during the core winter months and room temperatures of temporarily approximately 15 °C are considered sufficient in the living areas.

THE PROBLEM OF LONG-TERM STORAGE

The great challenge of electrically self-sufficient buildings is not on-site power generation via sun or wind, but instead long-term storage of electrical energy in relation to high seasonal variations of energy availability. While the previously discussed heat storage devices are indeed voluminous, yet technically simple, to date there exist no similarly convincing battery storage devices. Rechargeable batteries store only direct current, not the alternating current used in buildings today. Their integration into a building's power supply requires the coupling and decoupling of direct current via rectifiers and inverters [55]. Lead batteries similar to those in motor vehicles today are used as uninterruptible power supplies (UPS) for server centres and clinics and serve to bridge temporary failures of the electrical grid. Their energy density equals about 40 Wh/kg battery weight. This nominal capacity isn't available to full extent, because deep discharging strongly reduces battery life. Starting with a minimum charge of 20 % and a charging efficiency of 80 %, buffering the power supply of a single-family house for one day (10 kWh) requires a nearly 400-kg battery pack at a cost of approximately 1000 Euros. However, its life span lasts only a few years (Fig. A 1.25).

One of the central problems of batteries is self-discharge. Similar to thermal energy storage devices, they lose their stored energy more or less quickly depending on the technology used. In the context of research on electromobility, batteries with higher capacity and performance rates have already been developed. However, they still don't seem suitable for the seasonal storage of electric current in buildings in relation to life span, costs, environmental

A 1.19 Schematic diagram of a solar power system for hot water and space heating with a high solar load match as basis for simulations.

A 1.20 Additional heating demand and solar load match for a solar power system for hot water and space heating of a 150 m² single family house (space heating demand 25 kWh/m²a) and a four person household. The system configuration comprises flat collectors, a 300-litre water storage tank, and a variable-size buffer storage tank with 50 cm of insulation. Simulation results via TRNSYS using climate data for Freiburg, Germany.

A 1.21 Seasonal thermal energy storage transferred from the building into the ground (geothermal probes). Direct and indirect cooling (reversible heat pump operation) is also possible in addition to heat generation by a heat pump.

A 1.22 "Samer Mösl" residential development, Salzburg, Austria, 2006, Architect: sps-architekten, Energy planner: TB Stampfer.

A 1.23 Heat supply system schematic of the "Samer Mösl" project.

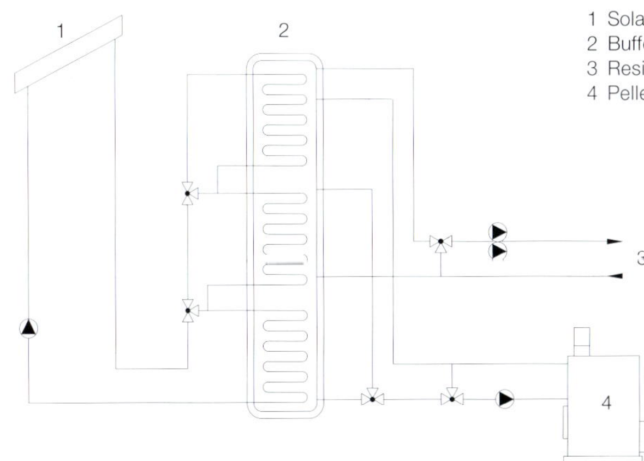

1 Solar collectors 197 m²
2 Buffer storage 21 000 l
3 Residence transfer station
4 Pellet boiler 100 kW

A 1.22
A 1.23

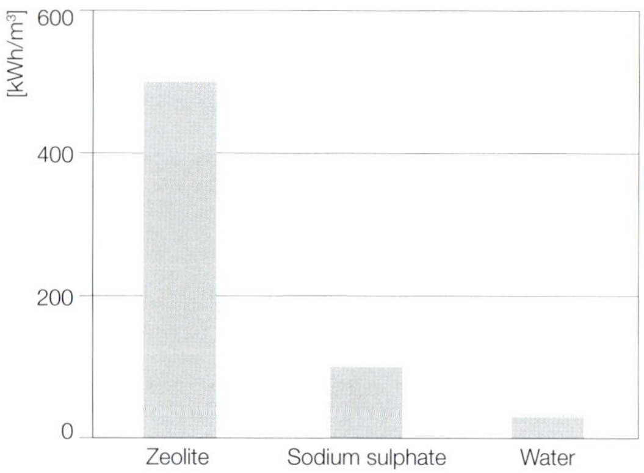

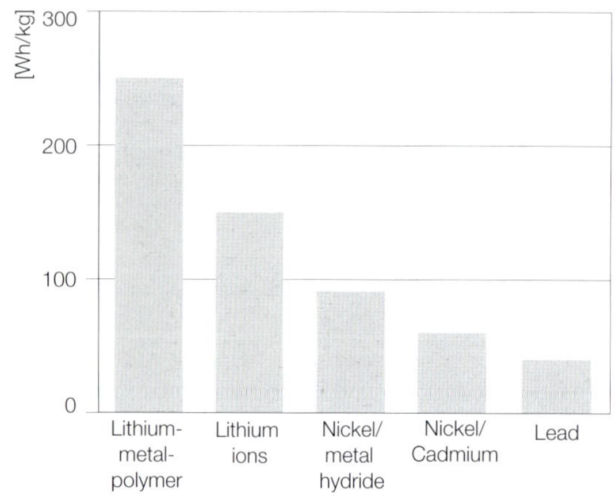

A 1.24 Typical energy density of different thermal-energy storage devices.
A 1.25 Energy density of different types of rechargeable electrical storage devices.
A 1.26 The new Monte Rosa Lodge above Zermatt, Switzerland, 2009, joint project of ETH Zürich, SAC, Lucerne University of Applied Sciences and Arts—Engineering & Architecture, and EMPA.
A 1.27 The energy diagram of the "energy self-sufficient solar house" experimental residential building emphasizes the complexity of gridless, yet comfortable life.
A 1.28 Experimental residential building "energy self-sufficient solar house," Freiburg, Germany, 1992, Architect: Planerwerkstatt Hölken Berghoff, Energy consultant: Fraunhofer ISE.

A 1.24
A 1.25
A 1.26

impact, and production energy. New approaches for small battery packs as temporary storage devices in buildings are discussed in the section "The Degree of Individual Demand Coverage as a Differentiation Characteristic" (p. 37f.). Yet, without grid coupling, small battery units aren't useful, since the high excess output of photovoltaic systems in summer would be disposed of unused.

HYDROGEN ECONOMY AS POSSIBLE SOLUTION

The experimental building "Energieautarkes Solarhaus" (energy self-sufficient solar house) was completed in 1992 in Freiburg [56]. As a single family residential house in the middle of the city and self-sufficient in terms of heating and electricity, it was occupied for several years and served as a case study for a decentralized solar hydrogen economy (Figs. A 1.27 and A 1.28).

The small 19 kWh battery pack in the building only serves as temporary power storage to prevent short-term use of the hydrogen system. During the summer months water is split into hydrogen and oxygen by an electrolyser (2 kW) powered by a 4.2 kW_p photovoltaic system. The two gases are stored almost lossless in pressure tanks (15 m^3 H_2, 27 bar) and recombined in a fuel cell (0.5 kW) for electrical power generation during winter, while also producing water. The waste heat released is an additional winter heat source and used to heat water. Similar to a rechargeable battery, a fuel cell is an electrochemical energy converter. However, when generat-

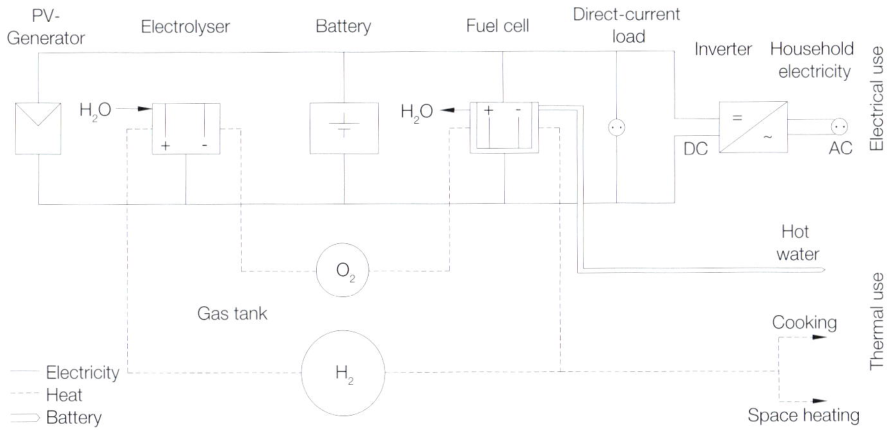

A 1.27
A 1.28

ing power, it is continuously fuelled by hydrogen. In addition to supplying power, hydrogen is also used for cooking (gas stove) and heating (air heating via ventilation system). More recent projects are also based on a hydrogen economy. The experimental box house "SELF" presented in 2010 by the Swiss research institute EMPA, Department of Building Technologies, uses hydrogen produced on-site by solar energy for heating and cooking [57].
Without the hydrogen system, the Freiburg solar house would require, as calculated at the time, a lead battery pack weighing four tons with a capacity of 160 kWh. That is just below half of the capacity of the Monte Rosa Lodge, which is seven times larger. The fact that the battery storage isn't much larger there is due to the combined heat and power unit fuelled by rapeseed oil. In the context of the discussion above, and considering a battery's short life span and production energy of approximately 60,000 kWh, the Energy Self-Sufficient Solar House project would have made no contribution to research with a battery unit alone. While acknowledging high labour expenditure for operation and maintenance, this emphasizes the basic feasibility of the solar hydrogen economy at a small scale with performance values that are, even under those circumstances, comparatively high. The high cost of the components used has changed little to date. Hydrogen economy models remain a possible future alternative, while combined energy solutions will continue to have significant advantages compared to singular systems.

PRACTICAL APPLICATIONS OF NET ZERO ENERGY BUILDINGS: MORE OR LESS GRID-BASED

What sounds contradictory at first has contributed to the most widely accepted understanding of zero energy buildings today. Because energy self-sufficient solutions – as shown above – result in extreme expenditure, applied concepts for climate-neutral buildings are based, at the very least, on their connection to an electrical grid. Depending on local conditions, additional grids or local use of biomass (such as in pellet stoves) or fossil fuels (such as in combined heat and power plants fuelled by natural gas) could be added to cover remaining energy demands. Such arrangements bridge the technology gaps in the area of power storage, eliminate thermal energy storage devices with extreme dimensions, and guarantee the customary reliability of energy supply.

GREEN POWER GRIDS ARE THE SIMPLEST PATH TOWARDS CLIMATE NEUTRALITY The designs that are easiest to build are based on energy supply through grids that exclusively use renewable energy, and thus, completely detach energy consumption from emissions (Fig. A 1.30, p. 24). The situation in Norway serves as an example: the standard there is the "all-electric building." Almost all electrical power generation is based on hydropower. Thus, it is free from fossil fuels or nuclear energy and produces no greenhouse gas emissions (2009: 96 %) [58]. Independent of the magnitude of consumption, all buildings are climate neutral. In this context and with low electricity costs, energy efficiency standards of buildings are low compared to Scandinavian neighbours and regarding the cold climate. This is changing, due to increasing domestic power demands, good export opportunities for green power, and limited capacity for expansion of domestic hydroelectric plants. In 2009 an extensive research program was started at Trondheim University with a new Centre for Zero-Emission Buildings [59]. Replacing electrical resistance heaters with heat pumps is actively promoted within building construction. This permits reducing electrical power consumption by a factor of about 3.

If there were an efficient undersea cable for relaying green power to central Europe, buildings there could also benefit from Norway's clean energy. This indicates that extremely high energy efficiency of buildings is required, but not to reduce excessive use of finite domestic green power resources, yet instead to market it globally. From this overall perspective, in an interconnected European power grid, the all-electric buildings in Norway would no longer actually be climate neutral: their high energy consumption prevents power from green networks from being available for climate neutral supply of houses in neighbouring countries. The seasonally strongly fluctuating power demand of all-electric buildings requires a grid with electricity generation from renewable energy sources that can be altered

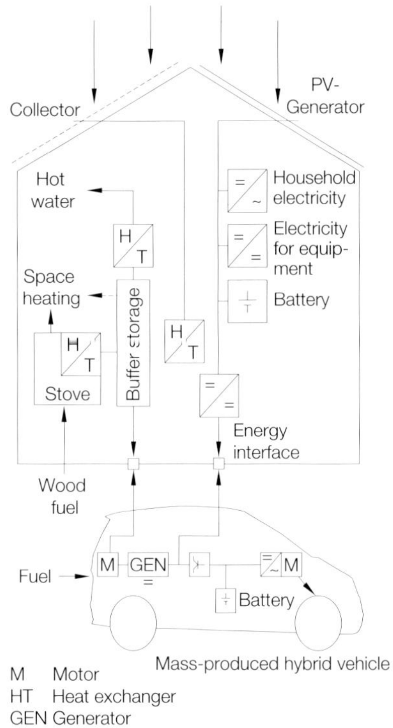

EXCOURSE MOBILE COMBINED HEAT AND POWER PLANT In the context of the "energy self-sufficient solar house" project (Fig. A 1.28, p. 23), a concept was developed for integrating an electric vehicle via its battery capacity, but also as a mobile combined heat and power unit (BCHP), into the electricity and space-heating supply of an off-grid building [60]. With a vehicle that comprises both an electric motor with a battery pack and an internal combustion engine – a so-called plug-in hybrid or a range-extension electric vehicle, surplus electricity can be transmitted from the building to the vehicle via an energy interface. The vehicle's internal combustion engine is normally used to produce electricity for its electrical motor with a generator. With a capacity of 5 kWh, the battery operates similar to a transmission that mediates between the motor's variable power demand and the motor generator's continuous electricity output at its operative energy optimum. Is a vehicle parked at a house and the building's energy reserves are depleted, the motor generator takes over the function of supplying the building with electricity and space-heating via an energy interface.
For off-grid buildings, this concept offers great advantages. The constraint, however, is lacking market availability of suitable hybrid or electric drive vehicles (EDV). Without the support of EDVs, buildings will need a stationary emergency-power unit.

accordingly, while possessing sufficient storage capability, for example in the form of pump storage power stations. The natural possibilities here vary greatly regionally and internationally. This also calls for a higher performance interconnection of grids than currently typical.

At sufficient building density, alternate grids are able to relieve green power grids (Fig. A 1.31). Heating or cooling grids could be supplied from central combined heat and power plants that are fuelled to 100 % by renewable energy in the form of wood or biomass. In light of higher grid costs compared to an electrical grid, mainly city grids or local micro-grids are suitable. Conditions look similar in gas distribution grids for biogas or hydrogen. The introduction of grids requires high-level decisionmaking in urban and infrastructure planning. Similar to the green power grid, it becomes clear quickly here that wood, biomass, biogas, etc. aren't available sustainably and sufficiently for long-term coverage of currently typical high building demands.

WITHOUT GREEN POWER GRIDS, ONLY THE BALANCING PRINCIPLE HELPS If electricity from grids or grid-based heating or cooling doesn't originate in renewable energy to 100 %, a building's

A 1.29
A 1.30
A 1.31

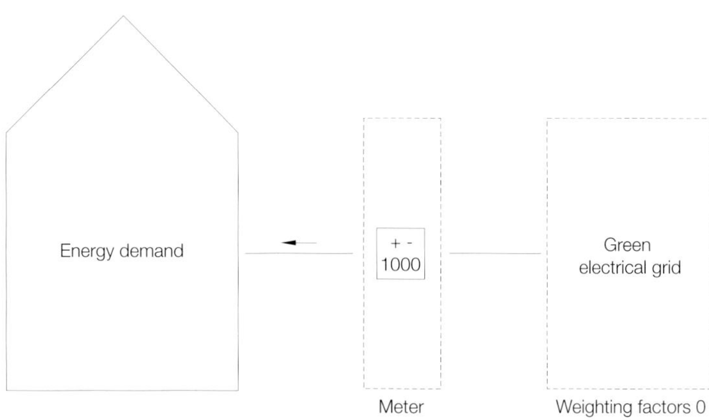

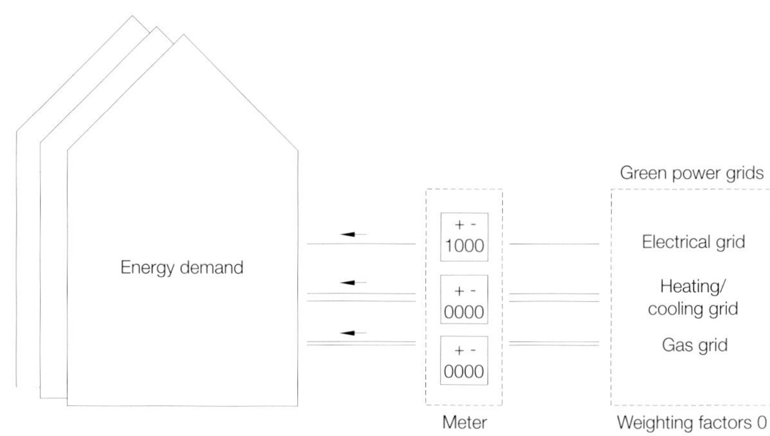

consumption impacts the environment. A "zero energy building" in the sense of "absolutely not reverting to fossil or nuclear resources" is impossible.

The buildings presented in the second half of this book should actually be called "net zero energy buildings." Their balanced energy or emissions are a result of calculations. This "net zero energy building" designation is common, however, only in English. The idea is based on the following approach:

Credits for locally feeding energy back into the grid offset energy consumption and climate gas emissions in an annual balance sheet. This method is similar to a business report. Fig. A 1.32 graphically indicates that the energy flow between meters and buildings takes place in both directions. Import and export can be balanced, since evaluation factors of the energy supplied are no longer zero. The goal is a zero balance (zero-energy building) or gaining credits (plus-energy building). This alternative has practical relevance, since exclusively green power grids are currently not available, with few exceptions. Therefore, nearly all buildings revert to fossil or nuclear sources when consuming energy. This also applies to buildings that are presented in the project section of this book.

A 1.29 Diagram, integration of an electrically self-sufficient building and a hybrid vehicle: electricity flows from the house to the vehicle and vice versa. The vehicle's motor generator can supply the house with exhaust heat from electricity generation.
A 1.30 What electrical self-sufficiency means is demonstrated figuratively by the ambitious project of circumnavigating the globe in five consecutive phases with an exclusively solar-powered electric glider [61]. The aircraft's wings span 63 m and are completely covered with high-performance solar cells. Batteries with a capacity of just under 100 kWh serve for buffering. Weighing 400 kg, the batteries comprise a fourth of the aircraft's total weight.
A 1.31 "Electricity-only buildings" connected to a green electrical grid are climate-neutral: environmental indicators of supplied electricity are equal to zero.
A 1.32 Buildings that use other green grids besides green electric power can remain climate-neutral even without their own on-site energy generation. They are 100 % grid-based.
A 1.33 Schematic representation of the balancing method for energy imported from and exported to a grid. Without 100 % green power, the zero-energy standard is only achieved if the annual balance is zero.

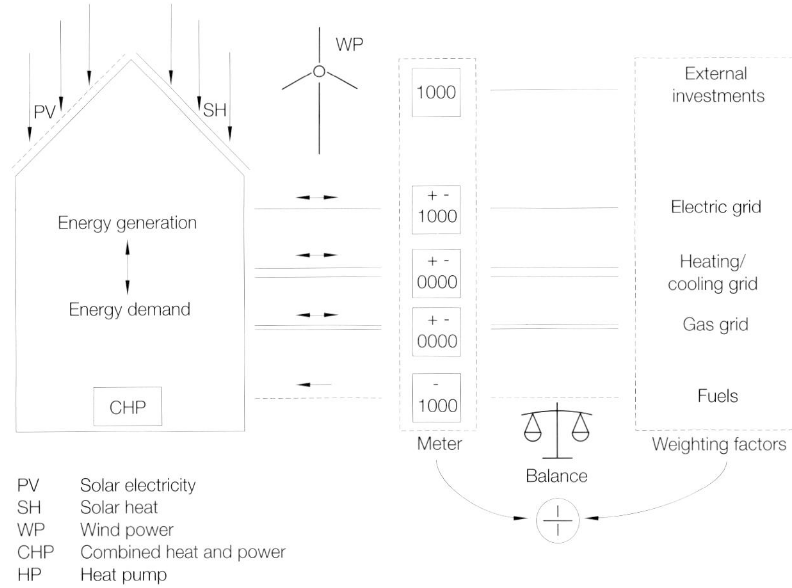

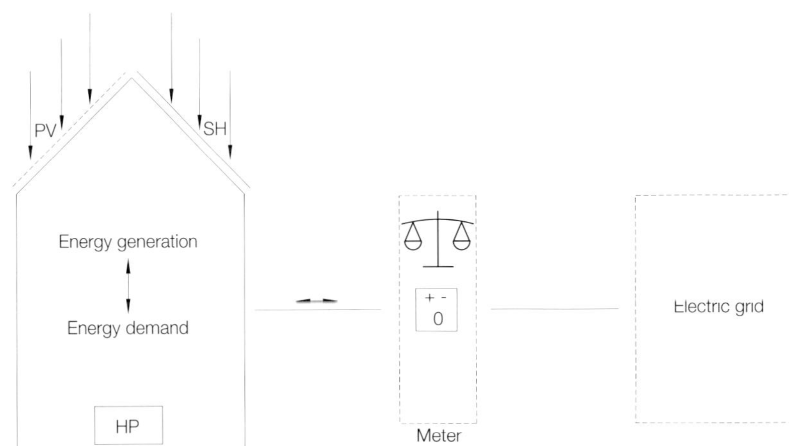

A 1.32
A 1.33

Possibilities for exporting energy into the power grid exist primarily for photovoltaic systems integrated into buildings and for combined heat and power plants. With solar collectors, excess heat can generally be exported from building systems to heating grids, if the required technical means are in place. Wind power systems are rarely found in direct connection to buildings. Many zero-energy buildings, however, include wind farms in their calculation and credit their power output to the building balance (designated in Fig. A 1.32, p. 25 as external investments). Even more far-reaching balancing scenarios include purchasing CO_2 credits from energy-savings measures in other buildings or, for example, from reforestation projects. These examples show how important a uniform, methodical understanding is for the term "zero-energy building" to gain significance. The following section therefore explains the methodical basis of balancing (p. 28ff.).

THE BALANCING PRINCIPLE IN "ALL-ELECTRIC BUILDINGS" The balancing principle is most easily understood in all-electric buildings. There is an undisputed balance at the power meter without additional evaluation. If power supply costs are equal to feed-in costs, then a net zero-energy building is also a "net zero energy cost building." Fig. A 1.33 (p. 33) shows the currently frequently used typical equipment including a heat pump and a photovoltaic system. The system is dimensioned so that the building's annual power consumption is balanced. Solar collectors can contribute to avoiding electricity consumption via the heat pump in summer. The significantly increased winter electricity consumption under conditions in central Europe causes higher demands on the grid and especially storage capacity and winter supply capacity, since the house's own solar power yield displays strong seasonal variation. The claim of climate neutrality is fulfilled in the strict sense only by means of suitable grids using a high percentage of renewable energy in winter.

A FIRST CONCLUSION According to the options illustrated above, it becomes clear that climate neutrality is produced in non-self-sufficient buildings only as a sum total of buildings, grids, and fuels. The reasonable expense for a local zero balance by means of energy generation and grid feed-in for a particular building is measured in economic and ecological terms in the context of an available alternative. This alternative is investing in a grid infrastructure based entirely on renewable energy. This may be advantageous in most cases, but difficult to influence, due to the individual decisions of individual building owners.

A 1.34

A 1.34 The balancing principle as applied to an "all-electric building" with heat pump and photovoltaic system: the photovoltaic system is sized to allow balanced annual electricity consumption. The optional solar thermal system reduces the heat pump's electricity consumption.

ENDNOTES

[1] McKinsey & Company: Kosten und Potenziale zur Vermeidung von Treibhausgasemissionen in Deutschland [Costs and potentials for avoiding greenhouse gas emissions in Germany]. Sektorperspektive Gebäude [Building Sector Perspectives]. Düsseldorf 2007.

[2] Institut Wohnen und Umwelt [Institute for Living and Environment]; Bremer Energieinstitut [Bremen Energy Institute]: Datenbasis Gebäudebestand [Existing buildings database]. Darmstadt 2010.

[3] Institut Wohnen und Umwelt [Institute for Housing and Environment]: Querschnittsbericht »Energieeffizienz im Wohngebäudebestand« [Cross-sectional report "energy efficiency in existing residential buildings]. Darmstadt 2007.

[4] Engelmann, Peter: Studentisches Wohnen im Passivhaus. Evaluierung energieeffizienter Studierendenwohnheime. [Student living in passive house construction. Evaluation of energy-efficient student residences] Dissertation, University of Wuppertal 2010.

[5] European Union (EU): Directive 2002/91/EG of the European Parliament and of the Advisory Board of December 16, 2002, on the total energy efficiency of buildings. http://eur-lex.europa.eu/LexUriServ/LexUriServ.do?uri=OJ:L:2003:001:0065:0065:DE:PDF. Accessed February 21, 2011

[6] European Union (EU): Directive 2010/31/EU of the European Parliament and of the Advisory Board of May 19, 2010, on the total energy efficiency of buildings (new version). http://eur-lex.europa.eu/LexUriServ/LexUriServ.do?uri=OJ:L:2010:153:0013:0035:DE:PDF. Accessed February 21, 2011

[7] Passivhaus Institut [Passive House Institute]: Passive House Planning Package PHPP. Version 2007. Darmstadt 2007.

[8] Federal Ministry of Economics and Labour (BMWA): Das 5. Energieforschungsprogramm der Bundesregierung [The 5th energy research program of the federal government]. Berlin 2005.

[9] Federal Ministry of Economics and Technology (BMWi): Entwurf zum Energiekonzept der Bundesregierung [Draft of the energy concept of the federal government]. Berlin 2010. Translated by the author

[10] ibid.

[11] http://www.enob.info. Accessed February 21, 2011

[12] ibid.
[13] SIA Media Information: Neues SIA Energieleitbild Bau [New SIA energy mission statement for construction]. Zurich 2009. http://www.sia.ch. Accessed February 21, 2011.
[14] SIA Merkblatt 2040: Energieeffizienz Pfad [Energy Efficiency Path]. August 2011.
[15] Preisig, Hansruedi; Pfäffli, Karin: Merkblatt SIA Effizienzpfad Energie [Efficiency Path for Energy]. TEC21 23/2010.
[16] Roland Stulz: Leichter Leben [Live Lighter]. Broschüre [Brochure] 2010. http://www.novatlantis.ch. Accessed February 21, 2011.
[17] Zimmermann, Mark; Althaus, Hans-Jörg; Haas, Anne: Benchmarks for sustainable construction. A contribution to develop a standard. In: Energy and Buildings 37/2005, pp. 1147–1157.
[18] cf. [14]
[19] In the following, the term CO2-emissions is used as an abbreviated form of equivalent CO_2-emissions.
[20] http://www.ipcc.ch. Accessed February 21, 2011
[21] cf. [13]
[22] Federal Ministry of Economy, Family, and Youth (BMWFJ); Federal Ministry of Agriculture, Forestry, Environment, and Water Management (BMLFUW): EnergieStrategie Österreich. [Energy Strategy for Austria] Vienna 2010. http://www.energiestrategie.at. Accessed February 21, 2011
[23] Federal Ministry of Transport, Innovation, and Technology (BMVIT): Haus der Zukunft Plus [House of the Future Plus]. 1. Ausschreibung [First Competition] 2008. Leitfaden für die Projekteinreichung [Guidelines for project submission]. Vienna 2008.
[24] Federal Ministry of Transport, Innovation, and Technology (BMVIT) (Ed.): Strategieprozess ENERGIE 2050 [[Strategy Process for ENERGIE 2050]. Themenfeld Energie in Gebäuden [About Energy in Buildings]. Vienna 2007. http://www.e2050.at. Accessed February 21, 2011
[25] Gesellschaft für Solarinstallation und Design, S.O.L.I.D.: Fernwärmeeinspeisung [District heating feed-in]. Wasserwerk Andritz [Andrix Waterworks]. Fernheizwerk/AEVG [AEVG district heating plant]. Wasserwerk Andritz. [Andritz Waterworks]. http://www.solid.at. Accessed February 9, 2011
[26] Forum Gas Wasser Wärme: Wärme aus der Ferne. [Forum on gas, water, heat: Heat from afar] In: FORUM SPECIAL 03/2009. Publisher Fachverband der Gas- und Wärmeversorgungsunternehmen (FGW) [Association of Gas and District-Heating Supply Companies] and Österreichische Vereinigung für das Gas- und Wasserfach [Austrian Association for Gas and Water] (ÖVGW). Vienna 2009, pp. 7–8.
[27] Federal Ministry of Transport, Innovation, and Technology (BMVIT): Forschungs- und Technologieprogramm »Haus der Zukunft« ["House of the Future" Research and Technology Program]. Vienna. http://www.hausderzukunft.at. Accessed February 21, 2011.
[28] United States Department Of Energy (US DOE): Building Technologies Program. Planned Program Activities for 2008-2012. Washington 2008. http://www1.eere.energy.gov/buildings/vision.html. Accessed February 10, 2011
[29] ibid.
[30] Pless, Shanti; Torcellini, Paul: Net-Zero Energy Buildings: A Classification System Based on Renewable Energy Supply Options. NREL Report/TP-550-44586, Golden/CO 2010. http://www.nrel.gov/docs/fy10osti/44586.pdf. Date: August 22, 2011
[31] Torcellini, Paul; Pless, Shanti; Deru, Michael; Crawley, Dru: Zero Energy Buildings: A Critical Look at the Definition. Paper No. 417, Proceedings (CD-ROM), ACEEE Summer Study on Energy Efficiency in Buildings, August 13–18, 2006, Pacific Grove, CA. National Renewable Energy Laboratory, Golden, pp. 12; www.nrel.gov/docs/fy06osti/39833.pdf.
[32] Deru, Michael: "Establishing Standard Source Energy and Emission Factors for Energy Use in Buildings." Proceedings of ES2007, June 27-30, 2007, Long Beach, CA, American Society of Mechanical Engineers.
[33] SIA 2039: Mobilität – Energiebedarf in Abhängigkeit vom Gebäudestandort [Mobility: Energy Demand as a Function of Building Location]. Mai 2011.
[34] http://www.cmhc.ca/en/en_001.cfm, Accessed May 17, 2011
[35] Griffith, Brend; et al.: Assessment of the Technical Potential for Achieving Net Zero-Energy Buildings in the Commercial Sector. Technical Report NREL/TP-550-41957, 2007.
[36] http://www.passiv.de. Accessed February 21, 2011
[37] http://www.minergie.ch. Accessed February 21, 2011
[38] In central European climate, a maximum heating output of 10 W/m² is equivalent to a target value of 15 kWh/m²a for annual space-heating demand [27]. Reference area is the heated net floor area within the thermal envelope; basement, storage, and infrastructure floor areas are included fractionally. In Austria and Switzerland, different reference areas are customary.
[39] MINERGIE-P: 30 kWh primary energy per m² of energy reference floor area for space heating, hot water, and ventilation including Swiss primary-energy factors. For household appliances, values per floor-area are not required, but rather the best efficiency class [26].
[40] cf. [7]
[41] cf. [5]
[42] SIA 380-4: Elektrische Energie im Hochbau [Electrical Energy in Building Construction]. Zurich 2006.
[43] Güttinger, Herbert; et al.: Energie-Detailbilanz des Eawag Forum Chriesbach [Detailed Energy Balance for the Eawag Building Forum Chriesbach]. Swiss Federal Office of Energy. Bern 2009.
[44] Lehmann, Beat; et al.: Eawag Forum Chriesbach – Simulation and measurement of energy performance and comfort in a sustainable office building. In: Energy & Buildings 42/2010, pp. 1958–1967.
[45] Voss, Karsten; et al.: Bürogebäude mit Zukunft [Office Buildings with a Future]. Berlin 2006.
[46] Lechner, Robert; et al.: PH Office. Standard für energieeffiziente Bürogebäude, Endbericht [Standard for energy-efficient office buildings, final report]. Publisher Federal Ministry for Transport, Innovation, and Technology (BMVIT). Vienna 2010.
[47] Lützkendorf, Thomas; et al.: Energiekonzepte und ihre Auswirkungen auf ausgewählte Nutzungskosten von EnOB-Bürogebäuden [Energy concepts and their effects on selected use costs of EnOB office buildings]. 3rd IFM Convention, Vienna 2010.
[48] Wagner, Waldemar; et al.: Energietechnische und baubiologische Begleituntersuchung Lehm-Bürogebäude Tattendorf. [Companion study on energy infrastructure and bioconstruction of the adobe passive office building in Tattendorf] In: Berichte aus Energie- und Umweltforschung [Reports on Energy and Environmental Research] 65/2009. Publisher: Federal Ministry for Transport, Innovation, and Technology (BMVIT). Vienna 2010. p. 43.
[49] Voss, Karsten: Experimentelle und theoretische Analyse des thermischen Gebäudeverhaltens für das Energieautarke Solarhaus Freiburg [Experimental and theoretical analysis of the thermal building behaviour of the energy self-sufficient solar house in Freiburg]. Dissertation, ETH Lausanne. Stuttgart 1997.
[50] http://www.jenni.ch. Accessed February 21, 2011
[51] http://www.nextroom.at/building.php?id=28858. Accessed February 10, 2011
[52] cf. [38]
[53] International Energy Agency: Solar Heating & Cooling Programme, Task 44. Solar and Heat Pump Systems. http://www.iea-shc.org/task44/index.html. Accessed February 21, 2011
[54] ETH Zürich (ed.): Neue Monte Rosa Hütte. Ein autarkes Bauwerk im hochalpinen Raum [New Monte Rosa Hut. A self-sufficient building in the high Alps]. Zurich 2010.
[55] A special direct-current grid in buildings isn't reasonable in most applications. Inverters today have very high efficiency rates. Only few energy-efficient appliances with direct-current models are available. Direct-current appliances also have different voltage levels requiring a variety of DC/DC converters.
[56] Voss, Karsten; et al.: The self-sufficient solar house in Freiburg. Results of 3 years of operation. In: Solar Energy 58, 1996, pp. 17–23.
[57] SELF. Das unabhängige Haus [The independent house]. http://www.empa.ch/self. Accessed April 19, 2011
[58] NVE Energiestatus 2010 [NVE energy status 2010]: http://www.nve.no/no/Energi1/Energistatus. Accessed February 21, 2011
[59] Norwegian University of Science and Technology NTU: The Research Centre on Zero Emission Building. http://www.sintef.no/Projectweb/ZEB/About-ZEB. Accessed February 21, 2011
[60] Goetzberger, Adolf; et al.: Mobile Blockheizkraftwerke [Mobile combined heat and power plants]. In: Erneuerbare Energie [Renewable Energy]. 04/1996, pp. 11–15.
[61] Solar impulse: Around the world in a solar airplane. http://www.solarimpulse.com. Accessed February 21, 2011

METHODICAL PRINCIPLES OF BALANCING

As simple as energy balancing appears to be for the previously described examples of all-electric buildings, its complexity is due to particular details. This is especially true for buildings that utilise multiple energy sources. However, the basic planning strategy on the level of individual buildings always comprises a dual concept:

- reduce energy expenditure
- optimise feeding energy into grids

INPUT/OUTPUT BALANCE Energy supply and feeding into grid infrastructure is calculated at the building's energy infrastructure (see Fig. A 1.33, p. 26). For a precise balance calculation, at least the following three provisions are required (Fig. A 2.02):

- one or more suitable indicators
- a balance boundary
- a balance period

Fig. A 2.02 displays the principle of an input/output balance. Selecting a suitable indicator, a balance boundary, and a period results in a particular balance [1, 2]. The right diagram exemplarily illustrates these provisions, i.e. primary energy as indicator (non-renewable fraction – PEI n.r.), heating, ventilation, air-conditioning (HVAC), hot water (HW), lighting and equipment as balance boundary and one year as period. In this context, the diagonal line in the diagram represents a net zero energy building. Within the annual balance, primary energy expenditure is compensated by a primary energy credit from feeding power into the grid. Anything above the diagonal line corresponds to net plus energy buildings. The illustration clearly shows that without prior reduction of demand, the path to zero energy buildings requires very large credits. Reasonable sizes of solar arrays or combined heat and power systems are dependent on significantly reducing energy demands.

In comparison, the concepts for planning-related energy balancing schemes employed in the projects described in this book vary and, therefore, lead to (more or less) different results. Typically, these schemes make use of calculation methods introduced in their respective countries in order to calculate a building's energy demand. Correspondingly, their balances for net zero energy buildings involve different balance boundaries. The data tables and diagrams in the project chapters serve to compare these concepts.

SOLAR CONCEPTS AND THEIR EXPANSION IN ENERGY-INTENSIVE BUILDINGS In all zero energy concepts, primary energy demand directly determines required grid feed-in credits. Against the background of currently preferred concepts of building-integrated solar power generation, size and

A 2.01
A 2.02

A 2.01 Rooftop wind turbines, five-story residential building in Chicago, Illinois, USA 2007, Murphy/Jahn.
A 2.02 Basic input/output principle, required balance criteria, and balancing example.

orientation of available exterior surfaces serve to define maximum permissible on-site building energy demands for an optimized balance. If the number of stories increases, balancing via solar power generation alone can no longer be achieved, since a building's energy demands increase more strongly than the exterior surface suitable for solar power generation. Here, the otherwise typically advantageous surface-to-volume ratio of tall and compact buildings comprises a disadvantage compared to small buildings.

Efficient building-mount wind power utilization is limited to a few unique cases, due to necessary wind speeds (at least 4 m/s annual average) and the unavoidable noise associated with wind turbines. Thus, wind power doesn't comprise a viable alternative. A five-story residential building in Chicago, Illinois, serves as example (Fig. A 2.01) [3]. Eight wind turbines with horizontal axis are located on its rooftop. They provide a total of 12 kW_p, equal to a photovoltaic system covering about 100 m^2. However, in combination with the building's energy efficiency, this is by far insufficient to attain zero energy standard (see Pixel Building, p. 134ff.).

In comparison, building-integrated combined heat and power (CHP) features certain advantages. Currently, CHP systems consist mainly of gas or diesel fuelled combustion engines directly axially coupled with a generator to produce electricity (Figs. A 2.03 and A 2.04, p. 30). The exhaust heat released by the engine is used to cover heating demands [4]. The electricity is used fractionally in the building or fed into the grid, similar to photovoltaic systems. In total, more than 90 % of the fuel's primary energy is utilised. A precondition for service, however, is that heat is used in the building without much delay. If heat isn't used, a CHP system becomes impractical in terms of energy efficiency. Typically, thermal storage devices are enlarged to increase their runtime. Technological developments are aimed at achieving the highest possible electricity yields (e.g. via fuel cells or micro-turbines) and introducing renewable energy as fuel type even for small output units (e.g. Stirling engines with external combustion). Integrating transient data for a building's expected electricity and heating demands as well as electrical supply and feed-in tariffs can serve to optimize the operating method in terms of energy and economy [5].
Such advances provide CHP systems with significant opportunities, even if future electrical grids comprise vastly higher percentages of renewable energy. Unlike photovoltaic systems, their electricity generation can be controlled. This is particularly valuable in a market with a high percentage of electricity generation from intermittent and variable wind and solar resources.

MANY DIFFERENCES BETWEEN ZERO ENERGY BUILDINGS ARE HIDDEN BY DETAIL Despite an annual net zero energy balance and correspondence between all three balance criteria, large differences emerge among net zero energy buildings. These differences are related mainly to:
- matching the timing of energy generation and demand (load-match)
- matching the timing of grid feed-in with grid demands (time-variable value of feed-in electricity)
- matching of feed-in and supplied energy (i.e. feeding in excess electricity in summer versus supplied natural gas in winter).

With the planned introduction of smart grids and daily and seasonally variable electricity metering, the grid-side cost of feed-in electricity or costs of electricity consumed will vary according to time. Thus, related cost differences will become increasingly relevant for building managers and operators. The more electricity costs vary according to time, the greater this effect will be, making costs a controlling factor. Zero energy concepts that mainly use grids for storage and power supply will become more expensive to operate than those that adapt their electricity generation and feed-in based on their own demand and the grid's tariff structure.

The tendency towards all-electricity buildings in residential and non-residential construction avoids the

BASIC PRINCIPLE

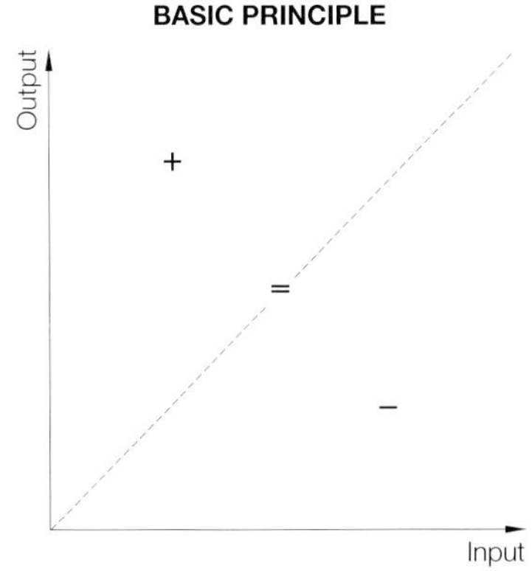

SPECIFICATIONS

INDICATORS
- Site energy
- Primary energy, non-renewable
- Primary energy, total
- Equivalent CO_2 emissions
- Energy costs

BALANCE BOUNDARY
- HVAC, HW, and lighting
- Equipment, tools, and centralised services
- Electromobility
- Construction components and materials
- External investments

BALANCE PERIOD
- Operating year
- Total use period
- Life cycle

EXAMPLE

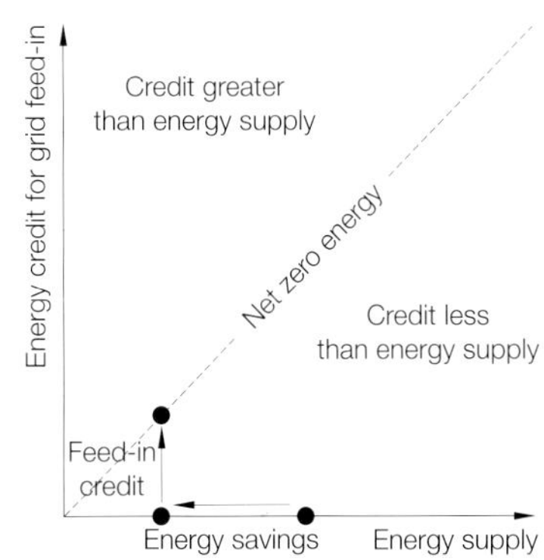

A 2.03 Schematic configuration of a combustion-engine CHP system and typical distribution of output energy into electricity and heat; 90 % of fuel energy content are utilized.
A 2.04 Typical CHP system.
A 2.05 Qualitative profile of electricity generation of a photovoltaic system and a CHP system compared to the electricity demand of a residential building in central European climate with generally high heating demands:
a without heat pump or PV, but with CHP
b with heat pump and PV
c without heat pump, but with CHP and PV

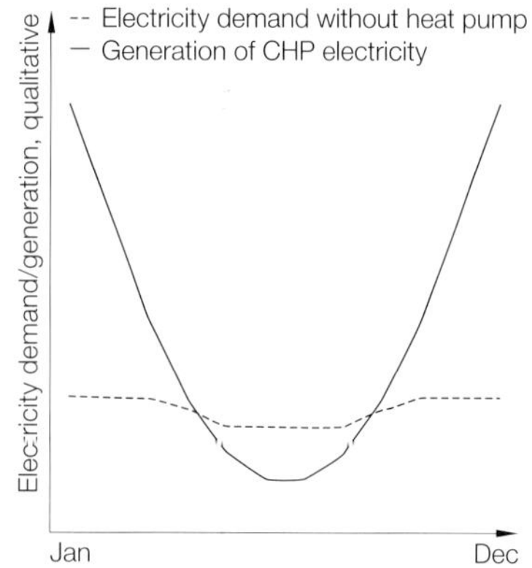

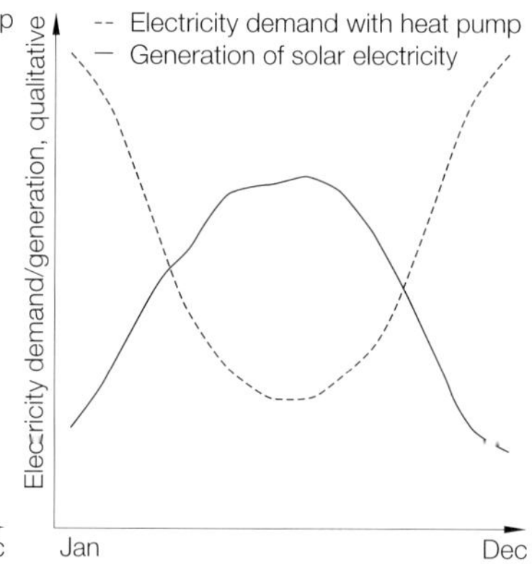

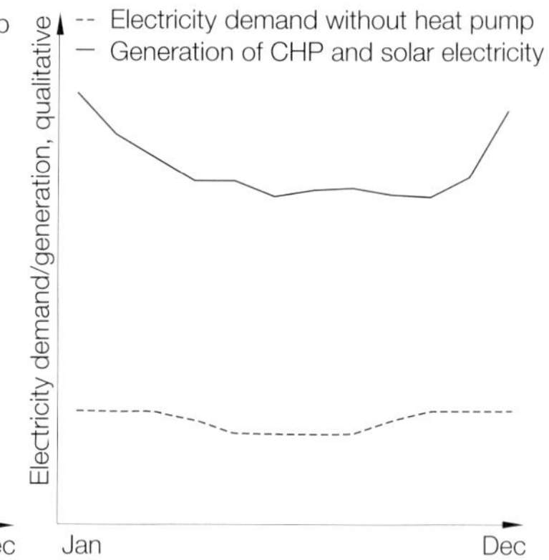

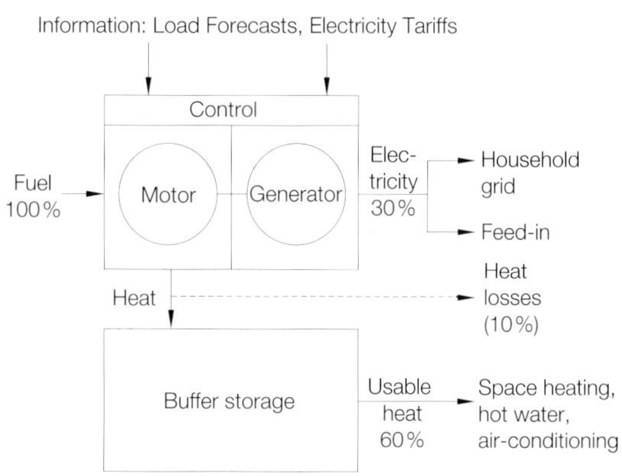

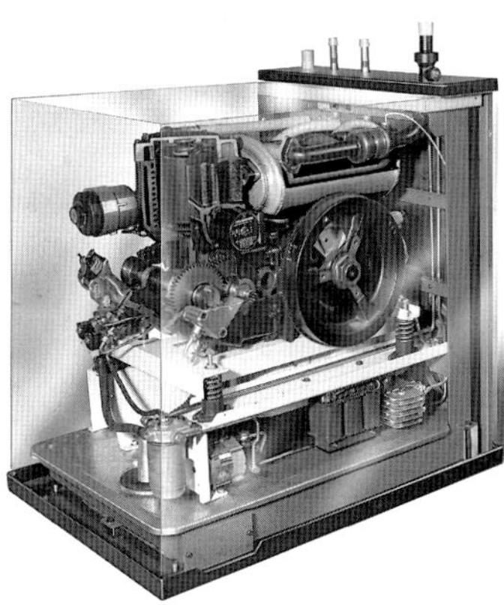

A 2.03
A 2.04
A 2.05 a
A 2.05 b
A 2.05 c

reliance on, for instance, natural gas during winter. However, electricity consumption of the required heat pumps increases the seasonal imbalance between electricity supply and potential of solar power feed-in. In winter, electricity consumption, and thus, the reliance on grids increases, while the load-match decreases. CHP-based solutions usually lead to a shortage of self-generated electricity supply in summer, because operating combined heat and power systems is typically influenced by heating requirements. When buildings aren't heated in summer and heat generation is only required for hot water, CHP systems will have short run times. Thus, a low degree of electricity is generated (Fig. A 2.05 a). This changes only in the case of systems that make use of summer waste heat to run a cooling system. With these systems, a demand for heat in summer remains. Such systems are complex and are, therefore, appropriate for large-scale projects, such as institutional buildings and hospitals [6].

THE BALANCING INDICATORS Since the overwhelming number of buildings today doesn't exclusively rely on electricity as energy source, balancing at the level of site energy is appropriate only in select cases (Figs. A 2.06 and A 2.08).

PRIMARY ENERGY Most countries, therefore, employ primary energy as a indicator for energy efficiency. The quantity of each energy source used in the building (site energy values) is converted into primary energy via primary energy factors and summed up. However, primary energy doesn't register on the meter or in the electricity bill. The case is similar for greenhouse gas emissions related to energy consumption. Both are calculated values.

The focus of primary energy calculation is typically on the non-renewable portion of supplied primary energy. This results in very low primary energy factors for biomass as fuel. While the procedure appears to be appropriate for using solar energy in buildings (primary energy factor 0), the limited availability of biomass from sustainable sources is becoming a mid-term problem [7]. In Europe, the potential of wood from sustainable forestry is still very high. In contrast, the current use of rapeseed oil or biogas already meets strict limits. In addition to calculating national primary energy factors, Switzerland employs policy-based weighting factors to create strategic incentives (national energy weighting factors) [8]. For example, within an energy weighting, wood pellets as an energy source are evaluated by a factor of 0.7, even though primary energy factors are calculated at only 0.3 [9].

GREENHOUSE GASES The transition to a CO_2-based weighting of buildings in the sense of a "zero emissions building" will turn climate change, rather than resource shortage, into the decisive issue.

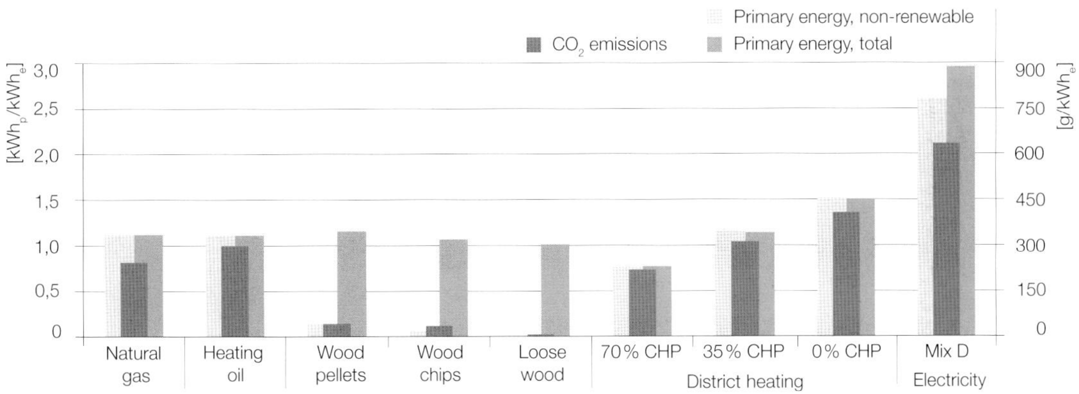

A 2.06 Calculated primary energy (non-renewable percentage and total cumulative energy expenditure in kWh primary energy per kWh site energy) and emission factors (CO_2 equivalent) following GEMIS 4.5, in the context of Germany's current electrical grid structure [10].

A 2.07 Primary energy factors and emission factors.

Large differences between weightings based on primary energy or emissions occur only in countries with an overwhelming share of nuclear power, e.g. France. Disaggregating greenhouse gases and energy consumption is otherwise only possible if the energy supply is based entirely on renewable energy sources. Most countries are still far away from this point.

The term zero emission isn't convincing in the case of buildings with central heating boilers and visible emissions from smokestacks, even if they employ biomass combustion, e.g. in a pellet stove, and its emissions are balanced by carbon credits. In addition to greenhouse gases, emissions also comprise health risks. In this context, the designation "climate neutral" is more accurate than "zero emission".

ENERGY COSTS The subject of energy costs and the zero energy cost house is adressed only in a limited number of cases [11]. Changes in energy costs according to time and political intervention in cost structures via tax rates and incentives influence balance calculations to a large extent. For example, a building with a primary energy balance deficit can neither incur energy costs nor a financial surplus if feed-in credits are very high. Conversely, a zero energy building may cause significant energy costs if feed-in credits are low.

DIFFERENT PRIMARY ENERGY FACTORS
Using established primary energy indicators, the question is raised which factors are applicable for evaluating the energy supply and grid feed-in of an actual building. Large differences occur especially in data on electrical grids, while data on other energy sources typically deviate to a smaller extent. An exception is Switzerland and its national energy weighting factors. In Europe, average factors for countries in the interconnected European electrical grid are often preferred when comparing buildings located in different countries (UCTE Mix, EN 15603). Depending on the respective country, these average factors deviate from the primary energy and emission factors of national grids. This leads to a different balance calculation (Fig. A 2.09, p. 32). Overall, the situation still appears very confusing and often leads to incorrect interpretations in building weightings, because data are not sufficiently transparent.

Concepts with different factors for supplied and feed-in energy are also possible. With the development of smart grids, time-related cost variations could also result in time-related primary energy and emission factors. These are already implemented in some American grids. Due to the plethora of possibilities, primary energy factors used for balance calculations of net zero energy buildings should always be itemized within the calculation.

INFOBOX **PRIMARY ENERGY AND EMISSION FACTORS**

The energy content of natural energy sources, such as natural gas, oil, or wood, is called primary energy. The energy that we use in buildings is called site energy. The non-dimensional primary energy factor expresses how much non-renewable primary energy is typically expended in order to supply a building with one unit of site energy. The lower the primary energy factor, the lower is single energy source dependency. The example of wood shows that this doesn't necessarily correlate with economic advantage. According to Fig. A 2.09, wood pellets are evaluated with a very low primary energy factor compared to natural gas. The factor for wood pellets is less than 20% of the factor for natural gas. Prices, however, typically reach 60% of natural gas prices. Conclusion: the magnitude of the primary energy parameter of a building does not mirror expected energy consumption costs. Climate gas emissions resulting from energy use are typically designated by the so called green house gas potential. In addition to CO_2 emissions, all other emissions are converted to a corresponding CO_2 quantity with equivalent greenhouse effect. The emission factor in units of g/kWh or kg/MWh expresses how much CO_2-equivalent emissions are generated by use of one unit of site energy. High emission factors are characteristic of electricity that isn't supplied by a grid that utilizes renewable energy sources.

A 2.06
A 2.07

A 2.08
A 2.09

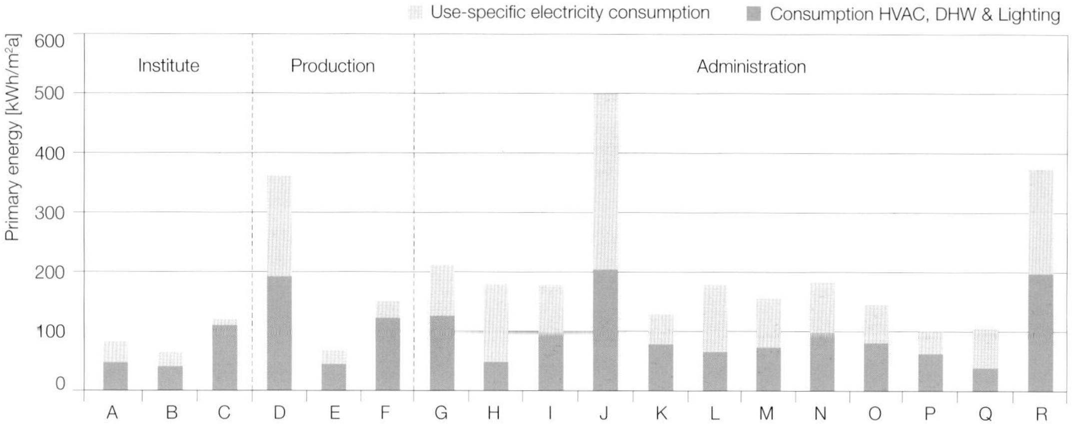

A 2.08 Summary of typical primary energy factors and emission factors for different countries and fuel sources. The differences are due to national electrical grid structure, calculation methods, and timestamps.

A 2.09 Measured primary energy parameters of selected buildings of the Energy Optimised Construction programme of the German Federal Ministry of Economics and Technology. Values are presented per net floor area (NFA) prior to climate-related generalisation, differentiated according to the normatively covered consumption sector of technical building equipment: space heating, hot water, ventilation, cooling, lighting (technical building equipment), and use-specific electricity consumption.

			Europe		**Austria**	**Denmark**	**Finland**		**Germany**[7]		**Italy**	**Norway**		**Spain**		**Sweden**		**Switzerland**	
			EN 15603 2008	PHPP 2007	Gemis Version 4.5[3)]	BR 2010 2010	BC 2012 2011	Gemis 2011	DIN V 18599/1 2007	GEMIS Version 4.5	UNI-TS -11300/4 draft 9/2009	NS 3700 2009	ZEB centre[10)] 2010-2050	I.D.A.E. 2010	CAL-ENER 2009	average[12)] 2008	pol. factors[13)] 2008	SIA 2031 2009	EnDK 2009
Electrical grid	PEI n.r.	kWh_p/kWh_s	3,14[1)]	2,70	1,3[4)]		1,70		2,60	2,61	2.18[8)]							2,53	2,00
	PEI total	kWh_p/kWh_s	3,31[1)]		1,91	2,50[5)]	1,70		3,00	2,96				2,28	2,60	1,50	2,50	2,97	
	CO_2 eq.	g/kWhs	617,00[1)]	680,00	389,00		329,62	331,00		633,00	531[9)]	395	132	350[11)]	649			154,00	
Natural gas	PEI n.r.	kWh_p/kWh_s	1,36	1,10	1,12		1,00		1,10	1,12	1,00							1,10	1,00
	PEI total	kWh_p/kWh_s	1,36		1,12	1,00	1,00		1,10	1,12				1,07	1,10			1,15	
	CO_2 eq.	g/kWhs	277,00	250,00	268,00		202[6)]	315,00		244,00		211		251[11)]	204,00			241,00	–
Heating oil	PEI n.r.	kWh_p/kWh_s	1,35	1,10	1,11		1,00		1,10	1,11	1,00							1,15	1,00
	PEI total	kWh_p/kWh_s	1,35		1,13	1,00	1,00		1,10	1,11				1,12	1,08	1,20	1,20	1,24	
	CO_2 eq.	g/kWh_s	330,00	310,00	302,00		279[6)]	381,00		302,00		284		342[11)]	287,00			295,00	
Timber	PEI n.r.	kWh_p/kWh_s	0,09[2)]	0,20	0,01		0,50		0,20	0,01	0,00							0,05	0,70
	PEI total	kWh_p/kWh_s	1,09[2)]		1,01	1,00	0,50		1,20	1,01				1,25		1,20	1,20	1,06	
	CO_2 eq.	g/kWh_s	14[2)]	50,00	6,00		32,40	17,00		6,00		14		0,00	0,00			11,00	
Wood pellets	PEI n.r.	kWh_p/kWh_s			0,14		0,50		0,20	0,14	0,00							0,30	0,70
	PEI total	kWh_p/kWh_s			1,16	1,00	0,50		1,20	1,16					0,00	1,20	1,20	1,22	
	CO_2 eq.	g/kWhs			41,00			19,00		41,00		14						36,00	
Disctrict heating	PEI n.r.	kWh_p/kWh_s		0,80	0,76				0,70	0,76	system specific							0,81[14)]	0,60
70% CHP	PEI total	kWh_p/kWh_s			0,77	1,00[5)]	0,70		0,70	0,77						0,90	1,00	0,8[14)]	
(fossil)	CO_2 eq.	g/kWh_s		240,00	219,00			230,00		219,00		231						162[14)]	

n.r.: non renewable
[1)] Electricity according to UCTE Mix 1996, CO_2-values for 2009: 432 g/kWh_{el}
[2)] Wood, general
[3)] Data from Environment Agency Austria
[4)] 60% hydro power in Austria's power generation, but 50% when considering imported power
[5)] 2015 requirements use 0,8; 2020 requirements use 0,6 for district heating and 1,8 for electricity
[6)] Based on Motiva report, 2004
[7)] The normative primary energy factors for the national building code are given in DIN V 18599, emission data are not listed; if emission data are applied, the most common source is GEMIS
[8)] Original source: AEEG – Autorità per l'energia elettrica e il gas – Aggiorrnamento del fattore di conversione 2008
[9)] Ministero dell'ambiente e della tutela del territorio e del mare
[10)] EU mix scenario for nearly carbon-free grid towards 2050 (in line with IPCC 450 ppm scenario): average 2010–2050
[11)] Carbon emissions only
[12)] Calculated according to EN 15316. For electricity: calculations based on Nordic electricity mix
[13)] Pol. factors are PEI promoted to/by the government as they believe that increased use of electricity means more fossil fuel use in condensing power plants.
[14)] Based on waste combustion

	CFA supply kWh/m²$_{CFA}$	NFA supply kWh/m²$_{NFA}$	Floor-area factor, CFA/NFA
Housing, multi-family house (MFH)	18,6	23,3	1,25 [1]
Housing, single-family house (SFH)	13,1	16,3	1,25 [1]
Administration	11,7	13,7	1,18
Schools	4,7	5,3	1,12
Retail	6,1	6,8	1,11 [1]
Restaurants	27,8	30,9	1,11 [1]
Community places	5,0	5,6	1,11 [1]
Hospitals	21,1	24,5	1,16
Industry	27,8	31,2	1,12 [2]
Storage	0,0	0,0	1,12
Sports buildings	1,1	1,2	1,10
Indoor swimming pools	4,4	5,1	1,14

[1] Estimated [2] Strongly dependent on type of production

A 2.10 Typical use-specific electricity consumption values (site energy). The values per unit of conditioned floor area (CFA) according to SIA 2031 [12] were converted to values per unit of net floor area (NFA) using floor-area factors [13].

A 2.11 Winning model of the Energy-Plus House with Electromobility 2010 competition of the German Federal Ministry of Transport, Building and Urban Development, by ILEK, University of Stuttgart.

THE BALANCE BOUNDARY Most normative energy balancing methods only include service technology energy demands and occasionally only parts of it. Typically, systems for heating and hot water as well as auxiliary energy for pumps and fans are included. As a result of the EU Energy Performance in Buildings Directive of 2002, ventilation, cooling, and lighting are also included for non-residential buildings [14]. The demands of use-specific devices (household appliances, information technology, production machinery) and centralised installations (server rooms, cold storage houses, escalators, elevators) are almost always omitted. Normative requirements mostly exclude photovoltaic systems on the generation side, because they are considered part of the electrical grid.

USE-SPECIFIC ENERGY CONSUMPTION In energy-efficient non-residential buildings, the examples in Fig. A 2.09 show that demands omitted by normative requirements average at 60 % of total primary energy consumption. Fig. A 2.10 shows typical values for use-specific electricity consumption based on standard data sets in SIA 2031 [15]. If these demands aren't represented in the balance, then two difficulties arise in particular:

- Calculated demand is difficult to compare with a building's metered consumption. Electrical meters register electricity without differentiating between types of consumption, but instead, based on electrical circuits or rental units. Thus, a normative balance is difficult to verify via measurement alone. The more successfully service technology energy demand is reduced, the greater the influence of use-specific demands will be. This effect applies not only to energy itself, but also to peak performance as important indicator for grid tasking.
- If the degree of electrical load-match is to be evaluated, limits oriented on the normative balance headroom result in calculating excess load-match, since significant demands aren't even included, even though they are part of the building feed-in. In a complete balance, surplus via self-generated electricity would be smaller, because either demand or consumption would be higher (see Load-Match as Criterion of Differentiation, p. 37).

Both aspects indicate that the balance boundary for net zero energy buildings should be expanded to all types of consumption particularly in the case of electrical applications. The SIA standards in Switzerland and the Passive House Planning Package (PPHP) have implemented this and orient planning on standard values (SIA) [16] or project-specific planning data (PHPP) [17].

ELECTROMOBILITY Initial projects already include electromobility as a component of building energy balance calculations. One example is the winning design for the Plus-Energy House with Electromobility competition [18]. The building is scheduled for a 2011 completion date as a showcase and temporary building in Berlin (Fig. A 2.11). The energy concept integrates buffer batteries in the building and in the electrical vehicles. Using the car as mobile CHP unit is not intended (see Excourse: Mobile Combined Heat and Power, p. 24). Electrical vehicles are included as demands similar to household appliances, because they are powered by the same grid when their batteries are charged at the house. Their electricity consumption is then metered by the building's typical electricity meter. In contrast, charging at electrical charger stations isn't taken into account. Their battery storage devices can be included in the energy management, making self-produced building electricity surplus economically usable when feed-in credits are low.

EXTERNAL ENERGY PRODUCTION SYSTEMS
In contrast, the expansion of balance boundary to systems external to the actual project (purchase of green power, shares in wind farms, and similar) doesn't seem convincing in the context of a building energy weighting, because such systems are part of the grid and don't prioritize coverage of a balanced building's demands. They feed power from beyond the building's internal grid and aren't registered by its meters (see Fig. A 1.33, p. 26). By doing so, they utilize the transport and storage capacity of grids and reduce the primary energy factor as well as

A 2.10
A 2.11

A 2.12

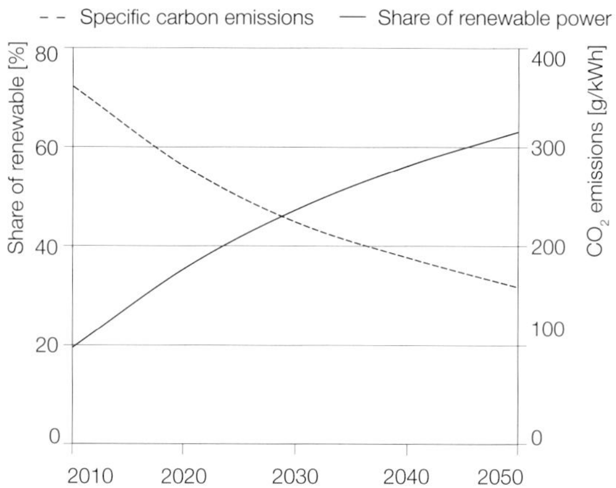

A 2.12 For two development scenarios of the interconnected European electrical grid, time dependencies of the primary energy factor and greenhouse gas emissions

A 2.13 Energy life cycle analysis for the University of Wuppertal house for the Solar Decathlon 2010 in Madrid. The sawtooth line for maintenance results from typical renovation cycles and influences the development of the total energy balance.

A 2.14 Life cycle analyses for three buildings featured in the project section of this book over a period of 60 years. Data were annualised for better comparison in relation to net floor area [19–21].

corresponding emissions. These effects are already taken into account on the expenditure side in the building's energy balance via related weighting factors for primary energy and emissions. Similarly, integrating a photovoltaic system in a building balance appears problematic if its goal isn't primarily covering the building demand, but instead prioritized grid feed-in. This applies, for example, to systems owned by energy providers or service associations on rented rooftops (see Renovation of Blaue Heimat, p. 78ff.).
The icons at the beginning of the "Projects and Lessons Learned" (p. 49) chapter of this book serve to point out which demands and generators are taken into account in each of the individual balance calculations.

THE BALANCING PERIOD In most cases, zero energy buildings are defined by an annual zero energy balance.

THE GRID OF THE FUTURE If net zero energy buildings become an issue within future scenarios, then the time variability of primary energy factors of grids will play a role. The growing percentage of electrical power from renewable sources and improved integration of infrastructure enable reduction of primary energy and emission factors of grids, but also credits for electricity or heating grid feed-in (Fig. A 2.12) [22]. While this effect doesn't change balances in all-electric buildings, it adversely affects balances in cases in which credits from electricity feed-in compensate the supply of a different energy source, such as wood pellets for a central heating boiler or natural gas for CHP. In the future, such a building would require greater energy efficiency for decreasing credits to remain sufficient for a zero balance in the case of equal feed-in. In the context of projections on future grids, such effects could be included in planning for net zero energy buildings in a progressive way.

THE INFLUENCE OF EMBODIED ENERGY ON THE LIFE CYCLE BALANCE Energy for construction, maintenance, and demolition of buildings, so-called embodied energy, isn't typically considered part of an energy balance [23]. Across the entire life cycle of a property, however, its percentage increases when service energy demands decrease and embodied energy increases, which happens in most cases [24, 25]. In combination with replacements and renovations that typically take place throughout a building's life cycle, embodied energy usually comprises 20–40 % of the total primary energy expense for energy efficient buildings for an assumed 60 years of service. Calculated for an annual figure and related to floor area, embodied energy comprises 20 to 50 kWh/m². Energy demands for maintenance and construction are approximately equal, because nearly all components must be replaced once or multiple times, with the exception of the building shell. The values for embodied energy vary considerably depending on construction type (wood or solid construction) and building-specific features (for example, with or without an underground garage). Comparative calculations show that these differences become greater than the additional primary energy expenditure required for a net zero energy building. This additional energy expenditure may result from increased insulation strength or installing a photovoltaic system.

If the zero energy standard is to be achieved in terms of a zero balance throughout the entire service cycle, the annual surplus of service energy balances must offset the energy expenses for construction and maintenance. This entails that the energy plus standard must be achieved in the annual service energy balance. The experimental houses for the Solar Decathlon in Madrid described in the chapter "Solar Decathlon Europe" (p. 168ff.) are capable of achieving this standard, since they feature relatively large systems for energy generation at simultaneously low energy demands. Fig. A 2.13 displays an analysis for the house developed for the Solar Decathlon by the University of Wuppertal [26]. After approximately twelve years of service and operation, the initial energy expenditure will be offset and the annual surplus will significantly exceed maintenance

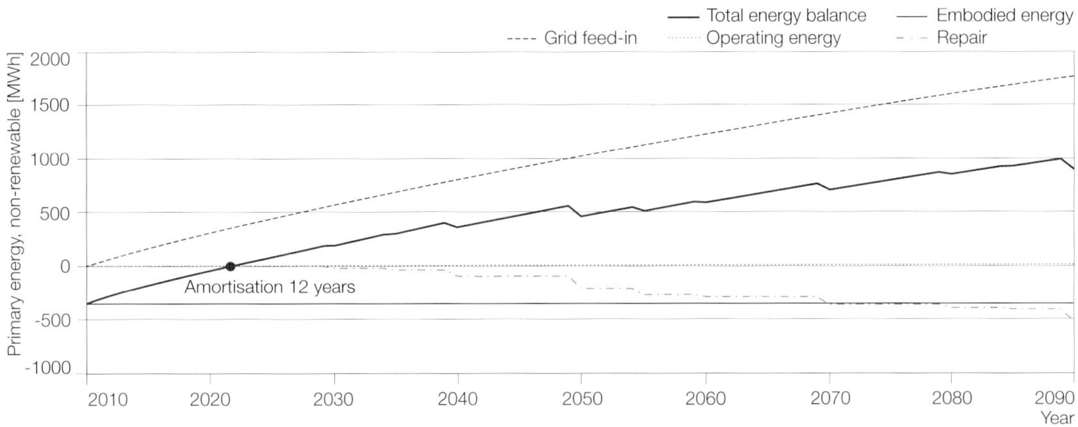

A 2.13

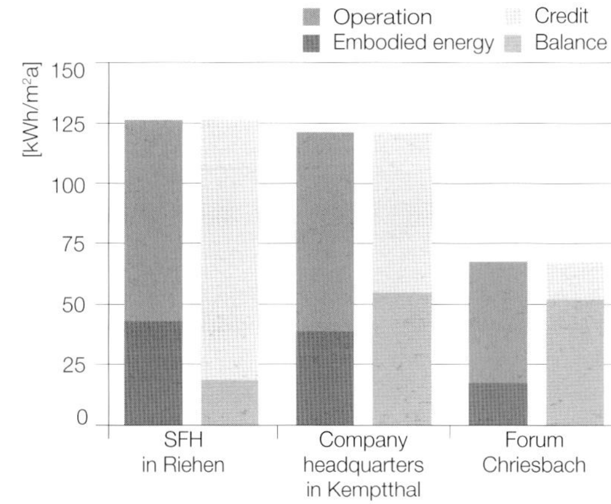

A 2.14

expenses. For a location in Germany, the balance wouldn't reach zero in the long term, and the net zero energy building status wouldn't be achieved throughout its life cycle. The small, energy-efficient buildings benefit from their relatively large photovoltaic systems (210–250 W_p/m^2_{NFA}). Such sizes, however, are rare in typical building geometries (see Fig. B 1.07, p. 53).

Fig. A 2.14 displays an overall calculation of expenditures for a life cycle of 60 years and differentiated annually for three Swiss buildings discussed in this book (see the Residention House in Riehen, p. 56ff.; Corporate Headquarters in Kemptthal, p. 120ff.; and Forum Chriesbach, p. 17). For the Corporate Headquarters in Kemptthal, credits compensate service energy, but not embodied energy. The solar power feed-in of Forum Chriesbach is also too low for balancing service energy. The building wasn't planned as a zero energy building. It primarily features low consumption values for both operations as well as embodied energy. The single-family house in Riehen generates a surplus of service energy. This surplus covers more than half of embodied energy, but is insufficient for a complete balance, primarily due to required periodic maintenance and renovation.
Because of their comparatively high embodied energy, the size of photovoltaic systems on buildings becomes a significant factor. Typical values for embodied energy today are on the order of 6 to 8 MWh primary energy per kW_p [27, 28]. For an annual power yield of approximately 950 kWh/kW_p in a suitable central European location, and with an assumed primary energy factor of 2.6 for electricity feed-in, calculations indicate an energy offset timeframe of 2.5 to 3.5 years. In this context, locations in southern Europe are less critical than those in northern Europe, because significantly higher yields offset corresponding embodied energy of photovoltaic systems. Critical countries are those with low primary energy factors, and thus, low credits forgrid electricity replaced by feed-in. Particular building certifications, such as the Swiss MINERGIE-A label [29] or the German DGNB label [30], already take embodied energy into account. Since collecting data requires significant efforts, these certifications are often based on generic values.
When renovating existing buildings, the high energy value of the building shell, interpreted as credit, considerably improves the result of a total energy balance calculation compared to new construction. Shell construction comprises approximately one-fourth of a building's embodied energy [31]. Therefore, in the mid-term, a balance calculation oriented on a building's life cycle appears expedient, in order to adequately reflect the decision between new construction or renovation. However, as the previously analysed buildings show, achieving zero energy building status within a life cycle balance calculation will, in most cases, be difficult to achieve.

LOAD-MATCH AS CRITERION OF DIFFERENTIATION Grid interconnection enables using decentralised energy provision in a net zero energy building to fractionally reduce external energy supply and for feed-in. To describe this differentiation process, a "load-match factor" (f) is applied within the following equation for each separate energy source in use:

$$f = \text{Minimum}\left[1, \frac{\text{self-generation}}{\text{self-consumption}}\right] \cdot 100\ [\%]$$

A maximum of 100 % is achieved in the observed period. Surplus isn't included, but instead calculated and added separately. The equation expresses this by comparing the calculated ratio to "1", while the lower of the two is considered valid [32]. This method is explained in detail in the chapter on "Energy Balancing: Practice, Standardisation, and Legislation" (pp. 40ff.). Strictly spoken, a separate addition of generation and consumption across a defined period leads to a "virtual" load-match. For example, if consumption and generation are equal for one particular month, then the load-match would be 100 %. In reality, however, electricity is occasionally supplied by a grid, for example at night, and surplus is fed into a grid, for example at noon. The formation of virtual load-match serves to simplify

A 2.15 Measured electricity consumption and generation values for the Solar XXI institute building in Lisbon, Portugal 2006, Architect: Pedro Cabrito e Isabel Diniz, Arquitectos. The building features 1100 m^2 area, has a 12 kW_p photovoltaic facade, and a 6 kW_p system that doubles as parking lot shading.
a Elevation
b Monthly values
c Cumulative values, 100 % = annual load

A 2.16 Measured electricity consumption and generation values for the Adam Joseph Lewis Center for Environmental Studies at Oberlin College, Ohio, USA 2001, William McDonough. Building I, with a net floor area of 1265 m^2, has a 160 kW_p photovoltaic system on the roof and above a parking deck.
a Elevation
b Monthly values
c Cumulative values, 100 % = annual generation

A 2.17 Calculated electricity consumption and generation profiles for a 1000 m2 office building with differing heating and cooling systems and a photovoltaic system that offers complete annual coverage of electricity demands at a central European location. Calculation with Ener-Calc (see Calculating Tool, p. 42)
a Gas heating, no cooling, photovoltaic system 25 kW_p: load-match 79 %
b Heat pump, no cooling, photovoltaic system 49 kW_p: load-match 60 %
c Heat pump, cooling, photovoltaic system 56 kW_p: load-match 69 %

A 2.15a
A 2.15b
A 2.15c

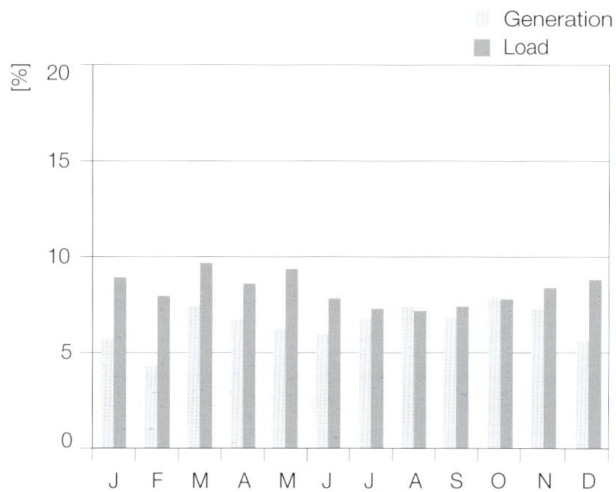

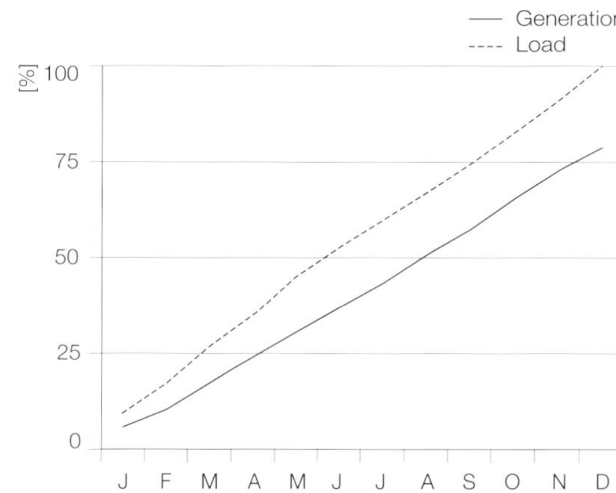

A 2.16a
A 2.16b
A 2.16c

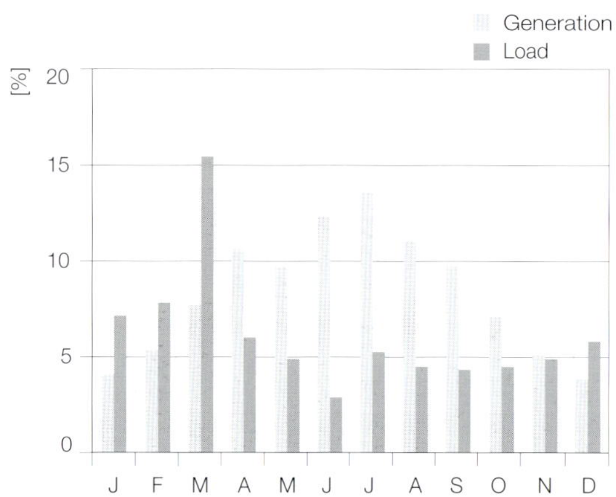

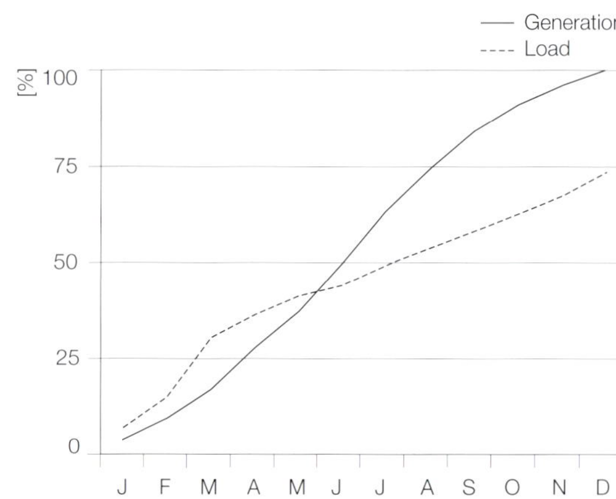

calculations during the planning phase. Demand and generation can be calculated separately and then compared. An integrated calculation requires extensive knowledge of time-related demand profiles. The load-match is influenced by three parameters:

- consumption profile and, thus, building use
- generation profile (photovoltaic system, CHP system)
- observation period.

Figs. A 2.15 and A 2.16 exemplarily illustrate monthly metered distribution of consumption and generation for a school in Ohio [33] and for an institutional building in Lisbon, Portugal [34]. Both are all-electric buildings. Consumption includes use-specific consumption. In the case of the school, annual generation exceeds consumption. The opposite occurs in the institutional building. The monthly load-match of the institutional building is 79 % on average, which equals the annual value. This is due to the seasonally balanced solar intake in Lisbon. The load-match for the school comprises 87 %. Load-match can be determined similarly for buildings with CHP units.

THE SELECTION OF BUILDING TECHNOLOGY INFLUENCES LOAD-MATCH The load-match for electrical power is determined by building technology choices, since these define the building's electrical energy demand as well as its seasonal profile. This is illustrated by the calculations shown in Fig. A 2.17 for an office building with different means of heating and air-conditioning. The photovoltaic system is designed precisely to enable a complete annual balance of electrical energy demands of HVAC, DHW, and lighting. The higher the consumption, the larger the size of the photovoltaic system. Use-specific consumption values aren't considered here.
By installing a heat pump, electricity demands increase significantly compared to a gas condensing-boiler heating system, especially during winter. In combination with geothermal energy, electricity replaces gas, and the photovoltaic system becomes bigger. The load-match decreases due to the seasonal imbalance of consumption and generation. For active cooling, however, the demand increases along with the load-match, because additional demand coincides with phases of high generation.

ACTUAL LOAD-MATCH IS SMALLER Even for weekly and daily balancing of generation and consumption, similar load-match factors are achieved for annual averages. When observing net metering values for specific times, the load-match for photovoltaic systems reaches approximately 30 % or lower (actual load-match – net metering). This result is essentially due to demand peaks that aren't covered by self-generation as well as nighttime total grid supply. Such effects are hidden in the observed time-related averages. Even larger size photovoltaic systems hardly change this [35].
Improving a building's capacity to cover its self-consumption is achieved by demand site management. Targeted activation and deactivation of devices enables shifting electricity consumption from peak-load times to low-load times, as long as certain devices aren't active during peak-load times. Another alternative is to buffer loads by internal electricity storage (see Grid Integration, below).
The introduction of a load-match leads to a differentiation of self-generated electricity. This permits identifying zero energy buildings that achieve net zero energy consumption to a large extent by self-generation coverage, as well as those that essentially utilize a grid's seasonal storage function. Because planning-related calculations of short-term data for generation and consumption aren't yet a standard procedure, and considering mostly minor differences between monthly, weekly, or daily load-matches, monthly differentiation seems appropriate. Monthly values indicate the demand on a grid's seasonal storage capacity for a building to obtain its annual net zero energy balance. In the Infobox: Definition of Zero- Energy Building (p. 41), this differentiation is introduced in the context of a standardised description of net zero energy buildings and later explained individually according to documented project examples.

GRID INTEGRATION As stated earlier, in addition to reducing energy consumption, optimising grid feed-in is a planning goal of net zero energy buildings. This applies not only to the quantity of energy

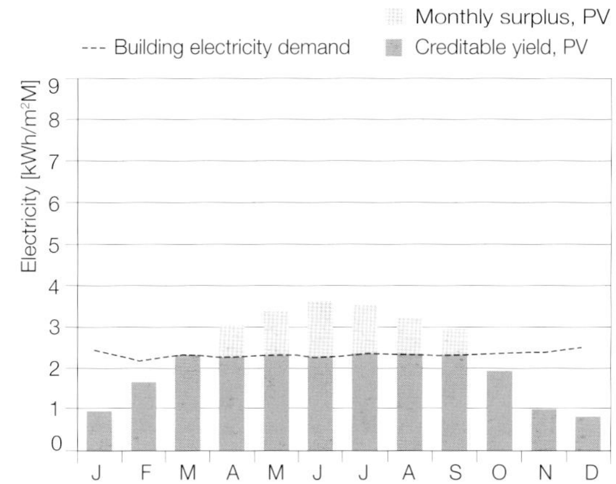

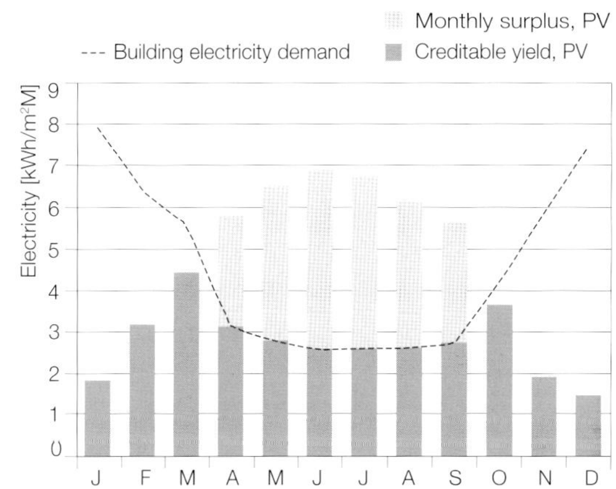

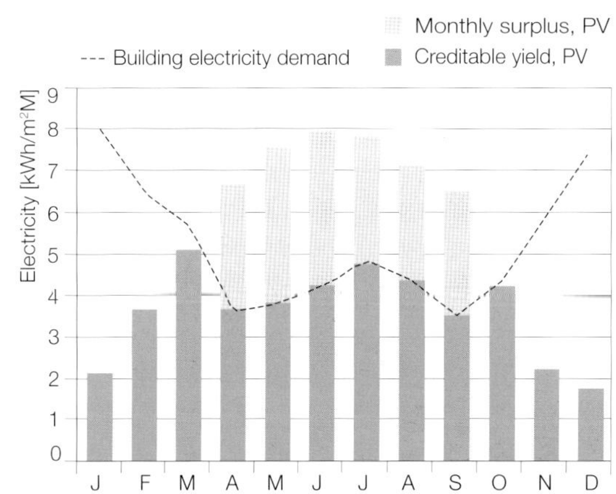

A 2.17a
A 2.17b
A 2.17c

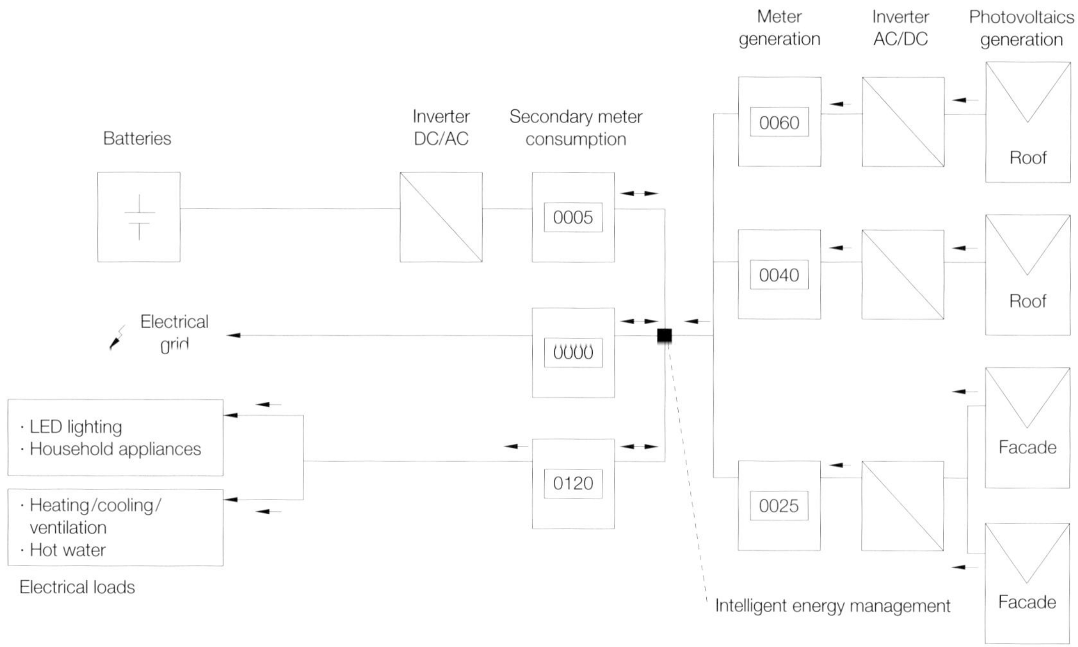

A 2.18
A 2.19

A 2.18 Schematic diagram of the electrical system for the University of Wuppertal house for the Solar Decathlon 2010

A 2.19 Calculated weekly load-match for electricity con-sumption of the Solar Decathlon building with and without battery buffering in
a Madrid
b Wuppertal

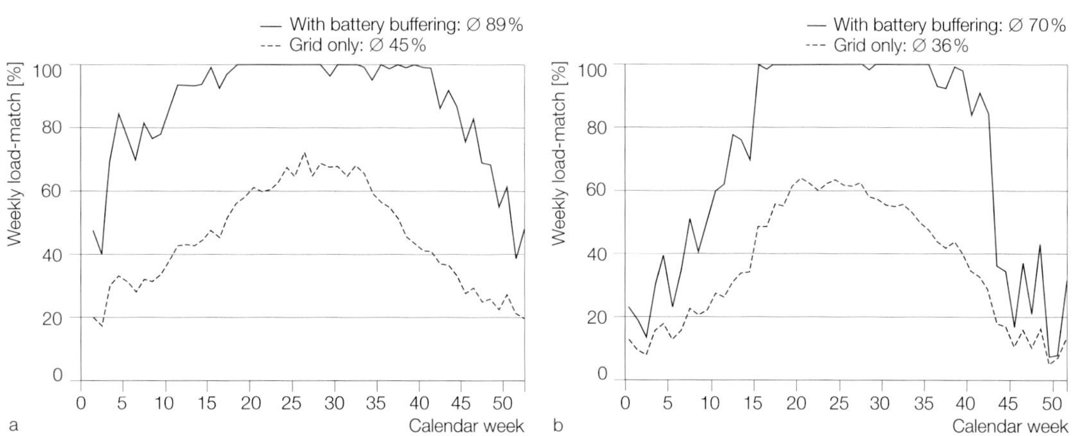

exported, but also timing of electricity or heat feed-in and demand. The fluctuation range of supply and generation values [36] comprises an initial indicator for grid feedback. However, this doesn't permit an authoritative differentiation, because weighting the effects that a building exerts on a grid also depends on local grid infrastructure and the mix of energy sources of each grid.
The required peak supply is a decisive factor, since it determines the grid's required capacity. Active load management and building-integrated energy storage can influence peak power supplies and time-related profiles of feed-in and consumption. While thermal energy storage in buildings exists and offers potential for expansion via new technologies (see "Zero Energy Extreme: Self-Sufficient Buildings", p. 18), the question is raised how decentralized power storage devices can serve for short-term storage of electrical energy. For this purpose, batteries in buildings as well as vehicles are currently subject to discussion and testing [37]. Weather forecasts as an element of a building's control system, combined with load and generation management, permit prediction of load and generation profiles.
Fig. A 2.18 features a schematic diagram of the electrical power system for the University of Wuppertal's Solar Decathlon 2010 building entry. A 7.2-kWh battery pack, in combination with an intelligent energy management system, enables prioritized self-demand coverage via self-generated solar energy. Calculations show that slightly less than half of electricity consumption would be covered by self-generated solar power without battery buffering. In Madrid, battery systems increase this percentage to around 90%, primarily because they supply electricity at night. In Wuppertal, the percentage can be increased from 36 to 70% (Fig. A 2.19). Similarly, grid feed-in can be shifted to times with the highest feed-in tariff, while electricity consumption can be shifted to times when tariffs are low [38].
Weighting which efforts invested in a building seem justified will only be possible once grids have been converted to supply a higher rate of renewable energy. In the current context, such solutions seem premature, due to the considerable effort and high costs involved.

ENDNOTES

[1] Sartori, Igor; et al.: Criteria for Definition of Net Zero Energy Buildings. Proceedings of EuroSun 2010. Graz 2010.
[2] Voss, Karsten; Musall, Eike; Lichtmeß, Markus: Vom -Niedrigenergie- zum Nullenergiehaus. Standortbestimmung und Entwicklungsperspektiven [From low-energy to zero-energy house. Siting and development perspectives]. In: Bauphysik [Construction Physics] 06/2010, pp. 24–34.
[3] http://greensource.construction.com/projects/0710_NorthApts.asp. Accessed February 17, 2011
[4] Lenz, Bernhard; Schreiber, Jürgen; Stark, Thomas: Nachhaltige Gebäudetechnik [Sustainable Building Technology]. Munich 2010.
[5] Musall, Eike et al.: Strom-wärmegeführter Betrieb. Voraussetzung für Wirtschaftlichkeit [Electricity-heat-led operation. Prerequisite for cost-effectiveness]. In: Euro Heat & Power 04/2010, pp. 50–56.
[6] ibid.
[7] Lücking, Rolf-Michael; Hauser, Gerd: Nachhaltige Energieversorgung von Gebäuden [Sustainable energy supply of buildings]. In: Technik am Bau [Technology in Construction] 10/2009, pp. 62–66.
[8] Institut Wohnen und Umwelt [Institute for Housing and Environment]: Kumulierter Energieaufwand und CO_2-Emissionsfaktoren verschiedener Energieträger und -versorgungen [Cumulative energy expenditure and CO_2 emission factors of different energy sources and supplies]. Darmstadt 2009.
[9] Federal Office of Energy: Gebäudeenergieausweise der Kantone [Building Energy Certifications of Administrative Districts]. Bern 2009.
[10] Schweizer Ingenieur- und Architektenverein: SIA 2031 [Swiss Association of Engineers and Architects]. Energieausweis für Gebäude [Energy Certification for Buildings], SIA 2040: Effizienzpfad Energie [Efficiency Path Energy]. Zurich 2009.
[11] Crawley, Dru; Pless, Shanti; Torcellini, Paul: Getting to Net Zero. In: ASHRAE Journal 51, 9/2009, pp. 18–25.
[12] European Union (EU): Directive 2002/91/EG of the European Parliament and of the Advisory Board of December 16, 2002, on the total energy efficiency of buildings. http://eur-lex.europa.eu/LexUriServ/LexUriServ.do?uri=OJ:L:2003:001:0065:0065:DE:PDF. Accessed March 2, 2011
[13] Federal Ministry of Transport, Building, and Urban Development (BMVBS): Bekanntmachung der Regeln für Energieverbrauchskennwerte und der Vergleichswerte im Nichtwohngebäudebestand [Announcement of Rules for Energy-Consumption Parameters and the Comparison Values in Existing Non-Residential Buildings]. Berlin 2009.
[14] cf. [10]
[15] cf. [10]
[16] cf. [10]
[17] Passivhaus Institut [Passive House Institute]: Passive House Planning Package PHPP. Version 2007. Darmstadt 2007.
[18] Güttinger, Herbert; et al.: Energie-Detailbilanz des Eawag Forum Chriesbach [Detailed Energy Balance for the Eawag Forum Chriesbach]. Swiss Federal Office of Energy. Dübendorf 2009.
[19] Primas, Alex: Berechnung der Grauen Energie [Calculation of Grey Energy]. Zurich 2010.
[20] Hall, Monika: University of Applied Sciences Northwestern Switzerland, Personal communication of building owners, 2010.
[21] Federal Ministry of Transport, Building, and Urban Development (BMVBS): Plusenergiehaus mit E-Mobilität. Auslobungstext zum Wettbewerbsverfahren im Rahmen der Forschungsinitiative »Zukunft Bau« [Energy-plus House with Electromobility. Text on Awarding Prizes for the Contest in the Scope of the "Future Construction" Research Initiative]. Berlin 2010.
[22] Graabak, Ingeborg; Feilberg, Nicolai: CO_2-Emissions in Different Scenarios of Electricity Generation in Europe, SINTEF report TRA 7058, 2011.
[23] Marszal, Anna; et al.: Zero Energy Building. A review of definitions and calculation methodologies. In: Energy & Buildings 43, 2011, pp. 971–979.
[24] König, Holger; et al.: Lebenszyklusanalyse in der Gebäudeplanung [Life cycle Analysis in Building Planning]. Munich 2009.
[25] Sartori, Igor; Hestnes, Anne-Grete: Energy use in the life cycle of conventional and low-energy buildings. A review article. In: Energy and Buildings 39, 2007, pp. 249–257.
[26] University of Wuppertal; Hochschule Rosenheim; Hochschule für Technik Stuttgart; Hochschule für Technik und Wirtschaft Berlin (Ed.): Solararchitektur 4 [Solar Architecture 4] Munich 2011.
[27] http://www.bauteilkatalog.ch. Accessed March 2, 2011
[28] Alsema, Erik; de Wild-Scholten, Mariska: Reduction on Environmental Impacts of Crystalline Silicon Photovoltaic Technology. http://www.ecn.nl/publicaties/default.aspx?nr=ECN-M--07-059. Accessed March 2, 2011
[29] MINERGIE-A: www.minergie.ch. Accessed March 2, 2011
[30] Ebert, Thilo; Eßig, Nathalie; Hauser, Gerd: Zertifizierungssysteme für Gebäude [Certification Systems for Buildings]. Munich 2010.
[31] Kugel, Daniela: Lebenszyklusbetrachtungen als ökonomisches und ökologisches Einsparpotential in der Gebäu-deplanung und -bewirtschaftung [Life Cycle Considerations as Economic and Ecological Saving Potential in Building Planning and Management]. Master Thesis, University of Wuppertal 2007.
[32] Voss, Karsten; et al.: Load Matching and Grid Interaction of Net Zero Energy Buildings. Proceedings of the EuroSun Conference. Graz 2010.
[33] Pless, Shanti; Torcellini, Paul: Energy Performance Evaluation of a Low-Energy Academic Building. ASHRAE Winter Meeting Chicago, Illinois 2006.
[34] Gonçalves, Helder et al.: Solar XXI. Towards Zero Energy. Lisbon 2010. http://www.lneg.pt/iedt/?unidade=17695. Accessed 2.03.2011
[35] Widén, Joakim; Wäckelgard, Ewa; Lund, Peter D.: Options for improving the load match capability of distributed photovoltaic. In: Solar Energy Journal 83, 2009, pp. 1954–1966.
[36] cf. [23]
[37] Multifunktionale Wechselrichter und Speicher für Solarstrom [Multifunctional Inverters and Storage for Solar Power]. In: BINE Projektinfo [BINE Project Info]. 10/2010. http://www.bine.info. Accessed March 2, 2011
[38] cf. [26]

ENERGY BALANCING: PRACTICE, STANDARDISATION, AND LEGISLATION

The following three sections describe methods for energy balancing according to examples of completed projects. They also indicate interrelations between respective normative methods in Germany, Switzerland, and Austria. By doing so, they illustrate current trends.

GERMANY The net zero energy buildings completed in Germany to date usually achieve an annual zero primary energy balance via normative primary energy factors. Complying with balance limits for normative energy demand calculations, most projects only include consumption values for space heating, domestic hot water (DHW), cooling, ventilation, and lighting. Selected projects already focus on total energy consumption, including use-specific consumption, or even integrate mobility within a balance. On-site electricity generation usually enters an energy balance independently of energy demand, regardless of coverage rate. This process doesn't correspond to normative requirements for issuing an energy certificate, but instead comprises a supplemental, voluntary calculation.
Apart from a few selected examples, energy balances aren't connected to fixed energy demand limits. The focus here is on passive houses or energy efficient buildings with energy values significantly lower than legally required. Their energy demand is typically covered by on-site generation. In exceptional cases, shares in wind power generation or investments in external technologies, such as biogas production or combined heat and power, supplement the on-site balance (see Projects and Lessons Learned, p. 56ff.).

LEGISLATION AND STANDARDISATION The basis for energy weighting of buildings in Germany is the Energy Savings Directive (EnEV) as amended on October 1, 2009 (Fig. A 3.04, p. 42) [1]. The requirement of cost-effectiveness in the energy savings law means that the directive can't demand energy savings measures if they cost too much. In this regard, the introduction of net zero energy buildings should be complemented by suitable incentives, increased energy taxes, or similar measures. The Law on Promoting Renewable Energies in the Heat Sector (EEWärmeG) demands a minimum share of renewable energy in the heating supply for new construction, but also permits compensation [2]. The Law for Prioritizing Electricity from Renewable Energy Sources (EEG) and in the Law for Continuation, Modernisation, and Expansion of Combined Heat and Power Installations (KWKG) regulates compensatory practices for feeding electricity into the supply grid. The Energy Savings Directive takes into account all relevant energy consumption covered by

A 3.01 This commercial property from 1963 was renovated for office use covering 1143 m² net floor area. After renovation, energy consumption is measured at 95 kWh/m²a for space heating and 23 kWh/m²a for electricity. With a 165-kW wood pellet boiler and a 44-kW_p rooftop-mount photovoltaic system, annual primary energy consumption is more than compensated. Freiburg, Germany 2006, Architect: hotz + architekten, Energy concept: Stahl & Weiss

A 3.01
A 3.02

A 3.02 Definition of a zero energy house

calculations following DIN V 18599 [3]. These include energy used for a building's space heating, DHW, cooling, ventilation, and lighting demands. Electricity consumption from electronic devices such as computers, household appliances, or building infrastructure such as elevators and escalators aren't included. Also, a building's embodied energy isn't taken into account. Calculations are based on monthly energy balances with results summed up as annual values. Primary energy requirement levels are determined by an analogous reference building that serves as basis for normatively defined reference construction and mechanical equipment [4]. Crediting electricity from renewable energies was introduced in 2009 in §5 of the Energy Savings Directive. Until then, photovoltaic systems were considered part of the electrical grid, due to their feeding generated electricity entirely into grids. Thus, they were omitted in a building's energy balance. Exported electricity reduces the use of fossil or nuclear fuels, and thus reduces the primary energy factor and CO_2 emissions of the building's grid electricity demand. The EEG [5] regulates purchase of and payments for electricity generated from renewable energy sources. The EEG sets the price for exported electricity and, in the most current version, also promotes using electricity for building self-consumption. §5 of the Energy Savings Directive states: "In buildings planned for construction, if electricity from renewable energy sources is used, then electricity may be deducted in calculations […] from the site energy demand if it

1. is generated in direct spatial relation to the building and
2. is primarily used in the building itself and only surplus energy is fed into a public grid.

No more than electricity according to requirement 1 that corresponds to a calculated electricity demand may be credited."

The German Institute for Civil Engineering (Deutsches Institut für Bautechnik) comments on this as follows [6]:

- A "direct spatial relation" exists if self-consumed electricity isn't carried via a public distribution grid. Installation of corresponding meters allows distinction between electricity "used in the building itself" and electricity "fed into a public grid." Installations on the level of entire developments or urban quarters, i.e. generation systems designed for multiple buildings, can then also be taken into account. In this regard, it is irrelevant whether the generation system is operated by the owners or a third party.
- The annual primary energy demand according to the Energy Savings Directive is calculated by month, as well as deducting self-generated electricity from renewable sources from the electricity demand. The maximum deductible electricity quantity results of calculated electricity demands.

Self-generation of electricity with integrated combined heat and power (CHP) systems is also taken into account in DIN V 18599. The central idea is that electricity generation coupled with CHP space heating supply is credited to the primary energy supply (in the form of fuel) used for operating the CHP. Crediting leads to deducting the primary energy equivalent of generated electricity from the primary energy equivalent of fuel for CHP operation. To simplify, crediting takes place in the context of an annual balance. Surprisingly, in contrast to the preceding procedure for photovoltaic electricity, monthly surplus in CHP electricity generation isn't limited according to a building's electricity demand. This rule applies regardless of whether fuel used for the CHP system comes from renewable sources such as rapeseed oil or wood pellets, or non-renewable sources. Payment for electricity exports from CHP systems is regulated by the Combined Heat and Power Act (KWKG 2010 [7]). The law is intended to promote new construction and modernisation of CHP systems with the goal of CHP-generated electricity

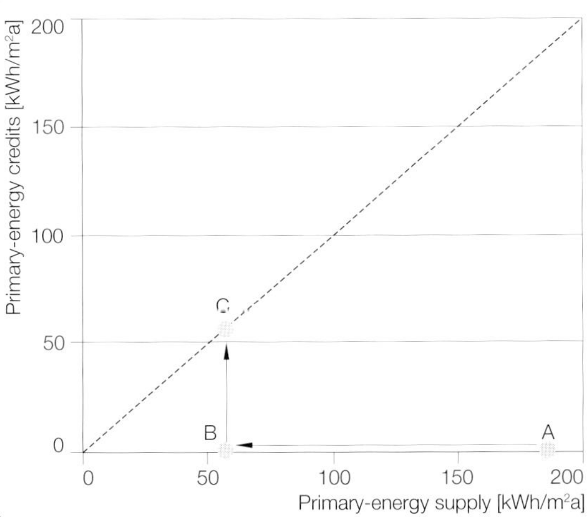

INFOBOX **DEFINITION OF A ZERO ENERGY BUILDING**
For this publication, a zero energy building is defined as an energy-efficient building that, within its annual balance sum, covers its entire annual primary energy demand in connection to the electrical grid and further grids if required, based on a monthly balance via primary energy credits for surplus energy feed-in. On-site energy provision is to prioritize coverage of building self-demand.

PROPOSAL FOR LABELING A zero energy building is designated by the acronym Net ZEB and by declaring the required or actual primary energy credits for a zero deficit in the annual energy balance. These are determined on the basis of a monthly balancing method that takes prioritized coverage of building self-consumption into account. This doesn't include annual surplus. The declaration on primary energy credits indicates that the designation is the result of a balance calculation.

EXAMPLE As a zero energy house, the 48 m^2 experimental house of the University of Wuppertal for the Solar Decathlon 2010 (see Solar Decathlon Europe, p. 168ff.) has a calculated annual total primary energy demand of 188 kWh/m^2 including household electricity, based on Wuppertal as location (A). The location choice explains the comparatively high balance value. The demand reduced according to EnEV § 5 by the photovoltaic system's monthly creditable yield is 131 kWh/m^2. As result, a demand of 57 kWh/m^2 remains (B). It is offset by the photovoltaic system's sum of monthly surplus fed into the grid and is used for seasonal compensation of the balance (C). The proposed correct designation for this house is: Net ZEB 57. The fact that the house generates higher surplus isn't included in this designation, since it is considered part of the grid, and thus, reduces its primary energy factor.

providing up to 25 % of total electricity supply in Germany by 2020.

As of 2010, there were no definitions or calculation methodologies for net zero energy buildings within related legislation or standards. Because summer solar electricity surplus isn't permitted in balancing calculated winter deficits, a net zero energy building can't actually be achieved in its strict normative sense in local climate. Even offsetting solar electricity grid feed-in against fossil fuel supply is excluded from the primary energy balance. Credits are only available for CHP electricity. However, the ratio of electricity demand versus space heating demand, for most buildings, is such that the electricity generation possible with CHP also doesn't cover the total demand in the annual balance.

CONCEPT FOR REDEFINING THE ZERO ENERGY BUILDING In the context of the monthly energy balancing method introduced in DIN V 18599, the authors propose a definition of zero energy building in this publication (Fig. A 3.02, p. 41). It is employed for comparison between numerous projects presented in the "Projects and Lessons Learned" (p. 56ff.) section in this book [8]. It remains to be seen if this definition will contribute to the development of future normative procedures. The definition is based on the balancing procedure according to Fig. A 2.02 (p. 29), i.e. a separate monthly meter balance for generation and consumption. Other than previous normative procedures, solar energy and combined heat and power are treated equally here. On-site electricity generation based on monthly balances is credited, but may not exceed electricity consumption. Monthly balance surplus is first added in the annual balance and then itemized separately. Crediting of other types of energy feed-in, for example heat or biogas produced on-site, is itemized via corresponding primary energy factors.

CALCULATION TOOL The past years have clearly demonstrated that, in practice, the calculation process for DIN V 18599 still evades control and is insufficiently supported by software. Broad acceptance has been prevented by the time-consuming nature as well as lacking transparency of calculation methods and results. In this context, dissertation research at the University of Wuppertal in 2010 produced a simplified method for balancing energy along with an Excel-based tool, EnerCalc [9]. EnerCalc already integrates simplified processing of photovoltaic and CHP systems within an energy balance for a net zero energy building following the de-finition presented here (download at www.enob.info, currently German edition only). Fig. A 3.03 exemplarily displays these results for an office building in a comprehensible way.

SWITZERLAND In 2010 in Switzerland there was neither a legislative basis nor a clear definition of net zero energy buildings. Individually planned and built examples are typically based on MINERGIE-P requirements, while including mechanical equipment such as solar collectors and photovoltaic systems [10]. For one, the term "zero energy building" is used for structures with the capacity to cover space heating, DHW, and electricity demands for ventilation systems via renewable energies according to annual average rates (Figs. A 3.06 and A 3.07).

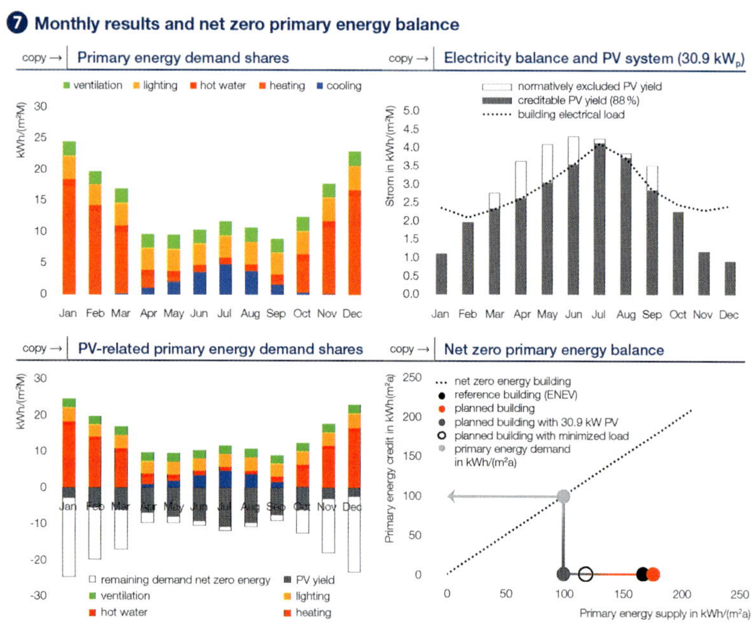

A 3.03

A 3.04

A 3.03 Illustration of energy calculation results with EnerCalc for an administrative building. A photovoltaic system is designed to offset annual total electricity consumption (in this case: a system with 30.9 kW_p peak power); natural gas is used for space heating. Summer surplus is weighted as grid electricity and initially not used for monthly building balances or seasonal compensation.

On the other hand, the term is used for structures that also balance total service electricity for equipment, household appliances, lighting, etc. (see the Residential House in Riehen, p. 56ff.; the Kraftwerk B in Bennau, p. 72ff; and the Corporate Headquarters in Kemptthal, p. 120ff.). Interpretation is usually at the level of net energy or site energy. Required energy is generated on-site.

LEGISLATION AND STANDARDISATION According to the Swiss constitution, municipalities have jurisdiction over setting the limits for energy consumption in buildings. Federal energy legislation covers additional sectors such as electricity supply, industry, and mobility, and gives the federal government supporting and coordinating roles in the building sector. As an important step towards implementing the 26 municipal energy laws, the Conference of Municipal Energy Directors drafted a model for directives, i.e. a comprehensive package of regulations on energy in the building sector. The draft regulations comprise a common denominator supported by the individual municipalities. They

Then, in a second step, an additional, fictitious photovoltaic system is defined that produces a zero primary energy balance on an annual basis, taking seasonal compensation into account (86.3 kWh/m²a primary energy credit for a 40-kW_p PV system).

A 3.04 The energy savings directive as central instrument in the building industry is an outcome of the energy savings law. The directive contains limit values and refers to norms, technical regulations, and official publications as calculation guidelines.

A 3.05 Framework conditions for energy regulations and standards for buildings in Switzerland

A 3.06 The zero energy residential building (MINERGIE-P-ECO) Eulachhof comprises 132 apartments and 8 retail units in Winterthur (CH) 2007, GlassX

A 3.07 Energy supply schematic with electricity grid coupling and Eulachhof's connection to district heating from waste incineration for energy peaks in winter and for redundancy purposes.

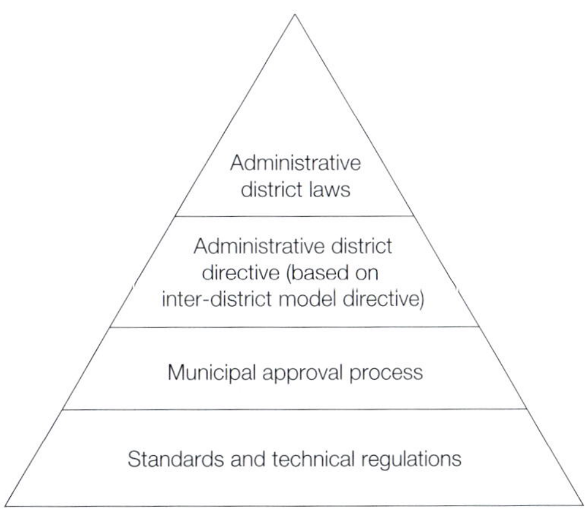

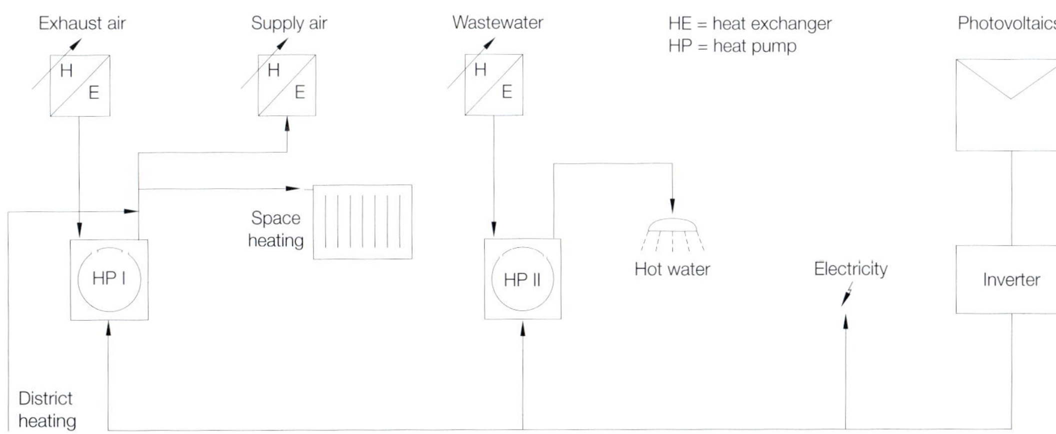

A 3.05
A 3.06
A 3.07

Building categories	MINERGIE	MINERGIE-P	MINERGIE-ECO	MINERGIE-P-ECO	MINERGIE-A
Housing I (multi-family houses)	•	•	•	•	•
Housing II (single-family houses)	•	•	•	•	•
Administrative buildings	•	•	•	•	
School buildings	•	•	•	•	
Retail	•	•			
Restaurants	•	•			
Community places	•	•			
Hospitals	•	•			
Industry	•	•			
Storage	•	•			
Sports buildings	•	•			
Indoor swimming pools	•				

A 3.08 Building categories: MINERGIE, MINERGIE-P, MINERGIE-ECO, MINERGIE-P-ECO, and MINERGIE-A

A 3.09 Sample energy balance for a 244 m² single-family house with a heat pump and geothermal heat exchanger and a 4.7-kW$_p$ photovoltaic system.

A 3.10 Sample energy balance for a 120 m² single family house with 18 m² solar collectors, 2 m³ buffer storage, and supplemental wood pellet heat generator.

A 3.08
A 3.09
A 3.10

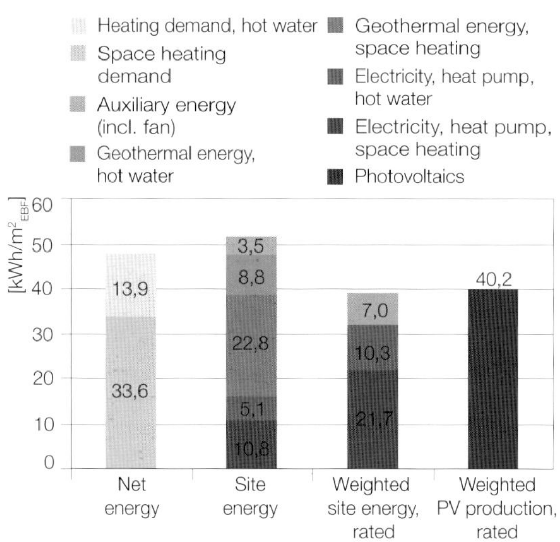

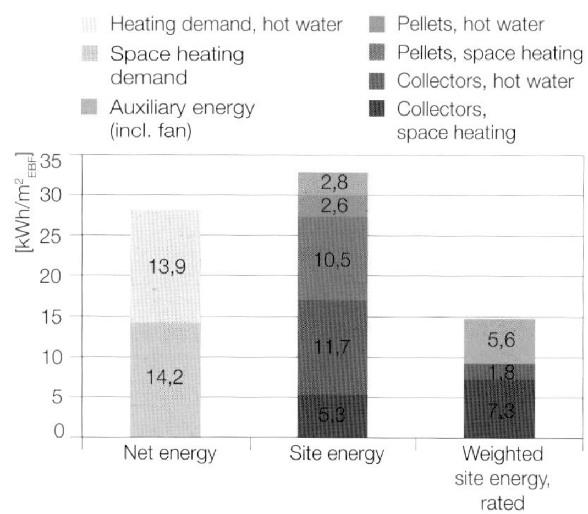

become part of legislation, and communities are responsible for their implementation to varying degrees.
Since early 2009 electricity from renewable energy sources has been promoted by the cost-covering feed-in credit program KEV ("kostendeckende Einspeisevergütung"). The KEV is financed by a surcharge to electricity costs per consumed kilowatt-hour. The entire KEV expenditure is distributed among hydropower, photovoltaics, and biomass, and features a maximum credit sum. Credit timeframes range from 20 to 25 years based on the technology used. The cost-covering feed-in can't be combined with simultaneously selling self-generated electrical power as green electricity on the free green power market (Fig. A 3.05, p. 43).

Only a normative proof of space heating demand according to the SIA 380/1 [11] is required for a building application. Space heating demands are determined monthly and summed up in an annual figure. Proof of energy demand for DHW, lighting, appliances, electronic devices, and building service equipment is currently not legally required. The SIA 380/4 includes recommendations on lighting, ventilation, and air-conditioning exclusively for commercial or public buildings with more than 2000 m² [12]. To support planning, the standard also provides specifications on service installations, mechanical equipment, and space heating. For air-conditioned buildings, the SIA 382/2 [13] is currently under revision. It is based on the SN EN ISO 13790 standard and calculates energy demands for heating and cooling via an included simplified dynamic per-hour weighting scheme.

THE MINERGIE LABEL The Swiss MINERGIE building label has been in effect since 1998. In 2004 the significantly tightened MINERGIE-P standard was introduced, supplementing the original version. In late 2010 a total of about 20,000 buildings were certified MINERGIE and about 1000 MINERGIE-P. With the addition of "ECO," an addendum aimed at healthy and ecological construction methods is now available. The MINERGIE-P standard corresponds roughly to the passive house standard and includes practically all possibilities for reduction of heat demands. MINERGIE and MINERGIE-P set requirements on energy demands for heating and DHW. Their limit values are 38 and 30 kWh/m² of weighted site energy. The EnDK Centre of Excellence for Energy of Municipalities provides weighting factors according to Fig. A 2.07 (Table of Primary Energy and Emission Factors, p. 31). Limits on heating demands are supplemental.

MINERGIE-A Certification according to the new MINERGIE-A building standard, which focuses on new residential construction, is available since March 2011 [14]. The goal of this new standard is for weighted site-energy demands to be zero on an annual basis for heating, cooling, and DHW, along with electricity demands for ventilation and circulating pumps. For buildings with large solar thermal systems, different balancing is provided. Additionally, there is an upper limit for the amount of embodied energy based on primary energy.
For the first time, this upper limit quantitatively regulates construction materials used and comprises a fundamental broadening of scope compared to the MINERGIE and MINERGIE-P standards.
It is conceivable that, at a later phase, these requirements will also be adopted by all other MINERGIE certification programs. For now, there is no limit on electricity demands for household uses, lighting, etc., since a satisfactory database isn't available yet. However, there are discussions on introducing related limit values at a later date (Fig. A 3.08).

MINERGIE-P AND MINERGIE-A These two standards represent two different strategies. While MINERGIE-P stresses optimisation of building envelopes and, therefore, specifies a strict upper limit on space heating demands, MINERGIE-A emphasises building equipment, including active solar energy use. The limit value for space heating demands for MINERGIE-A is at the basic level of MINERGIE. It corresponds to the requirements that apply to those Swiss municipalities that have set the strictest nation-wide regulatory requirements.
Fig. A 3.09 shows a sample energy balance for the heating demands of a single-family MINERGIE-A house. The demands for space heating and DHW correspond to the requirements of the MINERGIE basic standard. Heat is provided by a geothermal probe and heat pump via electrical power (here, projections for annual performance values are intentionally pessimistic). For illustration purposes, geothermal energy is also represented in the "Site Energy" column. The weighted geothermal energy becomes zero, while the electricity is multiplied by a factor of two. The weighted site energy is balanced by the photovoltaic system's annual electricity yield. Only photovoltaic systems that are installed on the building itself and that supply electricity for the building's own demand enter the balance calculation.
Fig. A 3.10 shows the calculated energy balance for a small single-family home with a large solar thermal collector. If the solar thermal system covers more than 50 % of the demand for space heating and DHW and requires additional heating fuelled by biomass (usually wood), then the limit value is 15 kWh/m² annually, and not zero. Obviously, this may be covered by enlarging the photovoltaic array.

AUSTRIA In Austria, the focus is on generating surplus energy, instead of merely consuming net zero energy. Despite the lack of normative legal frameworks, energy plus buildings have been developed and built for about ten years (see Energy Plus Community in Weiz, p. 100ff.; Office Building with Apartment in Villach, p. 129ff.). The focus is on generating electricity via photovoltaic systems, as well as producing heat from renewable sources (see Austria, p. 14f.).
The use of photovoltaic systems in Austria differs from the situation in Germany. The decisive factor are differing incentive practices in regard to green power generation. In Austria, incentives for systems above 5 kWp are regulated nationwide by the Green Electricity Act (ÖSG) [15]. An annual feed-in tariff budget of 2.1 million Euro is available nationally; the waiting list for this programme extends to 2019 [16]. Annual feed-in tariffs are regulated by the Green Power Directive (ÖSVO). Depending on system size and type of integration (integrated into the building or free-standing), credits range from 25 ct/kWh to 38 ct/kWh. Feed-in tariffs are guaranteed for a period of 13 years [17]. Private systems under 5 kW_p may take advantage of investment incentives from the Climate and Energy Fund, but don't receive higher feed-in tariffs. In 2009 and 2010 35 million Euro was available for this programme [18]. The limited budgets for these two incentive programmes have led to only a modest number of installations. In 2009 a capacity of 20.2 MW_p of new installations was documented, while 40 MW_p was predicted for 2010 [19, 20].

Balancing of previously existing energy plus buildings is performed independently of the current standard. In a first step, a building's energy quality is typically determined according to the Passive House Planning Package (PHPP) or the procedure of the Austrian Institute for Construction Engineering (Institut für Bautechnik). Consumption and generation aren't compared until a successive second step. This comparison is simple for buildings that only use electricity as an energy source. However, if other renewable energy sources are used, balancing may vary significantly. The main cause for these differences are the various data sources for primary energy factors in Austria (PHPP, GEMIS, EN 15603. See also Fig. A 2.07, p. 31). One the one hand, balances for buildings are based on PHPP primary energy factors. Others use the GEMIS database, which also includes total primary energy expenditure (see Office Building with Apartment in Villach, p. 129ff.). The climate:active (klima:aktiv) building declaration uses conversion factors according to EN 15603 [21], and climate:active criteria also evaluate building quality more comprehensively. In addition to energy demand and energy efficiency, ecological and convenience-related issues are also observed.
In the case of the "Sunlighthouse," balancing is based on the non-renewable primary energy share (Fig. A 3.11) [22]. CO_2 emissions are also considered. The Sunlighthouse is intended to compensate the CO_2 emissions caused by its construction and service with energy generated from photovoltaic and solar thermal systems within a timeframe of 30 years, and thus, become carbon neutral. Whether the Austrian success story of implementing passive house buildings can be repeated for energy plus buildings remains to be seen. Investment uncertainties associated with decentralised renewable ener-

A 3.11
A 3.12

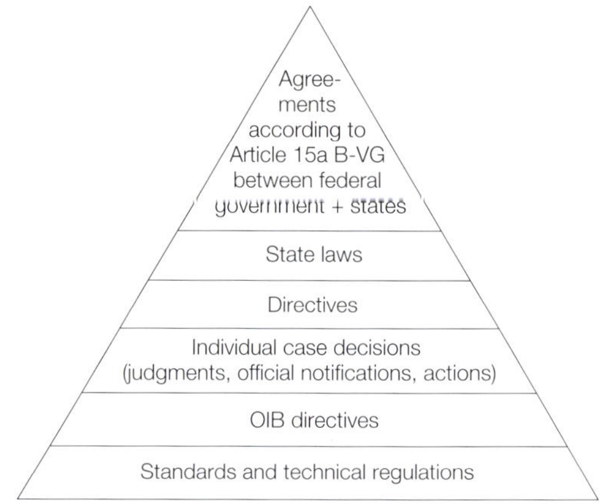

gy usage causes many potential clients to hesitate. Market trends are increasingly moving towards sustainable building concepts that promise an advantage in the real estate market through various certification options (ÖGNI [23], ÖGNB [24], climate:active [25], passive house) and that transcend energy weighting.

LEGISLATION AND STANDARDISATION In Austria, the federal states have authority over building legislation. Each state enacts its own building laws and codes. The instruments for implementing the EU building directive on energy performance of buildings consist of agreements between the federal government and the states [26]. The national Austrian Institute of Construction Engineering (OIB) deals with streamlining the parts relating to energy technology for the nine state's regulations, and therefore, also integrating the EU directives into the relevant OIB directives [27]. These directives specify technical details that are situated between legal regulations and norms (Ö-Normen) of the Austrian Standards Institute. Six core OIB directives were prepared to integrate and streamline building regulations nationwide. The individual federal states then enact laws or regulations ratifying the legal validity of directives [28].
Directive 6, which focuses on energy efficiency and thermal insulation [29], was declared legally binding by almost all states. It defines requirements and limits and refers to relevant norms as well as the OIB calculation guidelines on "Energy-Related Behaviour of Buildings" [30] for determining energy target values. Directive 6 also defines the contents and form of energy certificates. References to building-specific standards of the Austrian Standards Institute integrate directives and legislation. Within these complex constellations, the normative and legal implementation of the EU directive of 2002 on the energy performance of buildings has not been completed in Austria as these words are written in 2011 (Fig. A 3.12). One exception to state authority is issuing energy certificates in the case of building, selling, and leasing buildings or parts of buildings pursuant to the federal Energy Certificate Submission Law (EAVG) of 2006 is [31].
The current Austrian calculation procedures aren't suitable for balancing local energy supply and demand (status: early 2011). Calculating energy quality of residential and non-residential buildings follows the OIB calculation guidelines that supplement OIB Directive 6. The calculation is performed for monthly balancing; the results are typically presented only as an annual balance. The weighting focuses on space heating demands, but also identifies heating, cooling, and site energy demands. The fields for primary energy demands and greenhouse-gas emissions are, to date, blanks on the energy certificate. Space heating, DHW, and heating equipment energy demands for residential buildings are interpreted as site energy values according to OIB calculation guidelines. Cooling energy demands, similar to lighting and ventilation, are included only for non-residential buildings. Both procedures exclude use-specific consumption values from calculations.
The announced revision of OIB Directive 6 and the OIB calculation guidelines promises improved and in part new registration methods as well as uniform primary energy and CO_2 conversion factors for 2011. The next generation of energy balancing was announced for 2012 [33].

A 3.11 The "Sunlighthouse" is part of a set of eight experimental houses completed in different European countries within the Model Home 2020 project. A 7.6-kW$_p$ photovoltaic system and a solar thermal system are calculated to balance the entire annual energy consumption including household electricity of the building's 166 m^2 living area. Pressbaum, Austria 2010, Hein-Troy Architekten [32]

A 3.12 In Austria, building legislation falls under state jurisdiction. The decisive instruments for implementing the EU building directive are agreements according to Art. 15a B-VG between the federal government and the states. The OIB directives were drafted to streamline building-specific regulations and serve to integrate legal regulations and norms.

ENDNOTES

GERMANY

[1] EnEV Verordnung zur Änderung der Energieeinsparverordnung [Directive for Amending the Energy Savings Directive]. 2009.
[2] EEWärmeG Erneuerbare-Energien-Wärmegesetz [Renewable Energies for Heating Law]. 2009. http://www.enev-online.de/eewaermeg. Accessed March 3, 2011
[3] DIN V 18599:2007-02 Teil 1–10 Energetische Bewertung von Gebäuden [Energy Evaluation of Buildings]. 2007.
[4] Voss, Karsten; et al.: Energieeffizienz von Nichtwohngebäuden. Zielsetzungen und Bewertungsmaßstäbe [Energy efficiency of non-residential buildings. Goals and evaluation measures]. In: Deutsche Bauzeitung 07/2010, pp. 66–69.
[5] EEG Erneuerbare-Energien-Gesetz [Renewable Energies Law]. 2009. http://www.gesetze-im-internet.de/bundesrecht/eeg_2009/gesamt.pdf. Accessed March 3, 2011
[6] Achelis, Justus: Fachkommission Bautechnik der Bauministerkonferenz [Expert Commission on Civil Engineering of the Building Ministers' Conference]. Auslegungsfragen zur Energieeinsparverordnung. [Questions on Interpretation of the Energy Saving Directive]. Berlin 2009.
[7] KWK-Gesetz Kraft-Wärme-Kopplungs-Gesetz [Combined Heat and Power Law]. 2009. http://www.kwkg-novelle.de. Accessed March 3, 2011
[8] Voss, Karsten; Musall, Eike; Lichtmeß, Markus: Vom Niedrigenergie- zum Nullenergiehaus. Standortbestimmung und Entwicklungsperspektiven [From low-energy house to zero-energy house. Siting and development perspectives]. In: Bauphysik [Building Physics] 06/2010, pp. 424–434.
[9] Lichtmeß, Markus: Vereinfachungen für die energetische Bewertung von Gebäuden [Simplifications for the energy evaluation of buildings]. Dissertation, Bergische Universität Wuppertal [Bergische University Wuppertal]. 2010.

SWITZERLAND

[10] http://www.minergie.ch. Accessed March 3, 2011
[11] SIA 380/1 Thermische Energie im Hochbau [Thermal Energy in Building Construction]. 2009.
[12] SIA 380/4 Elektrische Energie im Hochbau [Electrical Energy in Building Construction]. 2006.
[13] SIA 382/2 Klimatisierte Gebäude – Leistungs- und Energiebedarf [Air-Conditioned Buildings—Power and Energy Demand]. 2011.
[14] cf. [10]

AUSTRIA

[15] ÖSG Ökostromgesetz [Green Electricity Law]. 2011. http://www.ris.bka.gv.at/GeltendeFassung.wxe?Abfrage=Bundesnormen&Gesetzesnummer=20002168. Accessed March 3, 2011
[16] Bundesverband Photovoltaic Austria [Federal Association of Photovoltaics in Austria]: Fördersituation Österreich. Bundesweite Tarifförderung [Situation of Promotions in Austria. Nationwide Tariff Promotion]. http://www.pv.austria.at. Accessed March 3, 2011
[17] ÖSVO Ökostromverordnung [Green Power Directive]. 2011. http://www.pvaustria.at/upload/2767_Oekostrom-VO%202011.pdf. Accessed March 3, 2011
[18] Klima- und Energiefonds [Climate and Energy Funds]: Photovoltaik Förderaktion 2010 [Action Promoting Photovoltaics 2010]. http://www.klimafonds.gv.at. Accessed March 3, 2011
[19] Biermayr, Peter; et al.: Innovative Energietechnologien in Österreich. Marktentwicklung 2009 [Innovative energy technologies in Austria. Market development in 2009]. In: Berichte aus Energie- und Umweltforschung [Reports on Energy and Environmental Research] 15/2010]. Ed.: Federal Ministry of Transport, Innovation, and Technology (BMVIT). Vienna 2010.
[20] Fechner, Hubert: 5 bzw. 8 % Photovoltaikstrom in Österreich bis 2020 [5 % or 8 % Photovoltaic Electricity in Austria by 2020]. FH Technikum Wien [Technical University of Applied Sciences of Vienna]. 2010.
[21] klima:aktiv. http://www.klimaaktiv.at. Accessed March 3, 2011
[22] Sunlighthouse. Model Home 2020. http://www.velux.at. Accessed March 3, 2011
[23] Österreichische Gesellschaft für Nachhaltige Immobilienwirtschaft [Austrian Society for Sustainable Real-Estate Management] (ÖGNI): Nachhaltige Architektur in Österreich [Sustainable Architecture in Austria]. http://www.ogni.at. Accessed March 3, 2011
[24] Österreichische Gesellschaft für nachhaltiges Bauen [Austrian Society for Sustainable Construction] (ÖGNB): TQB Bewertung [Total Quality Building Evaluation]. http://www.oegnb.net, http://www.eu-greenbuilding.org. Accessed March 3, 2011
[25] cf. [21]
[26] BGBl. II Nr. 251/2009 [Federal Law Gazette II No. 251/2009] Vereinbarung gemäß Art. 15a B-VG zwischen dem Bund und den Ländern über Maßnahmen im Gebäudesektor zum Zweck der Reduktion des Ausstoßes an Treibhausgasen [Agreement according to Art. 15a B-VG between the federal government and the provinces on measures in the building sector for the purpose of reducing greenhouse gas emissions]. 2009.
[27] Mikulits, Rainer: Die EU-Richtline über dieGesamtenergieeffizienz von Gebäuden [The EU Directive on the Total Energy Efficiency of Buildings]. Vienna 2003. http://www.fgw.at/publikationen/pdf/03/2003-4_Mikulits.pdf. Accessed March 3, 2011
[28] Österreichisches Institut für Bautechnik [Austrian Institute of Construction Engineering] (OIB): Harmonisierungskonzept [Streamlining Concept]. http://www.oib.or.at. Accessed March 3, 2011
[29] Österreichisches Institut für Bautechnik [Austrian Institute of Construction Engineering] (OIB): OIB-Richtlinie 6. Energieeinsparung und Wärmeschutz [OIB Directive 6. Energy Saving and Thermal Insulation]. 2007. http://www.oib.or.at. Accessed March 3, 2011
[30] Österreichisches Institut für Bautechnik [Austrian Institute of Construction Engineering] (OIB): Energietechnisches Verhalten von Gebäuden [Energy-Related Behaviour of Buildings]. 2007. http://www.oib.or.at. Accessed March 3, 2011
[31] EAVG Energieausweis-Vorlage-Gesetz [Energy Certification Submission Law]. 2006. http://www.ris.bka.gv.at/GeltendeFassung.wxe?Abfrage=Bundesnormen&Gesetzesnummer=20004886. Accessed March 3, 2011
[32] cf. [22]
[33] Pohn, Christian: Nearly Zero Energy Buildings und Plus-Energie-Gebäude. Definitionen für den urbanen Raum. [Nearly zero energy buildings and energy-plus buildings. Definitions for urban space]. Kongress für zukunftsfähiges Bauen [Congress for Future-Capable Building]. Vienna 2011.

PROJECTS AND LESSONS LEARNED

	Page	**BALANCED ENERGY CONSUMPTION** Technical building services	User-specific consumption	Electro-mobility	Embodied energy	**ENERGY SUPPLY** Solar electrical power	Solar-thermal power	CHP	Wind power	Heat pump	Heating grid	Biomass	Green power
PROJECTS AND LESSONS LEARNED – PART 1	50												
01 Residential House, Riehen (CH)	56	■	■		■	■	■			■			
02 ÉcoTerra Home, Eastman (CDN)	60	■	■			■	■			■			
03 Lighthouse, Watford (GB)	64	■	■			■	■					■	
04 Home for Life, Lystrup (DK)	68	■	■		■	■	■			■			
05 Kraftwerk B, Bennau (CH)	72	■	■			■	■			■	■	■	
06 Renovation Blaue Heimat, Heidelberg (D)	78	■				■		■			■		
07 Kleehäuser, Freiburg/Vauban (D)	84	■	■			■	■	■				■	■
08 Multi-Family Dwelling, Dübendorf (CH)	89	■			■	■	■			■			
09 Solar Community, Freiburg (D)	94	■	■			■					■	■	
10 Energy Plus Community, Weiz (A)	100	■	■			■				■			
11 BedZED Community, Sutton near London (GB)	103	■	■	■		■		■			■	■	
12 Masdar Urban Development Project, Masdar (UAE)	108	■	■	■	■	■	■	■	■	■	■	■	■
PROJECTS AND LESSONS LEARNED – PART 2	114												
13 Corporate Headquarters, Kemptthal (CH)	120	■	■		■	■				■			
14 WWF Headquarters, Zeist (NL)	125	■	■			■	■	■		■		■	
15 Office Building with Apartment, Villach (A)	129	■	■			■	■	■			■	■	
16 Pixel Building, Melbourne (AUS)	134	■	■		■	■			■	■		■	
17 Company Headquarters, Berlin (D)	138	■	■	■		■		■			■	■	
18 Zero Emissions Factory, Braunschweig (D)	144	■	■			■	■	■		■		■	
19 School Renovation, Wolfurt Mähdle (A)	150	■				■	■			■			
20 University Building, Saint-Pierre (F)	154	■	■			■							
21 Day Care Centre, Monheim (D)	158	■	■			■	■			■			
22 Elementary School, Hohen Neuendorf (D)	163	■				■		■				■	
23 Solar Decathlon Europe, Madrid (E)	168	■	■			■	■			■			

OVERVIEW OF PROJECTS AND THEIR CHARACTERISTICS – PART 1

B 1.01 Overview of the locations of the 291 projects identified worldwide
B 1.02 Overview of the worldwide completion of new net zero energy projects per year, broken down into residential and non-residential buildings, as well as new buildings and renovation projects
B 1.03 Plus Energiehaus, Kirchberg-Thening (A) 2001, Andreas Karlsreiter
B 1.04 Illustration of the relationship between main stakeholders and the motivation for the implementation of the net zero energy standard, broken down according to typologies

INTERNATIONAL DEVELOPMENTS The first zero energy buildings of the industrial era were scientifically initiated experimental projects as well as pioneering examples (see Figs. A 1.27 and A 1.28, p. 23). They were produced in the 1990s as energy autonomous buildings without a connection to a grid infrastructure. In the next couple of years they were followed by predominantly smaller residential buildings that generally focussed totally on covering heating needs as far as possible by means of solar energy or renewable energies with a connection to an electricity grid. Somewhat later the first projects emerged that achieved a complete balancing of all energy consumption, but these again used a connection to the electricity grid. Favoured by the solar electricity subsidies available in a number of countries, in the area of smaller residential buildings these projects represent a direct further development of the passive house or MINERGIE concepts. From 2000 onwards the number of projects completed each year has increased continuously (Fig. B 1.02). In many different climates, the availability of small combined heat and power units as well as those operated on biomass (▷ B 07, p. 84ff.; B 22, p. 163ff.) and more efficient photovoltaics has led to the development of different concepts, also for larger and more energy-intensive buildings, even including factories (▷ B 18, p. 144ff.; inverter factory or Solarfabrik, project list p. 178). Urban districts and entire cities are being developed on the basis of a net zero energy approach (▷ B 12, p. 108ff.). Interest in the concept is world-wide, and hence, in North America as well programs and concepts are being developed for "net zero energy buildings", "equilibrium buildings" (▷ B 02, p. 60ff.) or "carbon neutral buildings" (see Fig. A 1.12, p. 16).
2007 saw the beginnings of early renovations of existing residential buildings with the aim of turning them into net zero energy buildings. Only a few years later, this trend includes office buildings as well (▷ B 06, p. 78ff.; B 14, p. 125ff.; IdeasZ2, project list p. 178).

FOCUS ON CENTRAL EUROPE AND NORTH AMERICA International project research conducted for this book identified almost 300 building projects as equalised net zero energy or nearly zero emissions balance buildings.
These buildings were and still are erected mostly in north-western countries and climates (Fig. B 1.01). The clearly greater amount of building activity, technological advantages, and dependence on energy imports as well as awareness of climate change drive these projects, above all in central Europe. They are increasingly developed in climates domi-

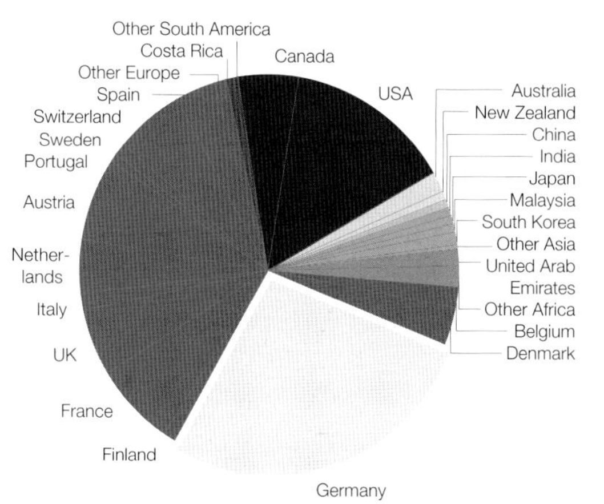

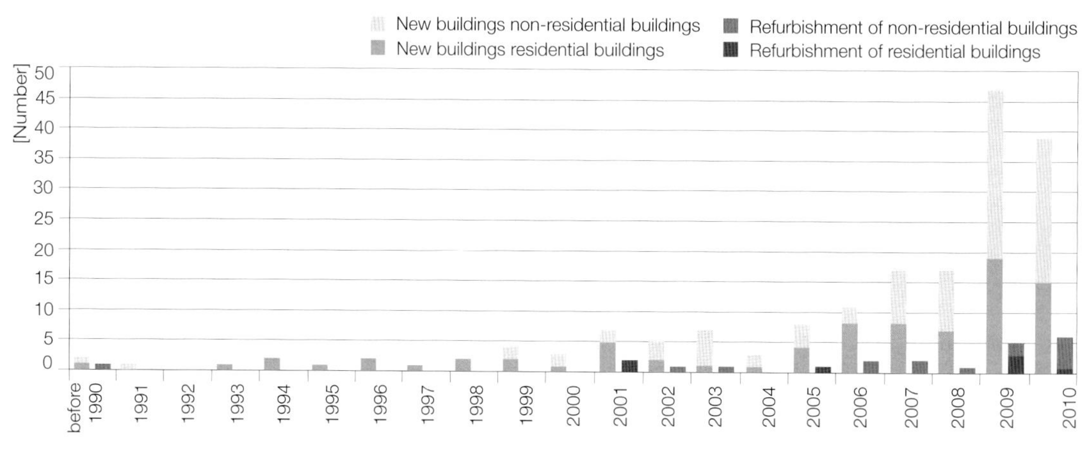

B 1.01
B 1.02

nated by a need for heating with marked seasonally changing energy demands. In such climatic zones, the lower amount of solar radiation, which also fluctuates considerably according to the season, makes it more difficult to cover these needs by active use of solar energy (see Figs. B 12.03 and 04, p. 109). Here, the primary goal is energy-efficiency. The passive house concepts that form the basis of a large number of the zero energy projects illustrate this fact.
The energy plus house of a family in Kirchberg-Thening is an early example. Since 2001 it has covered its consumption of energy, reduced by incorporating the passive house standard, by means of 17 m² of facade collectors, a heat pump in the ventilation system, and 10.4k kW_p photovoltaics (Fig. B 1.03).

THE STAKEHOLDER AND THEIR MOTIVATION
Whereas the first, mostly technically dominated demonstration projects were initiated as scientific studies, nowadays ecologically minded building clients, some within private construction partnerships, build net zero energy houses. The reasons for this range from the threat of finite resources and built climate protection to the desire for independence from energy supply companies and the intention to avoid rising energy costs (Fig. B 1.04). In a number of countries, subsidising solar energy encourages the use of photovoltaic arrays, and thus, promotes the development of energy efficient buildings into net zero energy projects (▷ B 01, p. 56ff.).
Alongside the construction of single-family houses, housing construction or real estate companies use the "green building image" conveyed by a net zero energy concept to increase the value and attractiveness of apartments in large residential buildings or small housing developments (▷ B 10, p. 100ff.). Such cases frequently go no further than covering the consumption of technical services by means of renewable energy (▷ B 06, p. 78ff.; SunnyWatt, project list p. 176). Solar electricity plants on the roof of a building or combined heat and power plants are not always owned by the apartment owner (▷ B 07, p. 84ff.; B 10, p. 100ff.; B 06, p. 78ff.).
A number of business companies, mostly from the field of renewable energy generation, carry out net zero energy factories for marketing reasons and in order to demonstrate their position with regard to sustainability (▷ inverter factory or Solarfabrik, project list p. 178). But already the first large business companies that are not primarily known for their position on sustainability are running their buildings with the aim of achieving a complete balance of the operating energy. As consequence, they aspire to gain a better reputation, among other benefits. In addition, the buildings carried out are increasingly certified in ways that exceed the purely quantifiable dimension "energy". Nowadays every tenth net zero energy building achieves an LEED, DGNB, MINERGIE-P, BREEAM, or similar certification. In such certification systems, further parameters are provided by sustainable materials, ecological concepts with regard to water consumption, local traffic, the projected duration of building occupancy and use or its adaptability to changing demands.
Architects employ the concept of net zero energy buildings for different typologies, in order to position themselves in this former niche and current boom branch of "green buildings" in accordance with their personal convictions (▷ B 07, p. 84ff.). Veterans of this branch are often drivers of many new projects and are consciously commissioned to carry out projects or are asked to adapt existing plans or ideas to achieve an equalised energy balance (▷ B 17, p. 138ff.; B 08, p. 89ff.). Since the council decision was passed by the European Union on "nearly zero energy buildings" (see European Union, p. 12f.), this requirement is more frequently found in competitions set up by governments or quasi-governmental institutions.

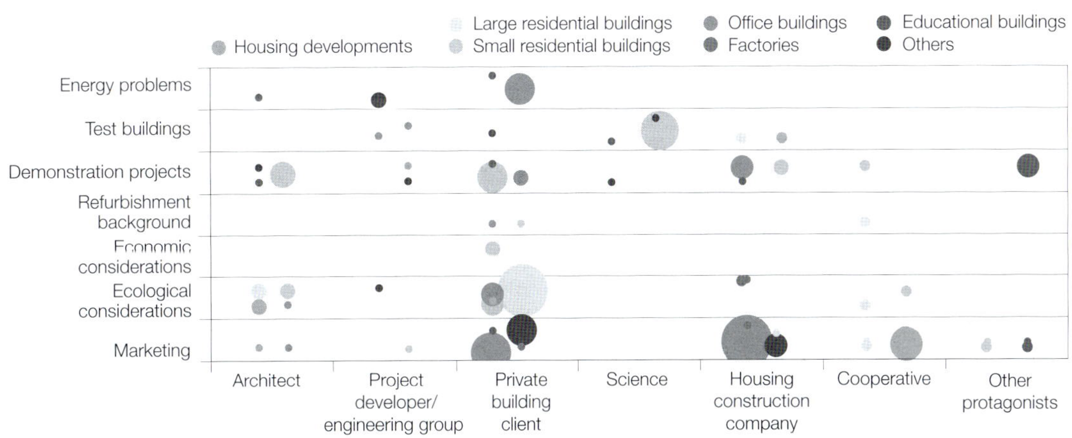

B 1.03
B 1.04

B 1.05
B 1.06

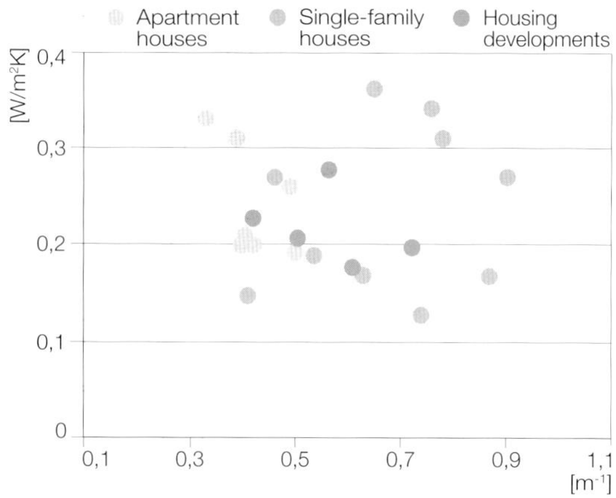

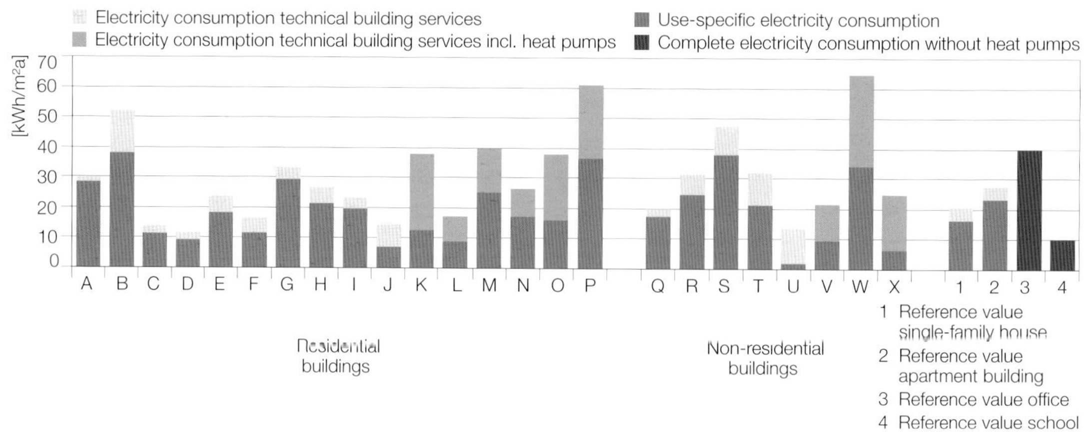

CHARACTERISTICS OF RESIDENTIAL BUILDINGS More than half of all the net zero energy projects carried out are residential buildings. This is due, on the one hand, to the fact that such buildings generally make up a large proportion of the total number of buildings. On the other hand, this is because related concepts in this field can be more easily realized by the protagonists. In most cases, all energy consumptions, including household consumption, are incorporated in the energy balance. In housing developments and urban planning projects, it is not the equalized balance of the individual houses, but the overall equalized balance that is the focus of attention.

ENERGY EFFICIENCY Most of the net zero energy buildings are located in climates where heating plays a dominant role. The concepts for lowering the demand for heating energy are well-known and integrated into various projects. For around one third of these buildings, certified MINERGIE or passive houses form the basis. A further 50 % pursue these ideas on the level of components and have highly insulated, very air-tight building envelopes with few thermal bridges. The average U-value (mean value for the entire building envelope) of the central European projects is 0.26 W/m²K which is very low (Fig. B 1.05). Buildings of all sizes aim at this standard. Projects in other climatic zones where heating does not play such a dominant role require considerably less expenditure (▷ project list p. 176ff.). Almost all the buildings researched have ventilation systems with heat recovery. All the same, maximum efficiency in the heat transfer medium (> 85 %) and low electricity consumption (below 0.5 W/[m³/h]) cannot yet be taken for granted. Earth tube heat exchangers are frequently used for pre-heating the air and keeping the intake air free of frost. For the most part, residents assess the air quality and the ventilation comfort positively and, strikingly often, mention the large amount of daylight and the visual relations between inside and outside. (▷ B 03, p. 64ff.; B 04, p. 68ff.; B 08, p. 89ff.). The ratio between the area of windows and the net floor area is on average 38 %, in those buildings with particularly large areas of glazing up to 50 %. A further reason for generously dimensioned openings in the houses is provided by the passive solar gains in winter. In contrast to non-residential buildings, all the residential buildings in central European climates have a markedly asymmetrical distribution of windows, with preference given to south-facing windows. These are generally shaded by externally mounted, movable shutters or blinds (▷ B 03, p. 64ff.; Fig. B 1.03, p. 51). Whereas in small single-family houses fixed overhangs or projections to provide shade in summer are rarely integrated in the architectural gesture (▷ B 04, S. 68ff.) or are placed as an additional element (▷ B 01, p. 56ff.; B 07, p. 84ff.), in large residential buildings they assume the form of roof or balcony overhangs or projections on which photovoltaic modules are fixed and are used to reduce large summer heat gains (▷ B 09, p. 94ff.; B 10, p. 100ff.; Sunny Woods, Fig B 1.10 and project list p. 176). As a rule, the deep window reveals that result from the thickness of the insulation do not offer sufficient protection from the sun (▷ B 08, S. 89ff.). Experience has shown that residents are generally critical in their assessment of room temperatures in summer. On the other hand, blinds are often used also in winter to provide privacy, which reduces the desirable winter heat gains (▷ B 05, p. 72ff.) and results in greater use of artificial light (▷ B 03, p. 64ff.). Building automation systems that can, for instance, control sun-shade systems, increasingly find use. Energy saving is not the top priority. Winter gardens as buffer spaces and a familiar attribute of solar building are no longer dominant today. All the examples of residential buildings in moderate climate zones operate without active cooling. Phase change materials or thick (loam) render layers (▷ B 03, S. 64ff.; B 02, p. 60ff.) introduce a further heat buffer. In addition, passive cooling strategies familiar from administration buildings, such as ventilation shafts (▷ B 03, p. 64ff.) or windows in lower facade areas or upper roof areas that open simultaneously (▷ B 04, p. 68ff.) are integrated. The energy efficiency measures described and the combination of a number of different measures reduce the energy demand of the residential buildings on the way to a net zero energy house by almost 60 % in comparison to standard buildings that were built according to the respective valid energy directives.

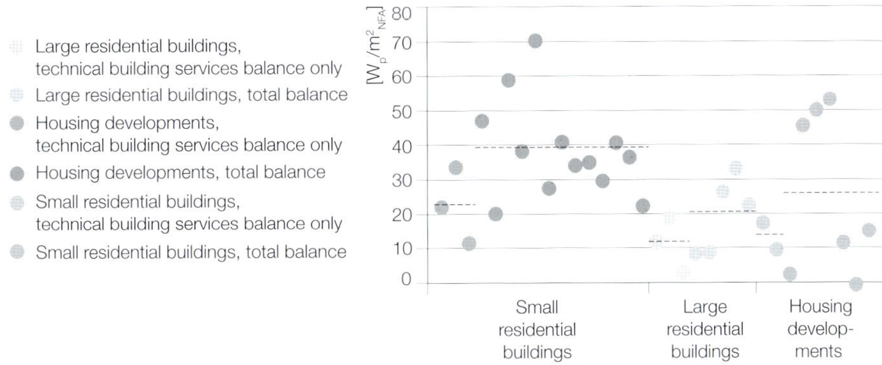

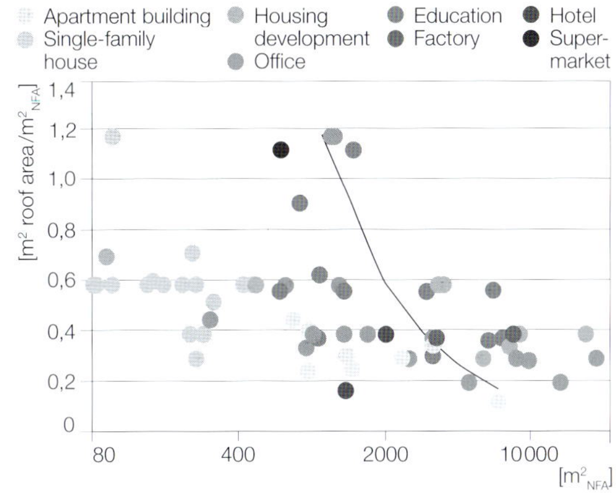

B 1.07
B 1.08

SAVING ELECTRICITY Even though household electricity consumption has, so far, not been included in normative energy balances (see The Boundaries of a Balance, p. 32f.), it is frequently employed in balances in building practice. The fact that this represents a decisive effort is shown by Fig B 1.06. In the case of residential buildings, user-specific electricity consumption is mostly by domestic appliances, and is, on average, four times the amount consumed by the building services. In electrically heated buildings, household electricity consumption still accounts for around 50 % of total consumption. Due to the limited local generation of energy (Fig. B 1.07), the residents use appliances with the highest energy efficiency ratings (▷ B 03, p. 64ff.), connect washing machines or dishwashers to the warm water network (▷ B 05, p. 72ff.; B 07, p. 84ff.), and use energy-saving lamps (▷ B 08, p. 89ff.). In rental projects, attempts are made to compensate for the lack of influence in this area by providing user information (▷ B 04, p. 68ff.; B 05, p. 72ff.; B 06, p. 78ff.). In operating the building it is shown that, with few exceptions, the measures do not yet achieve the desired success. The energy consumption figures are often close to the general consumption average (▷ B 06, p. 78ff.; B 09, p. 94ff., Fig. B 1.06).

ENERGY SUPPLY The lower the consumption, the lower the necessary self-generation by means of solar electricity plants or combined heat and power plants. Fig. B 1.08, p. 53, and Fig. B 2.12, p. 118

B 1.05 The mean U-value of the building envelope of residential buildings as a function of their compactness, illustrated for buildings at locations where heating plays a dominant role

B 1.06 Electricity consumption of residential buildings, broken down into household consumption and consumption for technical building services. Buildings with heat pumps are listed separately on account of their higher consumption; in the other buildings there additionally is a further energy carrier. The sources for the reference values are, for the residential buildings: SIA 2031 and the passive house projection package; for non-residential buildings: the publication of the rules for characteristic values for energy consumption; for non-residential existing buildings: the Federal Ministry of Transport, Building and Urban Development.

B 1.07 Installed output of the solar electricity plants per m$^2_{NFA}$ separated into buildings that aim at a completely equalised balance and those that balance only the primary energy demand for technical services.

B 1.08 Roof surfaces for the solar generation of electricity in relationship to the existing net floor area in the net zero energy buildings examined (assumed: flat roofs). The line shows the increase of usable floor area with an increasing number of stories and a constant roof area. For both one story with a usable floor area of 1000 m^2 as well as 7000 m$^2_{NFA}$ distributed on seven floors the roof area remains around 1200 m^2.

B 1.09 Renovated former military accommodation from the 1930s on what was once an American military base, Bad Aibling, district of Mietraching (D) 2010, SCHANKULA-Architekten / Diplomingenieure

B 1.10 Sunny Woods (1387 m$^2_{NFA}$), Zürich (CH) 2001, Beat Kämpfen

B 1.09
B 1.10

B 1.11 Dwelling house R128 (ca. 250 m^2_{NFA}), Stuttgart (D) 2001, Werner Sobek. A photovoltaic system measuring 52 m^2 in area offsets the electricity used for the heat pump and the household.

B 1.12 75 m^2 solar thermal collectors, 750 m^2 solar electricity modules and a ground source heat pump supply the net zero energy housing development (floor area 5900 m^2) with energy. Watt (CH) 2010, Beat Kämpfen

B 1.13 Energy flex house (216 m^2_{NFA}), Taastrup (DK) 2009, HENNING LARSEN ARCHITECTS

B 1.14 Characteristics of well-known urban and urban planning projects

B 1.11

show a cross-sectional evaluation in relation to all non-residential buildings. In addition to economic advantages, this also offers greater design freedom as the latter, depending on the relationship between living area and roof area, must not be completely covered with photovoltaic modules (▷ B 04, p. 68ff.) or indeed even specially expanded in the design process (Fig. B 1.08, p. 53).
Almost all net zero energy projects use solar electricity plants. On average, small residential buildings employ almost 40 W_p/m^2_{NFA} for the primary energy balance of their total energy consumption. If the energy expenditure for building services alone is considered, this figure is about halved. In renovation projects or those without solar thermal plants, larger photovoltaic plants are used (Fig. B 1.07, p. 53). Highly efficient buildings with low consumption figures also manage with considerably fewer photovoltaic arrays. Due to the low primary energy assessment, the use of biomass for heating has advantages with regard to the area of solar electricity arrays required to achieve an equalised balance.
In larger residential buildings or housing estates, the photovoltaic plants are, in relation to the living area, about half the size of those in single-family houses. Here, the dominant systems are those which are advantageous in primary energy terms, such as biomass heating systems or plants for combined heat and power, which allow the local generation of electricity by means other than solar electricity plants. The yields from the combined heat and power units exceed those from solar electricity plants. External wind power turbines are included in some of the balances (▷ B 07, p. 84ff.). “Kraftwerk B in Bennau” produces a sizable plus in solar heat and, to achieve an equalised balance, transfers this by means of a small heating grid to a neighbouring building (▷ B 05, p. 72ff.). At the scale of housing estates or residential housing developments, the related photovoltaic performance varies on account of very different approaches to supply. The passive house development erected in 1998 at Kronsberg in Hannover (▷ project list p. 176) covers the energy consumed entirely by means of external wind energy turbines. The “Energy Plus Community, Weiz” (▷ B 10, p. 100ff.) does not exploit the advantages offered by the scale of the housing development. Here, each individual house is understood as a single net zero energy house. In the Quartier Bad Aibling (Fig. B 1.09, p. 53) new buildings generate more energy than refurbished buildings. Therefore, the solar thermal and photovoltaic areas carried out differ significantly from each other. A number of the new buildings have a solar electricity output of 43 W_p/m^2_{NFA}. The example in Fig. B 1.09, p. 53 feeds heat into the development's own local heating grid by means of 2000 m^2 of solar collectors. The extremely large photovoltaic arrays of the “Solar Community, Freiburg” (▷ B 09, p. 94ff.) could be smaller, if all that was required was to achieve an equalised balance for the development itself.
A chronological view of smaller residential buildings reveals a tendency to reduce the size of PV systems. Improved efficiency levels of PV cells and increased energy efficiency of the buildings themselves will advance this trend in the years to come. This fact is shown in Fig. B 1.08, p. 53. The theoretically calculated potential of roof surfaces (here viewed as flat roofs for reasons of simplicity) for the generation of solar energy per m^2_{NFA} among known net zero energy buildings shows that, so far, construction hasn't reached particularly high densities yet. In the housing sector, buildings more than three stories tall are rare, as larger areas of living space mean higher consumption and proportionately, a smaller area of PVs. The same applies to administration buildings. In the net zero energy buildings carried out to date, the larger and higher-density (non-residential) buildings usually have fewer stories and the buildings, therefore, tend to be less tall.

COVERING HEATING DEMAND Three quarters of the net zero energy buildings use thermal solar plants. Here, 0.04 m^2 of collectors per m^2_{NFA} suffice for smaller residential buildings, or 0.03 m^2/m^2_{NFA} for larger residential buildings, to provide ca. 60 % solar coverage of domestic hot water. The areas of heating-aided solar thermal plants vary in size according to the integration in the heating system. In larger buildings, the systems use solar heat by means of buffer storage tanks or also supply neighbouring buildings. Here, the average area of the collectors rises significantly (0.27 m^2/m^2_{NFA}). In a number of small residential buildings, the collectors serve as heat source for the heat pump (▷ B 04, p. 68ff.; Energy flex house, Fig. B 1.13 and project list p. 176). In a striking number of cases, the design potential of

vacuum tube collectors is utilised (▷ B 08, p. 89ff.; Sunny Woods, Fig B 1.10 and project list p. 176). In contrast, the incorporation of facade collectors has, so far, played a minor role (▷ B 05, p. 72ff.; Sunny Woods and the Plus Energiehaus in Kirchberg, Thening, project list p. 176). There are already a number of pilot applications that use waste water for heat recovery (▷ B 05, p. 72ff.).
In recent years the number of net zero energy buildings with heat pumps has risen dramatically. Today they comprise 65 % of the generators of heat in net zero energy residential buildings. The improved technology and integration in systematic solutions, lower prices, as well as the opportunity to replace a gas connection with biomass delivery and storage areas all combined make heat pumps an attractive option (Fig. B 1.11). Generally, in the net zero energy buildings known, heat pumps with annual coefficients of performance of around three are used: an average thermal output of 22 W_{th}/m^2_{NFA} is thus, theoretically, contrasted by an electrical connected load of 6.5 W_{el}/m^2_{NFA}. The few measurements available in this area show that these figures, or indeed even more favourable relationships are generally achieved (▷ B 08, p. 89ff.). For reasons of cost in the housing sector, (exhaust) air-air heat pumps are most commonly used, in smaller buildings also those in conjunction with ground probes.

FROM THE HOUSING DEVELOPMENT VIA THE DISTRICT TO THE CITY In urban concepts it is not essential that every building should produce an equalised balance (▷ B 09, p. 94ff.), as they offer the possibility of a joint supply of heat through a local heating network. At the urban scale, networks in combination with individual solar electricity plants on the buildings take over the provision of renewable energy (▷ B 12, p. 108ff.). It will become possible to include renovated existing buildings with moderate energy efficiency in the energy balance and to compensate for a fixed, unchangeable orientation or unfavourable A/V-relationships (▷ Bad Aibling, project list p. 176). Here, the urban scale permits a rich variety of options (Fig. B 1.12). For approaches of this kind, an overall stakeholder or constellation of actors is required to formulate the ideas of a community and translate them into the energy concept (▷ B 09, p. 94ff.; B 11, p. 103ff.; B 12, p. 108ff.).

B 1.12
B 1.13
B 1.14

		Realised [Year]	Population [Persons]	Settlement area [ha]	Inhabitants /ha [Pers/ha]	FAR	Coverage heating/cooling energy consumption	Electricity generation
CITY	1st building phase Masdar, Masdar (VAE)	Building phase	8600	59	145	1.68	geothermal energy, absorption cooling, solar thermal energy, biomass-CHP plant	PV, geothermal energy, biomass-CHP plant
URBAN DISTRICT	Vauban, Freiburg (D)	2007	5000	38	132	1.20	local wood-fired CHP and boiler	partly PV
URBAN DISTRICT	Bad Aiblinger Mietraching (D)	2012		70		0.09	100 % renewable by local heating grid (CHP and solar thermal energy), de-centrally located heat pumps	PV, CHP, water power
URBAN DISTRICT	Bo01, Malmö (S)	2009	2000	22	91	1	100 % renewable, geothermal energy, biogas	PV
HOUSING DEVELOPMENT	Solar community, Freiburg (D)	2006	170	1.1	155	0.72	local, wood-fired boiler	PV
HOUSING DEVELOPMENT	Energy plus housing development, Ludmilla-Wohnpark, Landshut (D)	2011	180	2.2	84	1.05	de-centrally located heat pumps with ground collectors and local heating grid with CHP	CHP, PV
HOUSING DEVELOPMENT	Energy plus community, Weiz (A)	2008	66	0.7	94	0.30	de-centrally located heat pumps per dwelling unit	PV
HOUSING DEVELOPMENT	BedZED, Sutton (GB)	2002	220	1.2	183	0.74	central biomass CHP/boiler	biomass CHP, PV
HOUSING DEVELOPMENT	Eulachhof, Winterthur (CH)	2007	300	1.2	250	0.74	ground/district heat, de-central heat pumps	PV
HOUSING DEVELOPMENT	SunnyWatt, Watt (CH)	2010	50	0.4	125	0.68	semi-central heat pumps	PV

OVERVIEW OF PROJECTS AND THEIR CHARACTERISTICS – PART 1

RESIDENTIAL HOUSE

Riehen, CH 2007

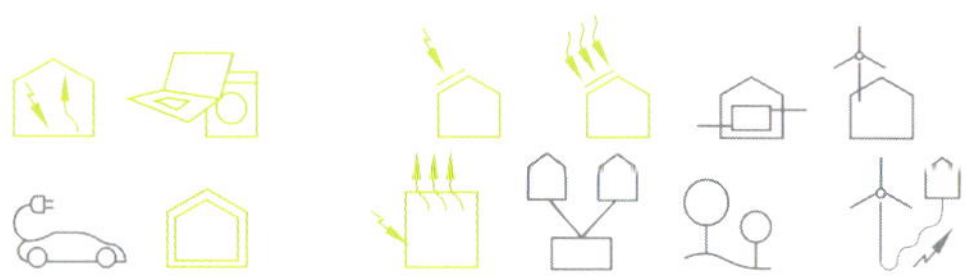

Client: Wenk-Furter family, Riehen
Architect: Setz Architektur, Ruperswil
Energy consultant: Otmar Spescha
Ingenieurbüro für energieeffizientes Bauen, Schwyz
Building services/photovoltaics: BE Netz
Bau und Energie, Lucern
Monitoring: Wenk-Furter family
Main stakeholder: client and architect

This wood-frame, two-family house in Riehen, Switzerland, was built with assistance from a grant program of the administrative district of Basel. The program for pilot and demonstration buildings for a 2000-watt society (see 2000-watt society, p. 14) in the Basel pilot region offers technical and financial support for demonstration projects that integrate ecological, economical, and social aspects of sustainable construction. By supplying a MINERGIE-P-certified building (see Switzerland, p. 43ff.) with solar and geothermal energy, an energy surplus is produced in the annual balance. Since no other energy sources than electricity are used, this house qualifies as a typical "all-electric house" (see zero energy building, p. 23ff). The architectural design and efficiency of the cube-like structure with its wood exterior and exposed concrete elements are representative of the more than 6000 Swiss small residential MINERGIE houses newly completed by late 2010. Its compact shape, moderately sized openings, and lack of both projecting elements or recesses reflect the design approach of efficiency maximisation. Other eco-friendly aspects include use of natural and non-hazardous building materials and rainwater harvesting.

DEVELOPMENT, DESIGN, AND STAKEHOLDERS

When planning began, the clients' only goal was to create a very efficient house based on the MINERGIE standard. However, the administrative

B 01.01
B 01.02

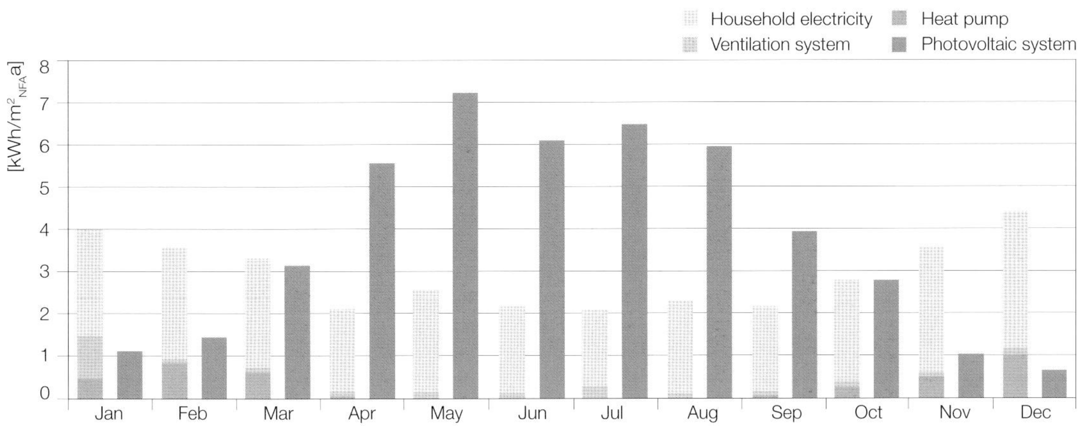

B 01.01 Site plan, scale 1:2000
B 01.02 View from south-west
B 01.03 Monthly electricity balance

B 01.04 Section, scale 1:500
B 01.05 Upper floor plan, scale 1:500
B 01.06 Ground floor plan, scale 1:500

district of Basel's "energy plus building incentive program" awakened their interest in a building that could generate a positive energy balance. The program subsidizes 40 % of the costs for a photovoltaic system. The system supplements the originally planned roof-mount solar thermal collectors. Thanks to the feed-in tariff of about 0.48 €/kWh guaranteed for twenty years the photovoltaic system will more than offset its costs within twelve years.

ARCHITECTURE Large windows along the southern garden facade permit ample daylight intake in interior spaces and support the use of passive solar energy (Fig. B 01.02). The northern facade has significantly fewer and smaller windows. External blinds and occasional roof overhangs provide required shading.
The ecological and economic aspects of the building design are of significant importance (Figs. B 01.04–06). A salient feature is the pre-weathered wood facade made of rough cut white pine cladding attached to the wood frame. Spaces between the wood framing are infilled with 38 cm of mineral wool insulation. A central ventilation system with heat recovery ensures that ventilation losses remain low and air quality adequate, which leads to meeting decisive criteria for attaining MINERGIE-P status. Airborne pollutant rates for formaldehyde as well as rates for total volatile organic compounds (TVOC) were analysed inside the house after the building envelope was completed and three months after the residents moved into the house. The levels are less than 50 % of those in the recommended guidelines of the Swiss Federal Office of Public Health.

ENERGY EFFICIENCY The energy concept and the positive energy balance are based on the building's efficiency as determined by the MINERGIE-P label. The low heating demands result from a number of factors: the extremely compact building design; an opaque building envelope that is sealed (air tightness: n_{50} = 0.5/h) and extremely well insulated by 38 cm of mineral wool; thermal windows with a heat transfer coefficient of 0.84 W/m²K; and a ventilation system with efficient heat recovery

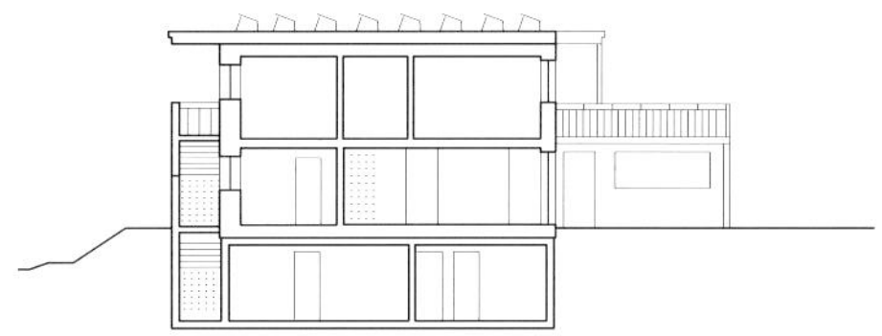

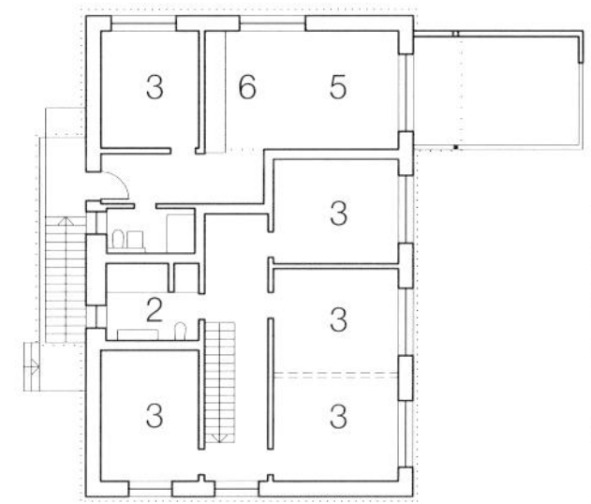

1 Office
2 Bathroom
3 Room
4 Living area
5 Dining area
6 Kitchen
7 Patio
8 Garage

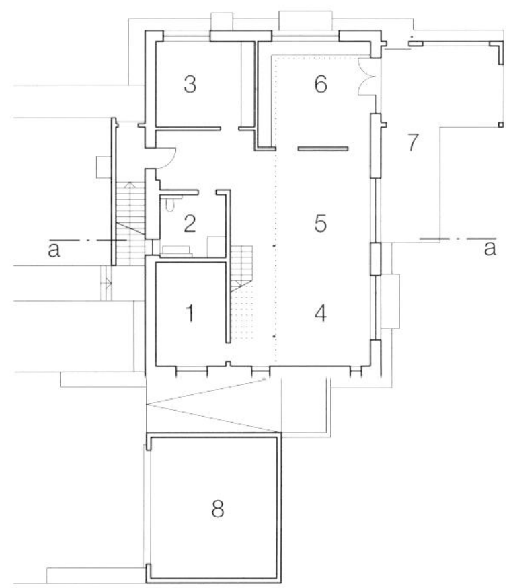

B 01.03
B 01.04
B 01.05
B 01.06

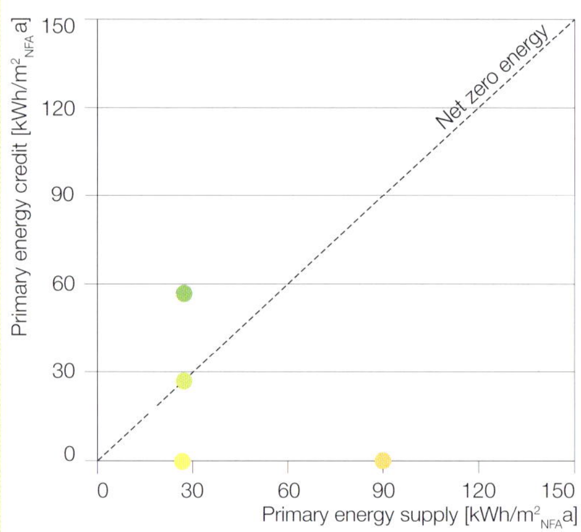

The Net-ZEB-29 standard comprises the following aspects:

- Metered annual total primary energy consumption including household electricity (87 kWh/m²a)
- Energy self-sufficiency from creditable monthly yields (60 kWh/m²a)
- Seasonal compensation of remaining consumption
- Annual energy surplus, public grid feed-in (28 kWh/m²)

Primary energy factors as per SIA 2031 (see Fig. A 2.07, p. 31)

B 01.07
B 01.08

and flat collectors for solar thermal energy. In addition, efficient household appliances reduce electricity consumption.

ENERGY SUPPLY Geothermal energy is extracted via a 120 meter deep borehole and serves as a heat source for a brine/water heat pump, as well as to pre-chill supply air in summer. The 14.4 kW_p photovoltaic system serves to cover remaining energy demands and is installed on the flat roof similar to the 7.5-m^2 solar collectors. The photovoltaic modules cover 84 m^2 and are inclined at 10 degrees, the flat thermal collectors are angled at about 30 degrees. The two systems are hardly visible from the street level and don't impair the cube-like appearance of the building.
The solar thermal collector provides about 60 % of annual hot water heating demands. The remaining heating demands are covered by the heat pump. In winter, the outside air supplied by the ventilation system is pre-heated via a U-shaped polyethylene pipe serving as a geothermal heat exchanger. Underfloor heating is also provided to control room temperature. During summer, air supply can be reversed to slightly chill the house and discharge some of the heat extracted during winter into the earth (Fig. B 01.09). The systems are kept simple to meet residential requirements. Simple, effective operation contributes to high comfort levels.

ENERGY BALANCE The building has a positive annual energy balance. In total, the solar power system produces about 30 % more electricity than consumed for heating, hot water, ventilation, and total household and operating power for the complete single family dwelling including a two-room apartment. The PV system is connected to the grid. Since electricity is the only energy source, consumption and generation can be offset directly without considering primary energy weighting factors. Self-generated electricity is directly compared to grid supply. Analysis of consumption data shows that operating the house requires more energy than was projected during planning (Figs. B 01.03, p. 57, and 10). By examining monthly progression, it becomes clear that the photovoltaic system was optimised to generate maximum electricity and not to prioritise coverage of self-consumption. The planners optimized the angle and orientation of the solar panels based on solar electricity feed-in credits and design freedom offered by the flat roof. This results in large amounts of excess energy between April and September, while there is a significant deficit during the winter months from November to February. Generation and consumption match only in March and October. The heat pump rarely operates between April and September because of higher temperatures and because the solar thermal collectors produce hot water. As result, the imbalance is amplified. Electricity gains are only offset by household and ventilation energy consumption.
The relatively high electricity consumption for a MINERGIE-P building implies neither particularly strong conservation measures nor unusually high power consumption. The positive energy balance partially compensates embodied energy across the house's life cycle. However, the surplus is insufficient to totally compensate embodied energy (Fig. B 01.07).

LESSONS LEARNED After living in the house for two years, the client evaluated it positively in all respects. The ventilation system produces a pleasant indoor climate, and chilled supply air combined with exterior shading during summer helps prevent overheating. No major alterations in lifestyle or energy consumption were necessary to generate more energy than is consumed.
The building was awarded the 2008 Swiss Solar Prize in the new construction category. Points were awarded for the careful integration of solar power systems into the restrained building design, as well as the simple, yet innovative energy concept.

B 01.07 Energy evaluation
B 01.08 View from north-west
B 01.09 Technical schematic of energy provision
B 01.10 Annual energy balance per m^2_{NFA} for 2009 (metered values)
B 01.11 Building and energy parameters (values refer to net floor area, NFA)

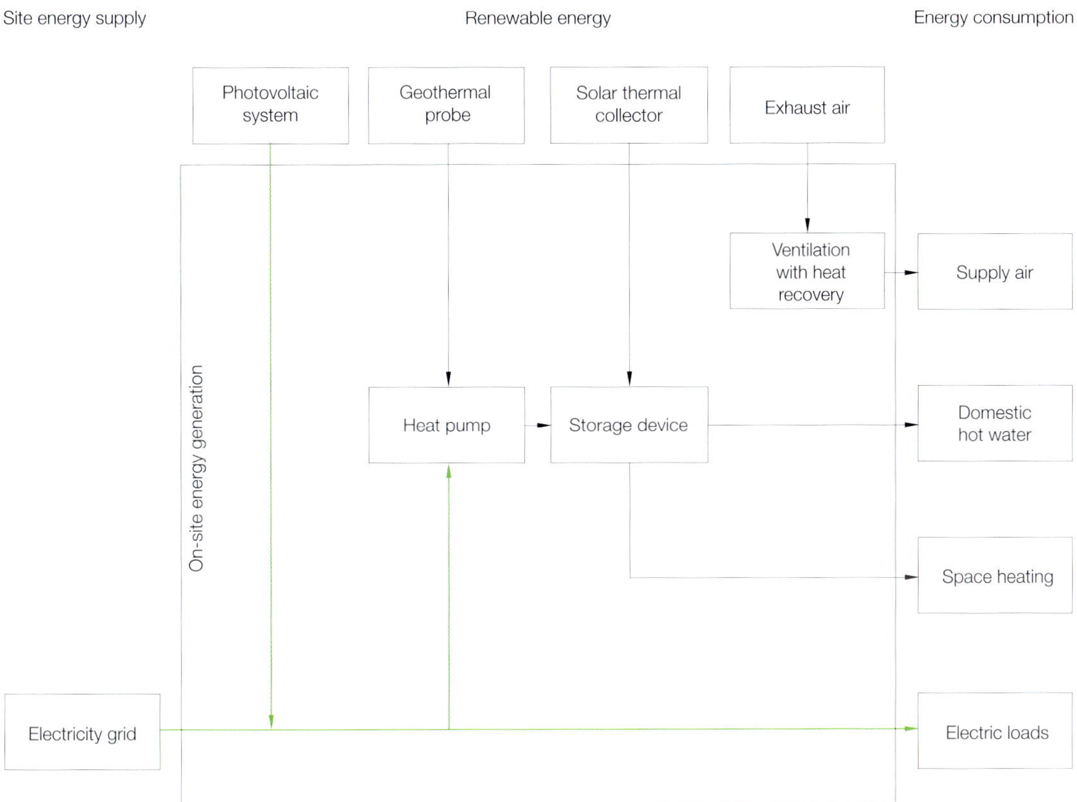

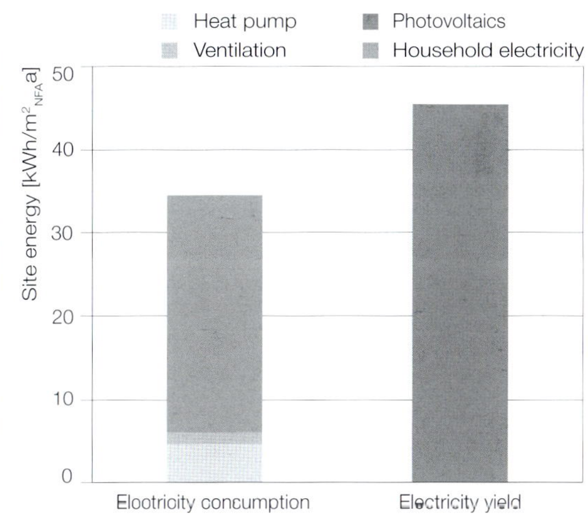

SITE	Riehen (CH)
Annual global radiation at site	1100 kWh/m²a
Annual mean temperature at site	10.1 °C
Context	suburban

BUILDING ENVELOPE PARAMETERS	W/m²K
U-value, exterior walls	0.12
U-value, windows (including frames)	0.84
U-value, roof area	0.11
U-value, ceiling above basement/floor slab	0.10
Mean U-value, building envelope	0.19

BUILDING EQUIPMENT PARAMETERS	
Solar collector area	8 m²
Area per m²	0.03 m²/m²
Photovoltaic system area	84 m²
Area per m²	0.36 m²/m²
Photovoltaic capacity	14 kW_p
Capacity per m²	47.60 W_p/m²

GRID INFRASTRUCTURE AND ENERGY SOURCES	
Supply infrastructure	electricity grid
Energy source supply	electricity
Feed-in infrastructure	electricity grid
Feed-in energy source	electricity

DESIGN STRATEGIES, CONCEPTUAL FOCUS
Passive house components, MINERGIE-P concept, mechanical ventilation with heat recovery, solar thermal collectors, brine/water heat pump, photovoltaics, ecologically sound materials

BUILDING PARAMETERS	
Net floor area, NFA	302 m²
Gross floor area, GFA	315 m²
Gross volume, V	1600 m³
Building envelope, A	858 m²
Surface to volume ratio, A/V	0.54 m²/m³
Building costs (net, construction/technical systems)	2020 €/m² (2007)
Number of units	2
Total number of users	7

CONSUMPTION PARAMETERS (2009)	kWh/m²a
Space heating consumption	11
Water heating consumption	14
Site energy consumption, heating (including hot water)	5
Electricity consumption	29
Total primary energy consumption	87
Total primary energy generation	115

B 01.09
B 01.10
B 01.11

RESIDENTIAL HOUSE IN RIEHEN

ÉCOTERRA HOME

Eastman, CDN 2007

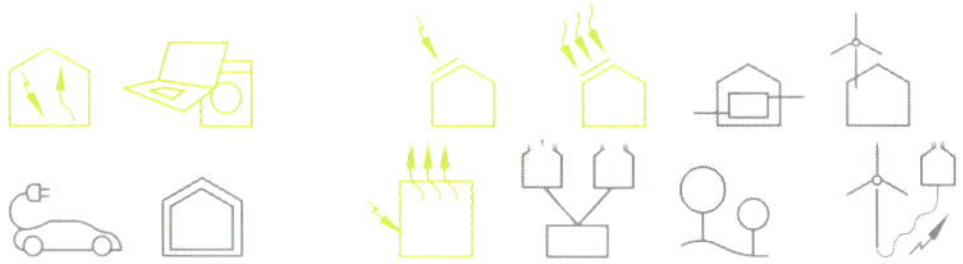

Client: Maisons Alouette, Saint-Alphonse-de-Granby
Architect: Masa Noguchi, Glasgow
Energy consultant: Concordia University Solar Laboratory, Montreal
Building services: Concordia University, NRCan, Régulvar, Montreal
Monitoring: Concordia University, Montreal
Main stakeholder: client

The two-story single-family house "ÉcoTerra Home" emerged in 2006 as one of twelve winners of the "EQuilibrium Initiative", a competition organised by the Canada Mortgage and Housing Corporation (CMHC). The house was built in late 2007 in Eastman, about 100 km to the east of Montréal, an area characterized by long, cold winters and relatively short, wet, and hot summers. The goal of the EQuilibrium Initiative is to promote demonstration buildings. While zero energy status is not an absolute requirement of the competition, 40 % of the possible points to be gained are awarded for overall energy performance, including credits for self-generated energy. Naturally, this motivated the teams taking part in the competition to achieve the zero energy goal. In addition to the characteristic timber construction method with low embodied energy and efficient appliances, the particular characteristics of the ÉcoTerra Home include a hybrid solar power system integrated in the roof. It comprises a combination of solar electricity modules with a thermal air collector known as BIPV/T (Building Integrated Photovoltaic/Thermal), as well as a ground water heat pump.

DEVELOPMENT, DESIGN, AND STAKEHOLDERS
In 2006 the CMHC started the EQuilibrium Initiative (E and Q stand for Environment and Quality). The goals of the subsidised new buildings included a healthy pleasant indoor climate, environmentally-friendly construction materials, economic use of energy sources and water, as well as achieving a zero energy balance. New technological approaches also receive support

ENERGUIDE RATING SYSTEM In Canada, the definition of a net zero energy building is based on a normalised local energy balance. The point-based "EnerGuide Rating System" (ERS) is used to categorise buildings. The energy generated is subtracted from the total energy consumption. This energy balance is then compared to a reference building and, using a key, the building in question is rated as an "Advanced House (Near Net Zero House)", "Net Zero House" or as a "Net Positive House". An ERS rating of 100 points represents a net zero energy building with a site energy balance of 0 kWh. Differences between local climate zones and heating systems are taken into account. Self-generated energy

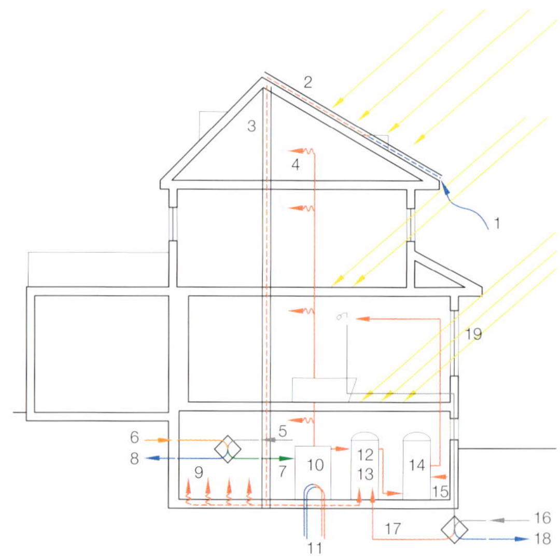

B 02.01 View from south
B 02.02 Cross-section showing building technical services concept, scale 1:250
1 Fresh air
2 BIPV/T system
3 Air duct
4 Heated supply air
5 Waste air
6 Fresh air
7 Heat transfer to heat pump
8 Exhaust air
9 Heat by thermal concrete core activation
10 Ground water heat pump
11 Geothermal probes
12 Storage tank
13 Heat exchanger (warm air from BIPV/T
14 Hot water storage
15 Electric heater
16 Well water supply
17 Hot water supply
18 Waste water
19 Fixed and flexible sun protection system
B 02.03 Ground floor plan, scale 1:250

B 02.01
B 02.02
B 02.03

used on-site is not considered in the building balance. The ERS calculation for the ÉcoTerra Home produced 98 points, resulting in a rating of the design as an "Advanced House", i.e. a near net zero house.

DEVELOPMENT OF THE BUILDING The planning of this house, which started in autumn 2006, involved architects and specialist engineers, as well as researchers from Concordia University in Montreal, which is part of SBRN, the Canadian Solar Buildings Research Network. The energy concept included a heat pump, heat storage, and a specially developed hybrid collector plant (the above mentioned BIPV/T). The selection of the insulation of the opaque envelope was based on building simulations and achieves U-values from 0.17 to 0.13 W/m²K, as it seemed more economical to cover slightly higher consumption by larger hybrid collectors.
The building was erected by a local manufacturer of prefabricated houses as client, with financial and technical support from the Canadian government and several industry partners. Following a test phase in August 2009 the all-electric house was occupied by two people and was comprehensively monitored from that time onwards.

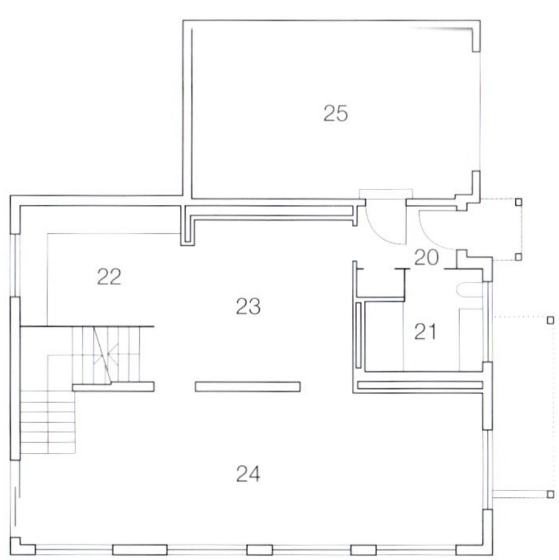

20 Entrance
21 Bathroom
22 Kitchen
23 Dining
24 Living
25 Garage

ARCHITECTURE The two-story single-family house is a timber frame construction except for the base plate, the foundations, and the basement shell. The prefabricated elements are insulated with polystyrene rigid foam panels (beneath the roof 33 cm, in the exterior walls 20 cm). The main part of the house was prefabricated in seven modules. It took only one day to erect the building on site.
The passive use of solar gains and a completely insulated, air-tight external envelope (air-tightness value 0.9 h^{-1}) form the basis for the zero energy approach. The floor plan has an aspect ratio of 1.4. More than 40 % of the main facade, oriented towards the south, are glazed. Overall, 15 % of the facade's 230 m² are glazed, providing a high amount of daylight intake in the rooms. The windows feature low-emission, argon-filled triple glazing, and in combination with the plastic frames have a U_w-value of 1.25 W/m²K.
In the hollow core concrete slab in the basement, hot air from the BIPV/T roof passes through several stainless steel ducts, thus distributing the heat in the slab. A solid concrete wall on the ground floor of the otherwise light-weight construction house serves as thermal storage. Overhanging roofs and automated external awnings shade the south-facing windows and prevent overheating in summer.

ENERGY EFFICIENCY A waste water heat exchanger connected to a heat accumulator reduces heating demands. A mechanical ventilation system with 70 % heat recovery reduces ventilation heating losses. To adjust the electricity demand of the ventilators, they can be operated at three different speeds. The average specific power consumption is 0.68 W/m³h. Efficient household appliances and economic fluorescent lamps are part of the standard fittings. By means of a building automation system the different energy consumption figures can be recorded and the operation of all systems regulated.

ENERGY SUPPLY An 11-kW_{th} ground water heat pump serves as the primary heating system. The heat source includes two 76-metre-deep boreholes containing U-probes back-filled with a brine mix. When required, heat is transferred via a heat exchanger to the supply air or the heat accumulator and supplements pre-heating by heat recovery from waste water. Via the 227 l heat accumulator, the heat pump also indirectly feeds a hot water storage tank of the same size. In addition, the heat pump has an electric auxiliary heater (9.6 kW) and a heating element in the hot water tank for DHW.
The grid-linked solar electricity plant, which is integrated flush in two pitched roofs, the larger part of which faces south at a 30° angle, consists of a total of 21 panels with amorphous silicon cells each with a nominal capacity of 136 W. The total capacity is 2.85 kW_p. Outdoor air enters a 3.8 cm cavity below the photovoltaic elements through openings in the eaves. Along the ridge, the warm air is collected by a distributor and directly used by means of a heat exchanger to preheat the process water. Thermal energy is further led to the hollow core of the concrete slab, then stored by means of the concrete's high thermal mass, and slowly released into the living space. This circulated air from the roof does not enter the living rooms and is, therefore, not used for conditioning or by the heat recovery system of the ventilation plant. The air temperature can rise up to 30 °C above the temperature at which it enters the roof (Fig. B 19.05). A variable speed fan controls the flow rate, and thus, the temperature lift. In summer night-time, air can also be used to cool the interior. Dissipating heat beneath the solar electricity modules also increases the efficiency of the latter.

ENERGY BALANCE The measured heating energy demand of the house at a low mean annual temperature of 6 °C amounts to about 43 $kWh/m^2_{NFA}a$, which represents about 12 % of the usual demand of a single-family house in this climate zone in Canada. Together with the consumption of the electric hot water heating, the electricity consumption of the heat pump is 18 $kWh/m^2_{NFA}a$. Thus, the measured values surpass the simulation data by around 7 %. The auxiliary electricity required for pumps and ventilators amounts to a figure of 15.90 $kWh/m^2_{NFA}a$. This can be largely attributed to the energy used by the BIPV/T fan and the air distribution system in the house. Comparable values for single-family houses with an

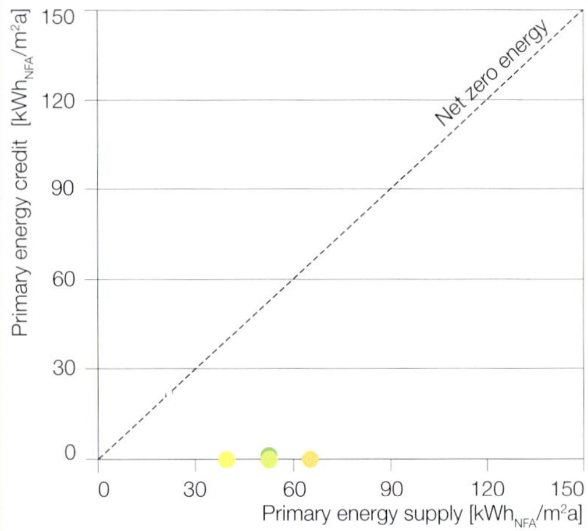

The zero energy goal is not achieved:

- Measured annual total primary energy consumption (65 $kWh/m^2_{NFA}a$)
- Building-specific primary energy consumption (41 $kWh/m^2_{NFA}a$)
- Consumption after monthly coverage of self-demands by self-generated solar electricity (51 $kWh/m^2_{NFA}a$)
- A feed-in above the coverage of self-demands does not take place. The consumption figures exceed the energy yields each month.

Primary energy factor for electricity (1.2), specific to Quebec.

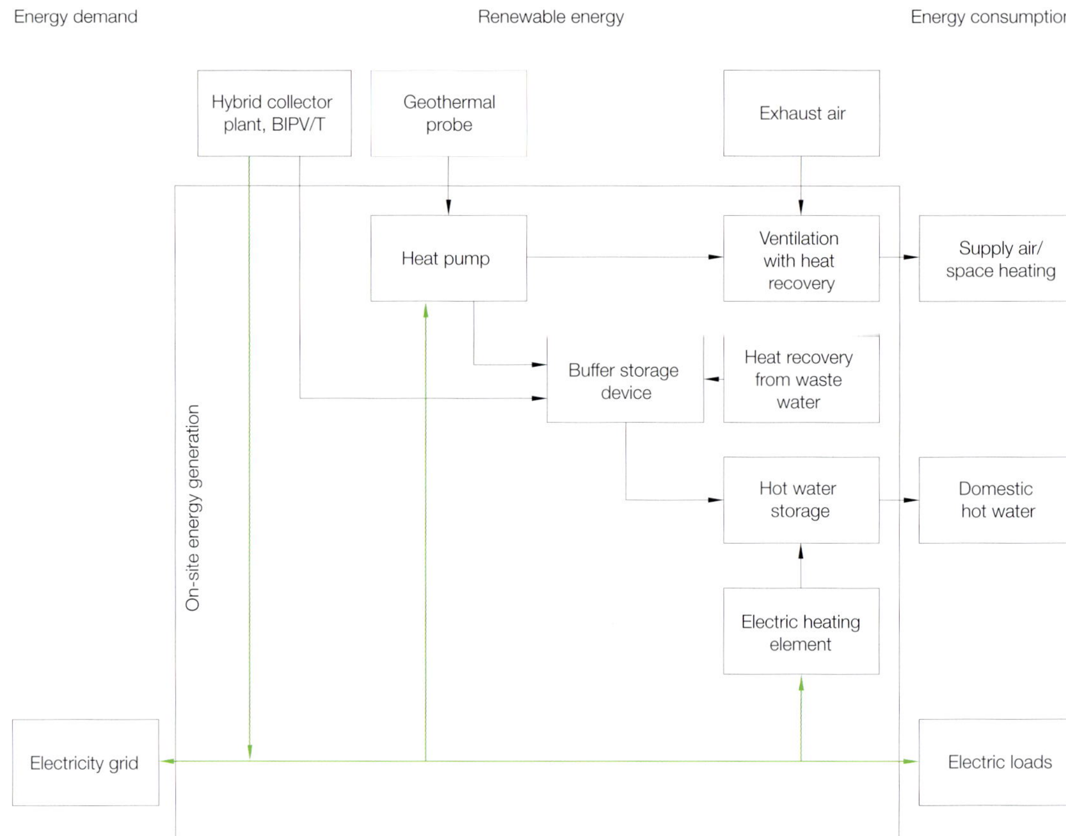

B 02.04 B 02.05

B 02.04 Energy evaluation
B 02.05 Technical schematic of energy provision
B 02.06 Monthly site energy yields contrasted with clearly larger energy consumption figures. In none of the twelve months does the energy generation exceed the measured energy consumption figures. The single-family house in Riehen, which is similar in size, (see Fig. B 01.03, S. 57) achieves clearly higher yields with slightly lower consumption.
B 02.07 Building and energy parameters (values refer to net floor area, NFA)

equalised energy balance are on average about 50 % lower. Much the same applies to the consumption of household electricity, which is relatively high with a figure of very nearly 20 $kWh/m^2_{NFA}a$ including lighting (c. 5 kWh/m^2a). However, compared to North America in general, 30 % less energy is used. The annual total electricity consumption, including all technical plant, lighting, and household appliances amounts to 54 kWh/m^2_{NFA} or 12 400 kWh and is, thus, clearly above the simulated demand values of 9800 kWh/a. However, this is still 75 % better than the Canadian average. The elaborate monitoring that was carried out with the aid of building automation allows uses not planned for to be identified. The users added light fixtures, set up a workshop in the garage that was originally planned to be unheated, and installed an electrical 5-kW-heating there that increased electricity consumption by more than 500 kWh in 2010.

The photovoltaic system measuring 55 m^2 lacks the capacity to cover the energy demand. In 2010 it produced only 2600 kWh, and thus, despite the high annual global radiation of 1270 kWh/m^2a, fell short of calculations that anticipated an electricity yield of ca. 3400 kWh/a. The difference here was caused by snow on the solar roofs and the shadows cast by the surrounding trees. To achieve a uniform appearance for the roof skin and to better integrate the photovoltaic modules, thin layer modules based on amorphous silicon were used, which are less effici-

ent. The use of mono-crystalline cells, which was originally considered, would have made possible an energy yield of approximately 7300 kWh/m²a. If, in addition to this, the shade problem had been solved, this would have balanced the measured energy consumption of the household systems, not including the garage heating.
An analysis of the energy values on the basis of primary energy produces, in comparison to other projects, clearly lower values (total primary energy consumption 65 $kWh/m^2_{NFA}a$), due to the low conversion factor of 1.2 (the generation of electricity in Quebec is based to a large extent on water power). Due to the use of electricity as sole energy source, this can be neglected (Figs. B 02.05 and 07).

LESSONS LEARNED Because it was prefabricated in the factory and then had to be transported to the site, the pitch of the BIPV/T roof is only 30°, which means that the solar electricity plant is not optimally used. The originally calculated optimal pitch of 40° would have led to snow sliding off the roof and also to higher yields.
Additional measures could bring the house closer to an equalised energy balance. For example: improving the thermal insulation of the walls to a U-value of 0.125 W/m²K would achieve savings of up to 500 kWh/a. There is similar potential for improvement by adapting the runtimes of the ventilator that distributes the heated air to match the times when there is no need for heating or cooling (ca. 700 kWh/a), dispensing with the air filter (442 kWh/a), and switching off the ventilation system when the residents leave the house (c. 50 kWh/a). Using weather forecast data to plan the loading and discharging of the thermal storage mass of the concrete slab or the use of the heat pump should, in the future, help avoid the generation of heat at peak load times and lead to lower energy use. The number of pipes and elbows in the system of ducts that distributes the heated air through the house from the BIPV/T roof leads to high pressure losses. In part, the heat gained is eventually not used, although it could, for example, replace the garage heating or serve as the heat source for an air-water heat pump. This could have made the drilling for the earth probes and the associated costs unnecessary, however, it would also have led to inferior performance by the heat pump. A larger heat accumulator would have allowed shifting the thermal loads away from peak load times. As this would reduce the network load, it may be of interest above all in regions where a high proportion of electricity is used for decentralised generation of heat. The additional consumption shows clearly that achieving a good to equalised energy balance is heavily dependent on the users.

SITE	Eastman (CND)
Annual global radiation at site	1270 kWh/m²a
Annual mean temperature at site	6.0 °C
Context	rural
BUILDING ENVELOPE PARAMETERS	W/m²K
U-value, exterior walls	0.16
U-value, windows (including frames)	1.25
U-value, roof surface	0.16
U-value, floor slab	0.66
Mean U-value, building envelope	0.27
BUILDING EQUIPMENT PARAMETERS	
Area of solar thermal collectors	55 m²
Area per m²	0,24 m²/m²
Thermal storage volume	54 l
Storage volume per m²	0,20 l/m²
Photovoltaic system area	55 m²
System area per m²	0,24 m²/m²
Photovoltaic capacity	2,84 kW_p
Capacity per m²	0,01 W_p/m^2
GRID INFRASTRUCTURE AND ENERGY SOURCES	
Supply infrastructure	electricity grid
Energy source supply	electricity
Feed-in infrastructure	electricity grid
Feed-in energy source	electricity

DESIGN STRATEGIES, CONCEPTUAL FOCUS
BIPV/T (solar electricity plant with combined thermal air collector), mechanical ventilation with heat recovery, ground water heat pump, increase and activation of thermal storage mass

BUILDING PARAMETERS	
Net floor area, NFA	230 m²
Gross floor area, GFA	250 m²
Gross volume, V	700 m³
Building envelope, A	544 m²
Surface to volume ratio, A/V	0,78 m²/m³
Building costs (net, construction/technical systems)	1100 €/m² (2007)
Number of units	1
Total number of users	2–4
CONSUMPTION PARAMETERS (2010)	kWh/m²a
Space heating consumption	42
Water heating consumption	26
Site energy consumption for heat (including hot water)	34
Electricity consumption	54
Total primary energy consumption	65
Total primary energy generation	13

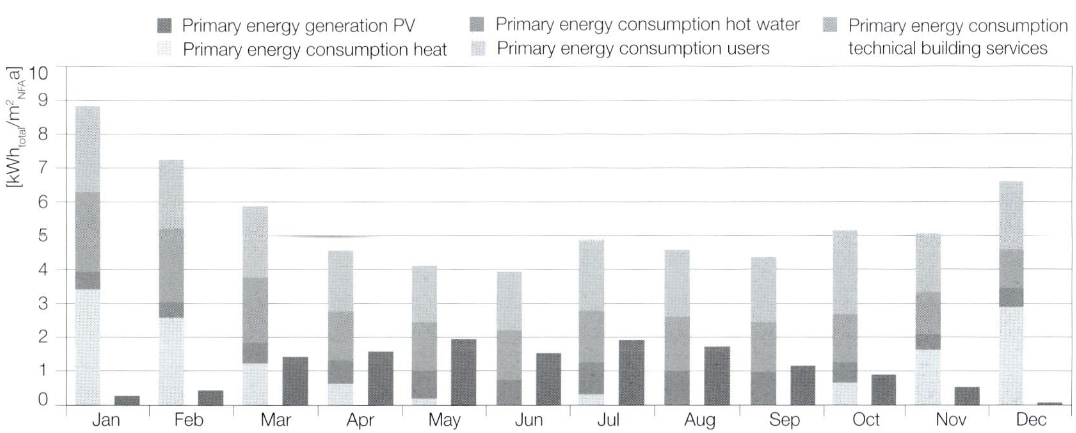

B 02.06
B 02.07

LIGHTHOUSE

Watford, GB 2008

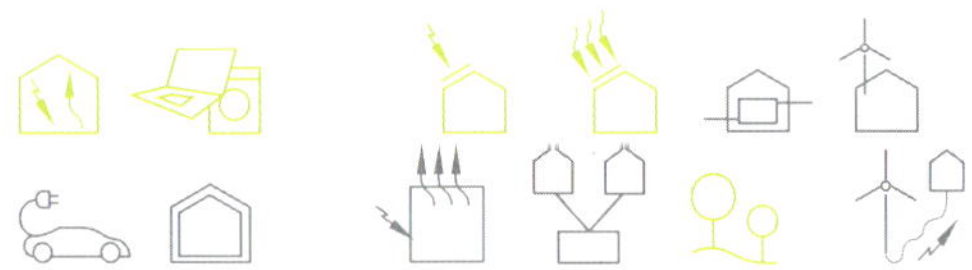

Client: Kingspan Off-Site, Kingscourt
Architect: Sheppard Robson, London
Energy consultant: Arup, London
Building services: Davis Langdon, London; CCB Evolution, Bristol
Monitoring: Kingspan Off-Site, Watford
Main stakeholder: client

The "Lighthouse", a demonstration housing project, was Great Britain's first zero carbon home, built to comply with the standards of the British "Code for Sustainable Homes" (CHS). A principal idea behind this small house in Watford near London is the emphasis on active use of solar energy by means of vacuum tube collectors and photovoltaic modules that are integrated in the roof skin. The shed roof slopes from north to south and smoothly transitions into the southern facade without change in materials, giving this building its characteristic appearance. This dynamic gesture, as well as the generous use of sweet chestnut wood in both primary and secondary structure, represents a departure from the rather restrained, typical British residential architectural design gesture. The building is intended symbolise the aim for a new start in terms of both architecture and energy in the UK (Fig. B 03.05).
Throughout the entire planning process, whether relating to energy consumption, construction, or materials, top priority was given to comprehensive and efficient use of resources. Energy demands were reduced by means of a concept similar to that of a passive house and heating demands are covered in part by either solar collectors or a wood pellet boiler. A photovoltaic system covers the complete primary energy use including the domestic power demands over the period of a year.

1 Entrance, utilities
2 Bedroom
3 Bathroom
4 Balcony
5 Living room
6 Kitchen
7 Dining room
8 Void
9 Study/work area

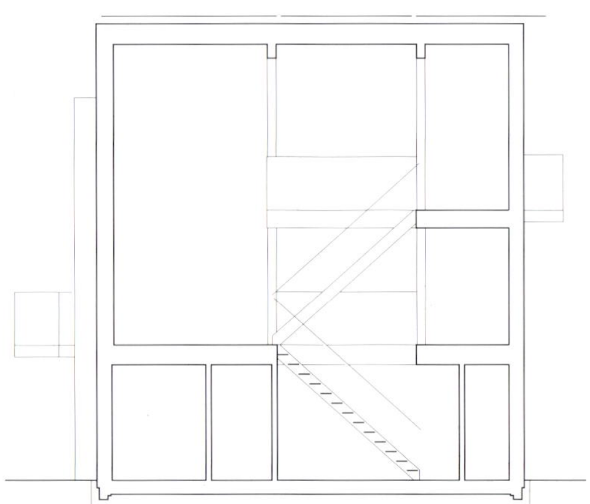

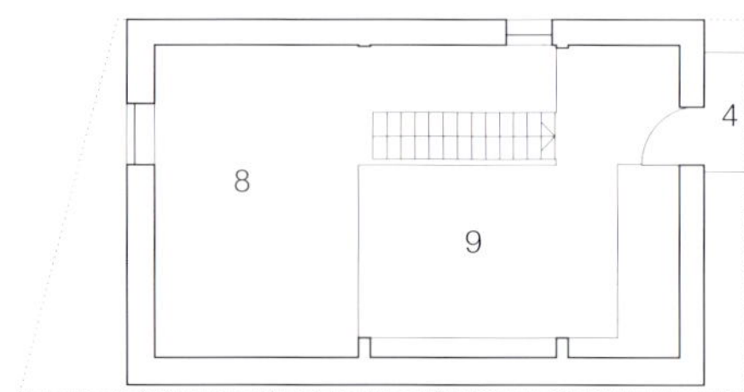

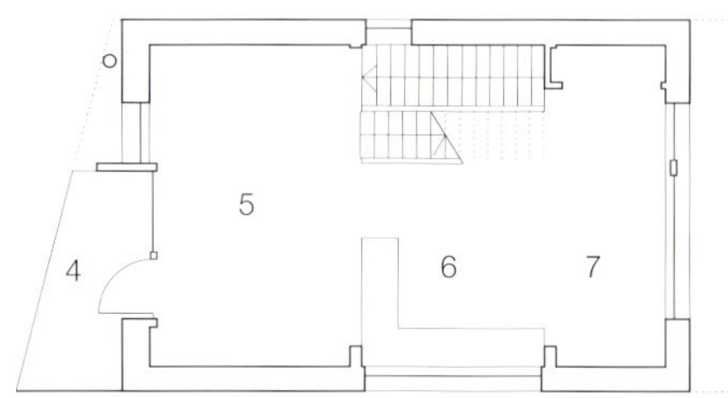

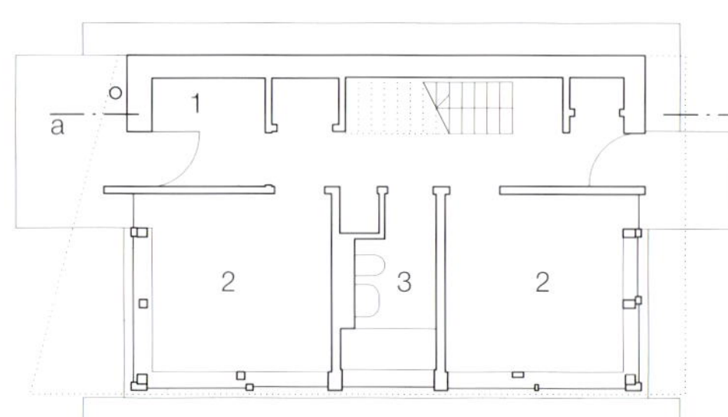

B 03.01
B 03.02
B 03.03
B 03.04
B 03.05

DEVELOPMENT, DESIGN, AND STAKEHOLDERS

The client – a company that develops and produces facade systems and solutions for sustainable building concepts – had the Lighthouse erected as a model project in the innovation park of the Building Research Establishment (BRE) in Watford. The BRE is an independent research-based business consultancy that provides sustainability and environmental expertise in the industrial and residential sectors. The park forms part of its research facilities.
The building for the Lighthouse is based on a design by architects Sheppard Robson, who aimed to implement the "Code for Sustainable Homes Level 6" (see Code for Sustainable Homes) in a new design, without compromising domestic comfort and quality. Close collaboration with the other engineers involved and a clearly defined goal in energy terms resulted in a design phase of only one month, while construction took three months.

ARCHITECTURE Because, as a demonstration project, this building is a temporary building, it doesn't feature a basement and is supported by screw pile foundations made of timber piles. The footings can be removed after demolition of the building. Instead of a standard concrete slab, a floating slab comprised of a massive concrete structure on ground floor level also serves to stiffen the building. The Lighthouse is a two-and-a-half story, free-standing house that, thanks to the flexibility of its layout, offers two to three (bed)rooms with a floor area of 90 m². The flexibility of the interior in all three dimensions enables diverse and long-term use by different residents, and thus encourages what is called social sustainability. The roof has a pitch of 40 degrees and leads seamlessly into the southern facade, creating, together with the solid north wall, a simple building shape. The orientation and pitch result from optimising not only the amount of useable interior space, but also the performance of the photovoltaic system and the solar thermal collectors on and in the south-facing roof, reserved for the active generation of energy. The 39-cm-thick prefabricated timber sandwich elements that make up the wall and roof have a 22 cm thick core of polyurethane hard foam insulation between two internal and two external OSB panels and achieve a U-value of 0.11 W/m²K. In constructing the external walls care was taken to reduce thermal bridges to an absolute minimum. Building corners and reveals along openings received additional insulation. Vestibules at both entrances serve as thermal buffer zones and contribute to a high level of air-tightness. A pressure test produced a n_{50} value of 0.8 h^{-1} for the building.
There are only few openings in the southern facade so as to avoid overheating of this house, which is built for the most part in lightweight construction. Only around 18 % of the entire facade area of the house is glazed, and a major part of the glazing is in the western entrance facade. Sliding shutters provide sun protection and visual privacy. They have a maximum reduction factor of 90 % but can be slid away completely to allow solar gains, when required. The windows in insulated timber frames have low-e coated triple glazing filled with argon gas with a Uw value of 0.7 W/m²K. The study/work area is on a gallery level that overlooks the living areas on the main (first) floor. A central open staircase intersects the various levels and terminates at the north-facing clerestory windows (Figs. B 03.01–03). As result, daylight falls from the mezzanine down to the rooms on the ground floor, where the bedrooms and a bathroom are located. A utility room, storage facilities, and a bike shed are also on this level (Fig. B 03.04). The storage room can be used for drying clothes, thanks to the waste heat from the boiler which is located there.
Almost all the materials used were chosen with sustainability in mind. Many components are recyclable or result from recycling processes.

ENERGY EFFICIENCY The energy concept of Lighthouse is based on the requirements of the British Code for Sustainable Homes.

CODE FOR SUSTAINABLE HOMES The CSH was introduced in April 2007 to assess the sustainability of new buildings, to establish standards of comfort, and to encourage a more sensible use of energy, materials, waste, and water. With a factor of 36 %, the category "energy and CO_2 emissions" is the most highly weighted of a total of nine categories.

B 03.01 Section, scale 1:200
B 03.02 Top floor plan, scale 1:200
B 03.03 Main floor plan, scale 1:200
B 03.04 Ground floor plan, scale: 1:200
B 03.05 West facade
B 03.06 Energy evaluation

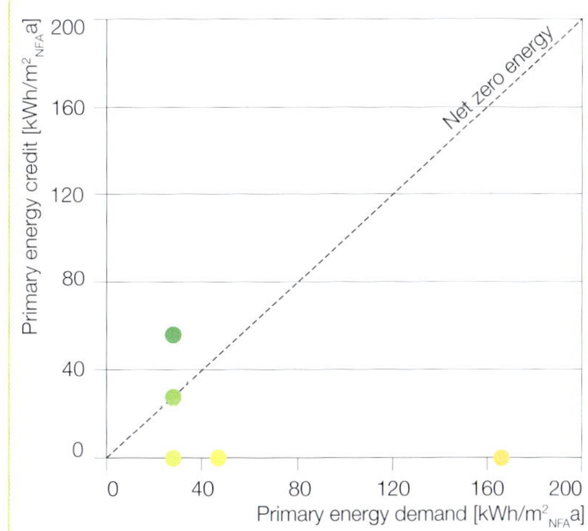

The energy values already partly calculated show that around 28 kWh/m²a of the high total energy consumption cannot be offset in the monthly balance. Standard Net-ZEB 28 results, as well as an energy plus over the entire year:

- Assumed annual total primary energy consumption (166 kWh/m²a)
- Building-related primary energy consumption (47 kWh/m²a)
- Reduction of consumption due to monthly electricity yields (138 kWh/m²a)
- Balance of remaining consumption
- Annual energy plus (26 kWh/m²)

Primary energy factors used according to GEMIS 4.5 and to energy directive EN 15603 (Fig. A 2.07, p. 31)

B 03.06

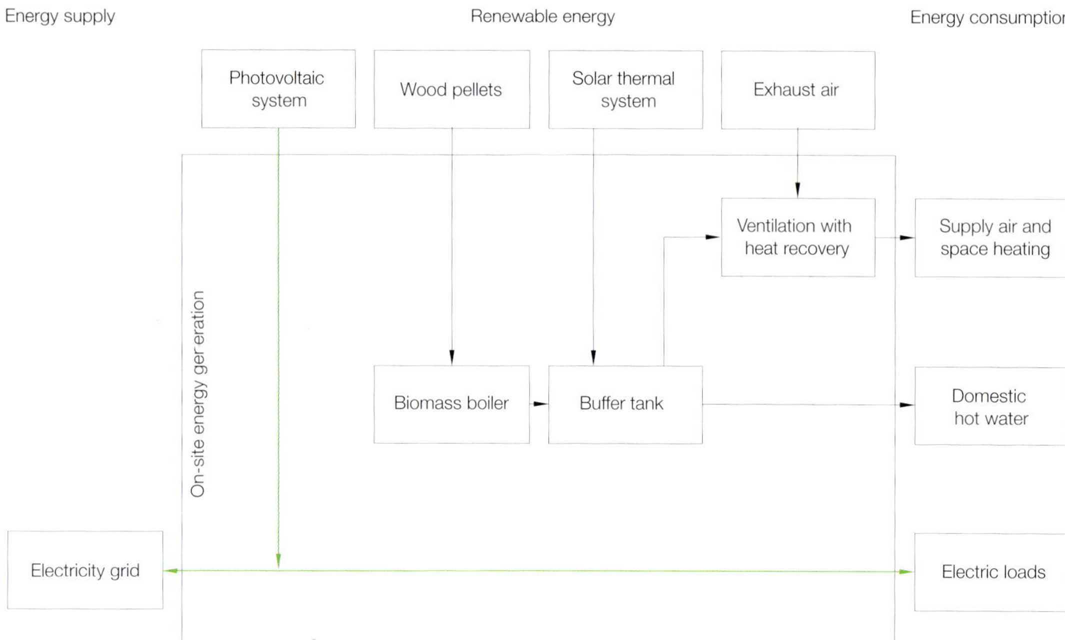

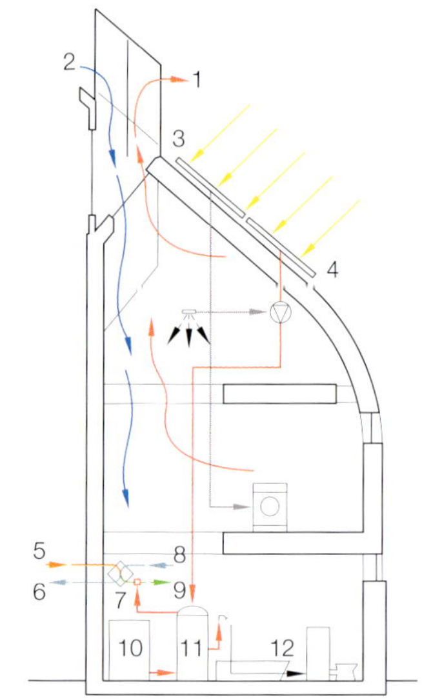

1 Passive extraction of heat
2 Passive cooling/ventilation
3 Photovoltaic system
4 Solar thermal collectors
5 Fresh air
6 Exhaust air
7 Heat recovery
8 Waste air
9 Supply air
10 Wood pellet boiler
11 Buffer tank
12 Grey water recycling

B 03.07
B 03.08

Depending on the year of completion, a particular number of points must be achieved; through a system of credits the building is given one of six different assessment ratings for sustainable houses. To meet Level 6, which is the most stringent, buildings must have net zero CO_2 emissions. Beginning 2016 this will be mandatory for residential buildings in England and Wales. This standard also requires that all consumption should be covered by local renewable energies as well as a mean U-value of 0.11 W/m²K of opaque building envelopes. The zero carbon building status includes all energy consumption sectors in a yearly balance. The energy content of building materials is not included in the energy balance, but is taken into account by sustainability certifications in the product labelling.

PASSIVE ENERGY CONCEPT The gypsum board and fibre cement panels used to line the internal walls contain phase change materials to improve room climate in summer (see phase change materials, p. 19). The daylight and ventilation concept as well as the layout of the interior are integrated. The central staircase is a light and air shaft which, together with the glazed “wind catcher” at roof level, creates a stack effect, and thus allows natural ventilation as well as passive cooling. When open, it collects cold outdoor air and leads it past the open living rooms down to the bedrooms on the ground floor. Warm air that spreads and rises can escape through controlled vents.
To reduce energy consumption, compact fluorescent lamps are used in the interior and LED lights on the exterior. All the household appliances have an A++ energy rating. Water-saving taps reduce the amount of water flowing through the shower and the washbasin to 8 l/min, which represents savings of 50 % in comparison to a conventional house. Rainwater is used for watering the garden and doing laundry, and with the help of a dual flushing system grey water is used for toilets. To make energy efficiency more easily understandable and to increase user awareness, energy monitors show all (energy) consumption, alongside general information.

The Lighthouse has a mechanical ventilation system with a heat recovery factor of 88 %. With a value of 3.31 W/(m³/h), the electrical energy efficiency of the entire plant lags behind that of standard plants (Fig. B 03.08).

ENERGY SUPPLY The energy demand for space heating and domestic hot water is covered by an automatically fed 10 kW wood pellet boiler in conjunction with the 4 m² vacuum tube collector system and a 210 l buffer tank. It is connected to the air-intake duct of the mechanical ventilation via the heat exchanger. Interiors are heated solely by means of space heating. The intention is that the pellet boiler should be used only for around six weeks of the year. The pellet storage area is integrated in the small house. Stored externally, the pellets could get damp, which would damage the automatic system of the boiler and would supply less heating energy. As a result of conceptually reducing space in the house, the pellet storage area is restricted to a volume of 0.25 m³.

In terms of primary energy, a rooftop-mount 4.7 kW_p photovoltaic system offsets the total annual energy supply (Fig. B 03.07).

ENERGY BALANCE A "smart meter system" records energy consumption and the resulting final energy debits or credits. This allows individual energy users to be identified and spotlights particularly high consumption or savings. The energy requirements of the Lighthouse were calculated during the design phase using the British SAP (Standard Assessment Procedure). The annual electricity demands for lighting, ventilation, and pumps as well as the other usual consumers of energy amount to around 52 $kWh/m^2_{NFA}a$. With a primary energy factor according to the energy directive EN 15603 this represents 163 $kWh_{prim}/m^2_{NFA}a$. This is balanced by an assumed energy output of 192 $kWh_{prim}/m^2_{NFA}a$ from the 4.7kW_p solar energy array. The surplus energy produced can be exported to the national grid and balances the primary energy equivalent for the pellets used, which amounts to approximately 3 $kWh_{prim}/m^2_{NFA}a$.
The consumption of energy is dominated by electricity. Due to the permanently operating ventilators and pumps, this figure is slightly higher than the average consumption in Great Britain. Due to lacking gas connections, a gas cooker (more common in the UK on account of high electricity prices) is not in use. The calculated annual energy costs with a wood pellet price of 2.5 Ct/kWh and an electricity price of 15 Ct/kWh amount to roughly 37 €/a, which represents around 10 % of the usual cost in a single-family house of comparable size built in 2006. The yield from solar power is calculated according to the price of the electricity supply. In the UK, crediting for solar electricity of 42 Ct/kWh and over a period of 25 years applies only to systems built after July 2009 (Fig. B 03.06, p. 65).

LESSONS LEARNED The house on the BRE research site is a demonstration building and, as such, not continuously inhabited. Consequently no real user experience can be reported, except for general observations: a number of windows on the first floor are placed too high, making it difficult to open, close, or clean them by hand. Due to the local weather conditions, the sweet chestnut wood used for the facade discolours more strongly than had been expected, leading to a deterioration in the appearance of the house over the course of time. It also negatively affects the sustainability assessment, due to the more frequent exchange of the materials used. The pellet storage area is incorporated in the small volume of the house to avoid the pellets becoming damp. The concept-related limitation of space for this house results in a very small storage area, and therefore up to three deliveries a year are required.
The high degree of prefabrication leads to the conclusion that this residential house, originally built for demonstration purposes, could be carried out in different forms, ranging from a free-standing house to a row house or as part of a housing estate with external heat supply from a biomass combined heat and power station or external wind power turbines. However, so far this model hasn't been replicated on a broad scale.

SITE	Watford (GB)
Annual global radiation at site	950 kWh/m^2a
Annual mean temperature at site	10.4 °C
Context	suburban
BUILDING ENVELOPE PARAMETERS	W/m^2K
U-value, exterior walls	0.11
U-value, windows (incl. frames)	0.70
U-value, roof surface	0.11
U-value, skylights (incl. frames)	0.72
U-value, floor slab	0.11
Mean U-value, building envelope	0.17
BUILDING EQUIPMENT PARAMETERS	
Area of solar thermal collectors	4 m^2
Area per m^2	0.05 m^2/m^2
Thermal storage volume	210 l
Storage volume per m^2	2.70 l/m^2
Photovoltaic system area	47 m^2
System area per m^2	0.59 m^2/m^2
Photovoltaic capacity	5 kW_p
Capacity per m^2	59.50 W_p/m^2
GRID INFRASTRUCTURE AND ENERGY SOURCES	
Supply infrastructure	electricity grid, delivery of wood pellets
Energy source supply	wood pellets, electricity
Feed-in infrastructure	electricity grid
Feed-in energy source	electricity

DESIGN STRATEGIES, CONCEPTUAL FOCUS
Passive house components; mechanical ventilation with heat recovery, photovoltaic system, solar thermal energy, biomass boiler, ecological building materials

BUILDING PARAMETERS	
Net floor area, NFA	79 m^2
Gross floor area, GFA	93 m^2
Gross volume, V	432 m^3
Building envelope, A	375 m^2
Surface to volume ratio, A/V	0.87 m^2/m^3
Building costs (net, construction/ technical systems)	2260 €/m^2 (2008)
Number of units	1
Total number of users	2
CONSUMPTION PARAMETERS (calc.)	kWh/m^2a
Heating energy consumption	22
Domestic hot water heating (consumption)	16
Electricity consumption	52
Total primary energy consumption	166
Total primary energy generation	192

B 03.09

B 03.07 Technical schematic of energy provision
B 03.08 Technical concept in section, scale 1:200
B 03.09 Building and energy parameters (values refer to net floor area, NFA)

HOME FOR LIFE

Lystrup, DK 2009

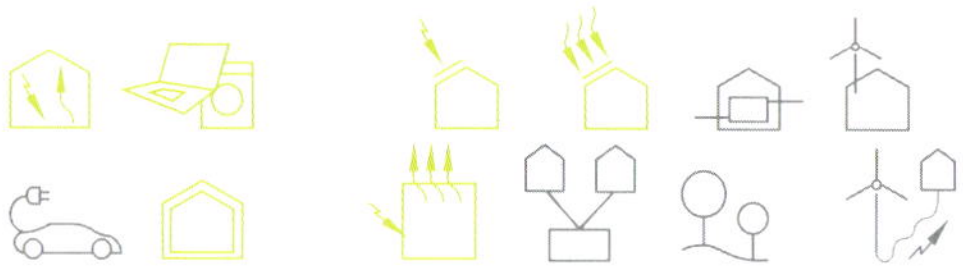

Client: VKR Holding, Hørsholm
Architect: Aart Architekten, Aarhus
Energy consultant: Esbensen Rådgivende Ingeniører, Aarhus
Building services: KFS Boligbyg, Støvring
Monitoring: University of Engineering, Aarhus, and WindowMaster, Aarhus
Main stakeholder: client

In 2009 the single-family house “home for life” (in Danish: Bolig for livet) was built on the outskirts of the Danish town of Lystrup, north of Aarhus. It is the first of six demonstration and test buildings erected throughout Europe in the framework of the programme “Model Home 2020”. Its aim is to present solutions and strategies for building energy-efficient homes without any reduction of user comfort, and additionally, to produce knowledge of the combination of different technologies for saving energy. An important aim and core aspect of the project “home for life” is the CO_2 neutral operation of the building. Thermal solar collectors and a heat pump cover the heating requirements for space heating and domestic hot water. In the annual balance, a photovoltaic system more than offsets the complete consumption of this all-electric house. Over a period of 40 years the calculated electricity surplus of 11 $kWh/m^2_{NFA}a$ is supposed to balance the energy used to produce the building.

DEVELOPMENT, DESIGN, AND STAKEHOLDERS

In the course of the programme “activehouse” the Danish VKR Holding with its subsidiaries Velux and Velfac, as well as regional partners, architects, engineers, and scientists contributed to the realisation of eight residential and non-residential buildings in five different European countries (Fig. A 3.11, p. 46). Two of these buildings were erected between 2009 and 2011 as part of the associated programme “Model Home 2020”, one of which was the “home for life”. The aims and contents of these two programmes include achieving the highest standards in terms of comfort, energy efficiency, and aesthetics, combined with a positive balance of operating energy over the life cycle to balance the embodied energy used in producing the building materials. Alongside a low space heating demand, the solutions and strategies comprise high exploitation of daylight, providing a healthy indoor climate by means of regulated introduction of fresh air, a building automation system, as well as the use

B 04.01 Site plan, scale 1:750
B 04.02 View from north-east
B 04.03 Technical schematic of energy provision
B 04.04 1st floor plan, scale 1:500
B 04.05 Ground floor plan, scale 1:500

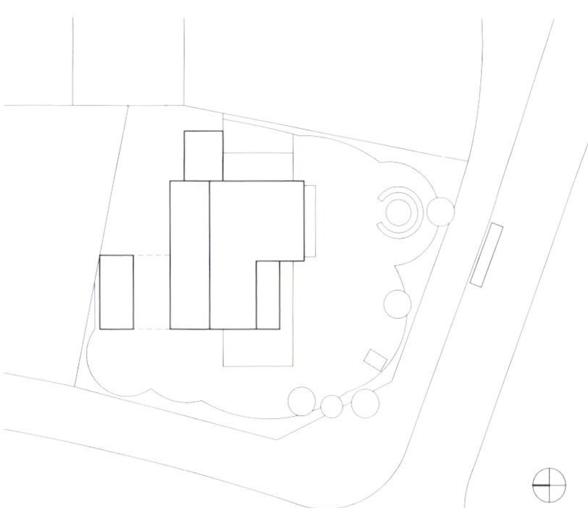

B 04.01

B 04.02

of renewable energy sources, with the focus on thermal solar energy. The buildings erected, intended to be climate neutral, should more than meet the goals of the European Union for new buildings from 2020 onwards (see European Union, p. 12f.).
Even before the completion of the “home for life” the decision was made that, following a six-month occupancy-free test phase, a four-person family should move into the house. Assisted by technical, scientific, and sociological experts, the residents tested the house, its systems, as well as its integrated concept over the period of one year.

ARCHITECTURE The design of the one-and-a-half story pitch-roofed house is a re-interpretation of a standard Danish single-family house. Through its great openness, it establishes clear relations between inside and outside. Energy consumption is reduced by solar gains through the numerous windows, which also allow the interior to interact with nature (Figs. B 04.04–06). Thanks to its pitched roof, this single-family house with a net floor area of 157 m^2 fits into the area of new buildings in the bay of Aarhus.
A large open living area on the ground floor and private rooms on the upper level offer the family members opportunities to be together as well as spaces to withdraw to. Terrace doors and full-height windows open the living room and kitchen towards the south to the garden, where a terrace serves to connect the building with the landscape. The continuous slate floor underscores the connection between inside and outside. The large number of openings, some of them skylights, create brightly lit spaces. From the utility room there is a direct connection to a roofed and partly insulated space known as the “multi-house” where vegetables can be stored, clothes dried, and bicycles kept. The rooms on the upper floor have two balconies that offer views across the bay of Aarhus.

CONTSTRUCTION AND INSULATION The house consists of a light timber frame structure with load-bearing I-beams and glued laminated timber beams. Steel beams support the cantilevered construction of the gallery above the living room and kitchen. Load-bearing laminated wood is used around the balcony and storage space on the first floor. Mineral wool behind the stiffening plywood panels in the roof and the external walls, 54 and 40 cm respectively, insulate the building. The U-value of the facade is 0.1 W/m^2K. The roof and the reinforced concrete floor slab, insulated by 50 cm of mineral wool, have a low U-value of only 0.07 W/m^2K. The natural slate tile facade determines the external appearance of the house. Little energy is required to produce the slate tile, which has a long life-span and requires no maintenance. Their dark grey colour is reminiscent of the dark surfaces of the photovoltaic array, solar collectors, and skylights, and gives the building a monolithic overall impression. The wooden facade of the entrance area provides a pleasant contrast (Fig. B 04.02).

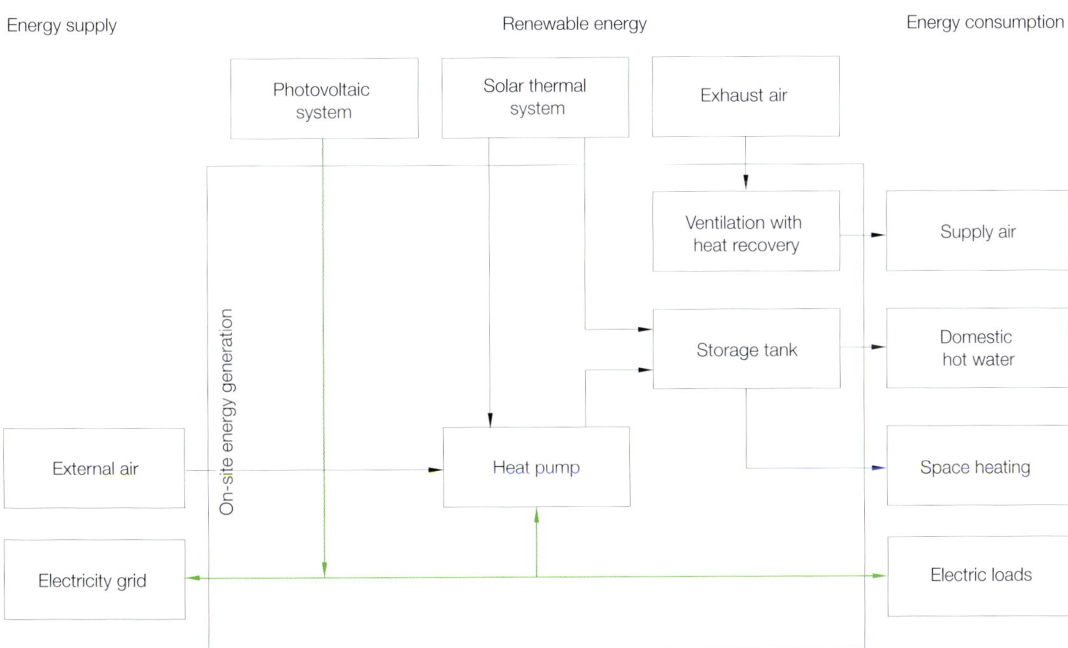

1 “Multi-House”
2 Utility room
3 Kitchen/living
4 Bathroom/WC
5 Bedroom
6 Terrace
7 Pergola
8 Carport
9 Garden shed
10 Corridor
11 Gallery
12 Balcony

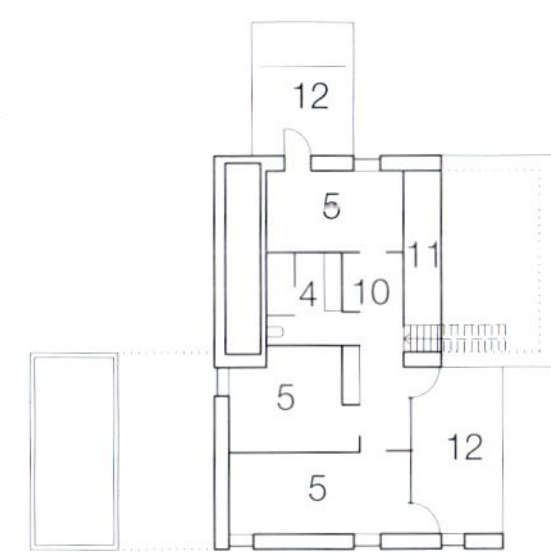

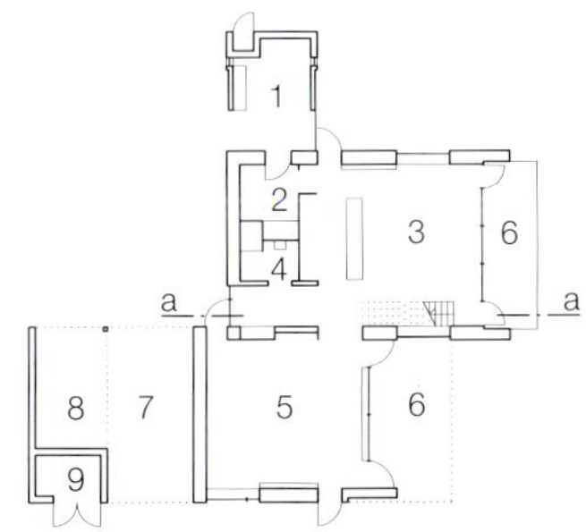

B 04.03
B 04.04
B 04.05

ENERGY EFFICIENCY The numerous windows provide plenty of daylight and solar energy gains. The ratio of the 75 m^2 of windows to the heated floor area is nearly 50 %, or more than twice the figure of a typical Danish single-family house. Twelve skylights in total are integrated in the roof surfaces. Thus, daylight enters almost every room from at least two directions. To reduce transmission heat loss, the triple-glazed windows have a low U-value of 1.0 W/m^2K in the facade and 1.1 W/m^2K in the roof, whereby the large frame proportion worsens the value of the glazing filled with argon gas. To reduce summer overheating, the overall energy transmission factor (of the windows) amounts to 45 %, with a light transmittance level of 67 %. Compared to a calculation that doesn't include solar radiation, the annual heating energy demand is covered up to 50 % by passive solar gains.

External shade elements avoid increased heat gains in summer, thus obviating the need for cooling. In the south the roof projects by almost one metre and protects the large windows from the high summer sun. In addition, sliding wooden shutters as well as internal blinds protect against glare and provide privacy. The skylights feature external blinds. PCM materials were not used. Following the initial experiences of the house, however, the latter are recommended to provide a more stable interior climate.

BUILDING AUTOMATION Many of the functions in the house are controlled by a building automation system and can be operated by remote control. When radiation increases, the inner (winter) or outer (summer) shading elements are activated automatically and separately, according to orientation and without depriving the rooms of daylight entering indirectly from each respective opposite side. On cold nights the shutters close in conjunction with the windows to reduce loss of heat.
Daylight and movement sensors switch on the artificial lighting and avoid excessive energy consumption in rooms that aren't occupied.
In summer the house is ventilated exclusively through the windows. A stack effect ensures that the skylights permit a controllable and high degree of venting of air from the ground floor to the roof (Fig. B 04.07). The windows open and close automatically in relation to CO_2 levels and room temperature. Venting stale warm air and fresh air intake through the lower windows in the facade makes passive cooling of the house possible. During colder seasons, a mechanical ventilation system with heat recovery provides adequate air exchange without great ventilation heat losses.

ENERGY SUPPLY A solar collector array covering almost 7 m^2 primarily feeds a buffer tank for the supply of space heating and hot water. An air-water heat pump connected to it uses both the heat from the solar thermal collector and environmental heat as a heat source and also feeds the joint buffer storage. From here, the building is supplied with heat by means of underfloor heating and also provided with hot water. A 50 m^2 photovoltaic system that is connected to the grid offsets the electricity consumption of the heat pump, the technical building services, and the household electricity (Fig. B 04.03, p. 69).

ENERGY BALANCE Following planning-based figures, a surplus of 1730 kWh/a is generated based on an annual solar electricity yield of 5500 kWh and by deducting 1270 kWh/a of energy used for the heat pump and the other technical equipment and household consumption of 2500 kWh/a. This surplus is fed into the national grid. This energy surplus is large enough to offset, over a period of about 40 years, the energy used to produce building construction materials, which amounts to around 500 kWh/m^2_{NFA}. The amount of energy consumed for the building structure was calculated using the software "Beat 2002" from the Danish Building Research Institute (SBi) and was based on information provided by manufacturers.

LESSONS LEARNED In the course of the project "Model Home 2020" the building underwent extensive monitoring. Following six months during which the house was open to the public, a four-

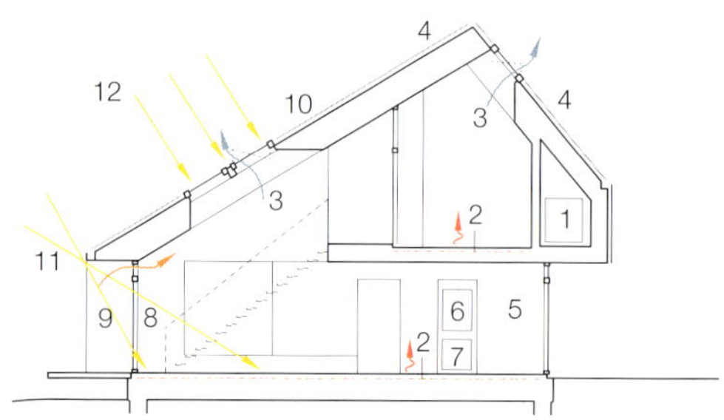

1 Ventilation/Heat recovery
2 Underfloor heating
3 Natural ventilation
4 Photovoltaic array
5 Control system (lighting, shade, ventilation)
6 Heat pump
7 Water tank
8 Internal sun shading/ night-time insulation
9 External sun shading
10 Solar thermal collectors
11 Sun position in winter
12 Sun position in summer

B 04.06
B 04.07

B 04.06 Bright living area with dark slate floor
B 04.07 Section with energy concept, scale 1:250
B 04.08 Energy evaluation
B 04.09 Building and energy parameters (values refer to net floor area, NFA)

person family lived in it between July 2009 and June 2010 in order to test it under real-life conditions. A team of engineers from the Aarhus University of Technology and anthropologists from the Alexandra Institute in Aarhus consulted with the residents and also recorded data on energy consumption, electricity production, and outdoor and internal climate, as well as the influence of the users on the building automation and the way they used it.

The recorded data and resident comfort information initially revealed a number of problems with the linked ventilation and shade components and called for optimisations of the building services. Windows remained open even on cool summer days, although they had been closed shortly beforehand by remote control. On the other hand, blinds closed and reduced heat gains as well as daylight intake. In addition, the underfloor heating initially reacted rather slowly to the dramatic weather changes typical in this area, which meant that it remained cold in the house. The movement-controlled light switch system requires that those working at a computer remain "active". If you sit motionless, the light is switched off. All in all, however, the internal room climate is good. During a heat wave at the beginning of July 2009 the shade system and the clerestory windows ensured that the house remained pleasantly cool. Comfort measurements produced ratings in the highest categories according to EN 15 251. The energy consumption during the year of measuring exceeded the values obtained from simulations by about 15 $kWh_{end}/m^2_{NFA}a$. In addition to the lack of fine tuning of the building automation system, which was frequently influenced by user intervention, the arrival of new family members resulted in greater consumption of heat and hot water. The night-time reduction of heating was omitted, as activities were continued late into the night. The average interior temperature was 23 °C, considerably higher than the assumed temperature of 20 °C. In addition, the sun protection systems were often kept closed during the day to prevent people looking into the house from outside. As result, solar heat gains were lower than originally assumed.
The touchscreen connected to the building automation led the family to constantly check their energy consumption and the generation of electricity. A kind of internal competition ensued about avoiding excessive use of energy surplus built up over the summer. When the test family moved back to their bungalow from the 1970s, located just a few hundred metres away, what they missed most of all about their temporary home was the plentiful daylight and that air in the rooms was always fresh and preheated. After they moved out, the house was sold at a normal market price. The recording of the energy and comfort data has been continued since that time.

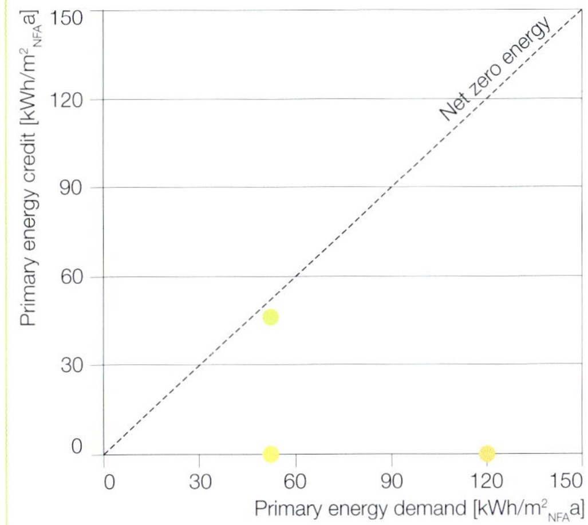

In the first year of measurements, achieving an equalised zero energy balance narrowly failed. Reasons include lack of harmonisation, partial exchange of systems, as well as changed patterns of use due to an increase in family size.

- Measured annual total primary energy consumption including household electricity (120 kWh/m^2a)
- Self-demand coverage by monthly chargeable solar power yields (66 kWh/m^2a)
- Feed-in of generated electricity not entered in the monthly balance (44 kWh/m^2a) and shortfall (9 kWh/m^2a)

Primary energy factors according to energy directive EN 15603 (Fig. A 2.07, p. 31), synthetic processing of monthly data

SITE	
Site	Lystrup (DK)
Annual global radiation at site	980 kWh/m^2a
Annual mean temperature at site	7.6 °C
Context	suburban

BUILDING ENVELOPE PARAMETERS	W/m^2K
U-value, exterior walls	0.10
U-value, windows (including frames)	1.00
U-value, roof surface	0.07
U-value, skylights (incl. frames)	1.10
U-value, floor slab	0.07
Mean U-value, building envelope	0.45

BUILDING EQUIPMENT PARAMETERS	
Area of solar thermal collectors	7 m^2
Area per m^2	0.04 m^2/m^2
Thermal storage volume	800 l
Storage volume per m^2	5.1 l/m^2
Photovoltaic system area	50 m^2
System area per m^2	0.26 m^2/m^2
Photovoltaic capacity	6 kW_p
Capacity per m^2	39 W_p/m^2

GRID INFRASTRUCTURE AND ENERGY SOURCES	
Supply infrastructure	electricity grid
Energy source supply	electricity
Feed-in infrastructure	electricity grid
Feed-in energy source	electricity

DESIGN STRATEGIES, CONCEPTUAL FOCUS
Passive house components; mechanical ventilation with heat recovery, controlled window ventilation, building automation, photovoltaic system, solar thermal energy

BUILDING PARAMETERS	
Net floor area, NFA	157 m^2
Gross floor area, GFA	190 m^2
Gross volume, V	520 m^3
Building envelope, A	469 m^2
Surface to volume ratio, A/V	0.90 m^2/m^3
Number of units	1
Total number of users	4–5

CONSUMPTION PARAMETERS (2010)	kWh/m^2a
Space heating consumption	30
Water heating consumption	18
Site energy consumption for heat (including hot water)	25
Electricity consumption	38
Total primary energy consumption	120
Total primary energy feed-in	111

B 04.08
B 04.09

KRAFTWERK B

Bennau, CH 2009

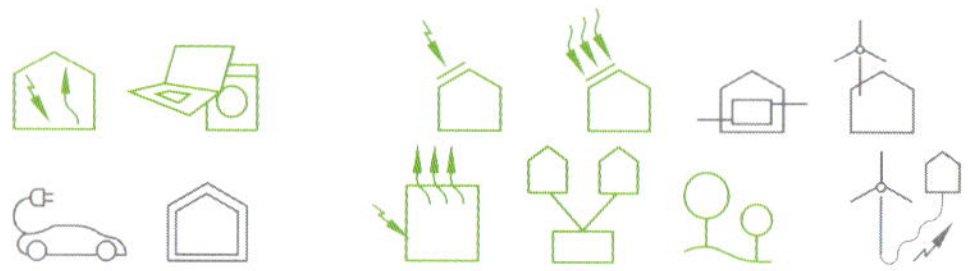

Client: Sanjo Immobilien, Altendorf
Architect: Grab Architekten, Altendorf
Energy consultant: Amena, Winterthur; Intep – Integrale Planung, Munich
Monitoring: Amena, Winterthur
Building services: Planforum, Winterthur
Main stakeholders: client and architect

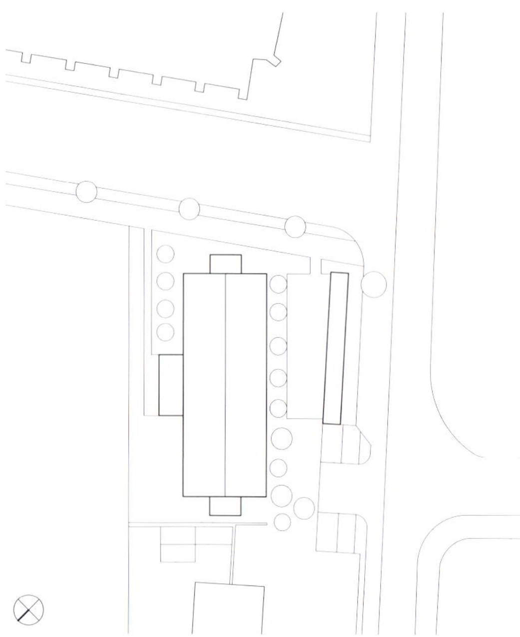

B 05.01
B 05.02

This apartment building in Bennau in the Swiss administrative district of Schwyz contains six 4-room rental apartments, each with a floor area of 140 m^2, and a loft apartment at roof level measuring 240 m^2. The programmatic name "Kraftwerk B" (Power Station B) is derived from the goals of the design: to reduce energy consumption, to cover more than the total annually required operating energy by means of active and passive solar systems, to achieve maximum heat recovery from waste air and water, and finally, to transfer heat surpluses to a neighbouring building. After its completion in 2009 Kraftwerk B was one the first buildings to be certified according to the Swiss label MINERGIE-P-ECO (see Switzerland, p. 43ff.).

DEVELOPMENT, DESIGN, AND STAKEHOLDERS

The architects, as both client and general contractor, initially set up a design competition between four teams in their own office in order to arrive at a concept for the building. This concept proposed a building with a flat roof, but the administrative district conservation authorities voted against this. The authorities' demand for a pitched roof, among other reasons due to the neighbouring church, turned out to be an advantage in the end (Figs. B 05.01 and 02): the pitch of 43 % allowed the integration of a solar electricity plant and better energy exploitation than a flat roof. The use of solar systems integrated in the facade and the roof, which determines the character of the design, called for a considerable amount of

B 05.01 Site plan, scale 1:1000
B 05.02 View from west
B 05.03 Longitudinal section, scale 1:500
B 05.04 Attic level plan, scale 1:500
B 05.05 Ground floor plan, scale 1:500

1 Entrance
2 Bathroom
3 Kitchen
4 Dining/living
5 Bedroom
6 Balcony

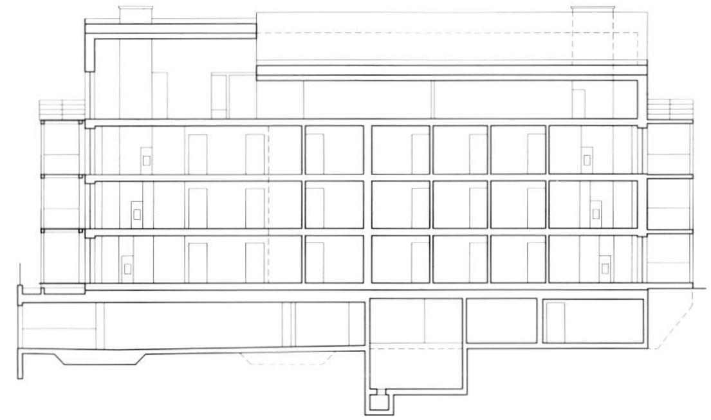

design work. With the intensive coordination of all specialists involved (for instance on questions of building physics), this extended over a period of four years. Ultimately, the building costs were around 15 % higher than those of cantonal standard construction methods. However, this figure does not consider the increased expenditure that this project demanded, or the reduced energy costs, increased value of the real estate, and the payments for electricity to be fed into the national grid, or heat to be transferred to other buildings.

ARCHITECTURE The building is comprised of a load-bearing reinforced concrete core with facades and roof made of highly insulated prefabricated wood elements. The massive building core is not only a load-bearing element, but in conjunction with adobe render walls, serves as a heat and moisture buffer and regulates fluctuations in room temperature and humidity. The south-western orientation of the building, which resulted from the line of the street and the direction in which the site faces, allowed optimal exploitation of solar radiation by means of the photovoltaic array integrated in the roof and the solar thermal system incorporated in the facade. The elements for passive and active use of solar energy, thus, determine the appearance of the building on the street. A total area of 150 m² of flat collectors alternates with story-height windows on the south-western facade. The collectors aren't separated by a ventilation gap from the facade construction and are part of the completely prefabricated curtain-wall facade elements. Thus, they form part of the building's thermal insulation and are the equivalent of an additional 10cm layer of thermal insulation.

Unheated rooms, such as the staircase on the north-western side and the basement, are completely thermally separated from the apartments. The decision was made not to use skylights, as the top apartment obtains light from both gable ends and from a glass wall facing the staircase (Figs. B 05.03–05).

The use of natural construction materials as well as the selection of materials that, when the building is eventually demolished, can be separated and recycled, meet the stringent ecological criteria of the MINERGIE-P-ECO label. In the interiors, untreated oak boards and adobe render alternate with exposed concrete surfaces and unadorned ventilation ducts. Two 10,000 litre rainwater tanks in the basement serve the bathrooms and for irrigating the garden.

ENERGY EFFICIENCY The triple-glazed, krypton-filled passive house windows with stainless steel edge spacers achieve a U_w-value of 0.57 W/m²K. They reach from floor to ceiling and allow a high exploitation of passive solar heat gains in the living rooms. To restrict ventilation heat losses, the windows can only be fully opened, not tilted. Consequently, the only way to ventilate by means of the

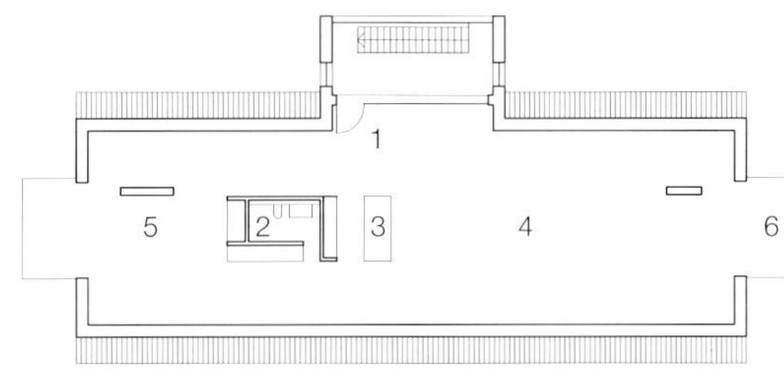

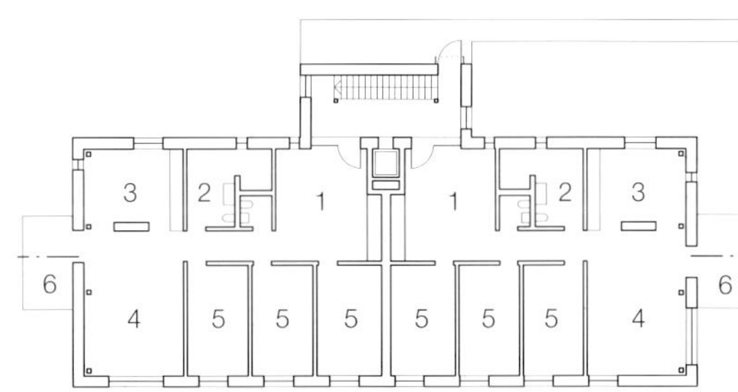

B 05.03
B 05.04
B 05.05

windows is to open them for short periods of time. The construction, which has few thermal bridges, has 44 cm of cellulose thermal insulation in the roof structure and 36 cm in the prefabricated timber element facade, which reduces the average U-value of the entire envelope to 0.2 W/m²K (Fig. B 05.11, p. 77). Possible thermal bridges at window reveals or blind stores are practically excluded by completely covering them with 6 to 8 cm of insulation. A 24-cm thick layer of insulation between the ceiling above the basement and the ground floor slab separates the basement thermally. The air tightness of the building envelope, which was proven by a blower door test, is below 0.6 h^{-1}. All the vertical shafts housing the heating pipework to the closed timber stoves are thermally insulated and air-tight. External louver shutters on all windows serve to darken rooms and provide shade in summer. The aim is to achieve, together with the ventilation heat recovery, an annual heating energy demand of 13.8 $kWh/m^2_{NFA}a$ (see passive house 15.0 kWh/m^2a).

ELECTRICAL APPLIANCES The kitchen appliances supplied by the landlord, such as fridge, cooker, and stove have an energy standard of A+ or A++ according to the MINERGIE-P-ECO system, and thus reduce the consumption of household electricity. Each apartment has a washing machine and clothes dryer with heat recovery. The washing machines and dishwashers are, in addition, connected to the hot water network. The hot water for these appliances is not heated by electricity but centrally, and therefore more efficiently, by means of solar collectors or heat recovery. In typical residential developments, such measures are rarely implemented.

BONUS-MALUS-SYSTEM Access via an internet browser to a visualisation of the data of the individual apartments allows energy consumption to be dealt with in a transparent way. Touchscreens beside the entrance doors to the individual apartments inform the tenants at any time about their energy consumption and the energy credits as monthly and annual balances. A bonus-malus system for consumption of hot water and heating energy encourages tenants to monitor their consumption of energy and thus influence the amount of rent they pay. The monthly basic rent includes an energy credit. If consumption exceeds this credit, the tenants are charged. Electricity consumption is also shown on a display, but is directly paid to the electricity provider: in this case, there is no energy credit.

ENERGY SUPPLY The ventilated 32-kW_p-photovoltaic array integrated in the roof completely covers the roof skin of the 43 % pitched roof, as well as the roof of a shed that separates the site from the road.
The 150 m² of solar collectors in the south-western facade generate all the heat required for space heating and hot water in the summer months, while covering more than 60 % of annual demands. The decentralized small storage stoves are equipped

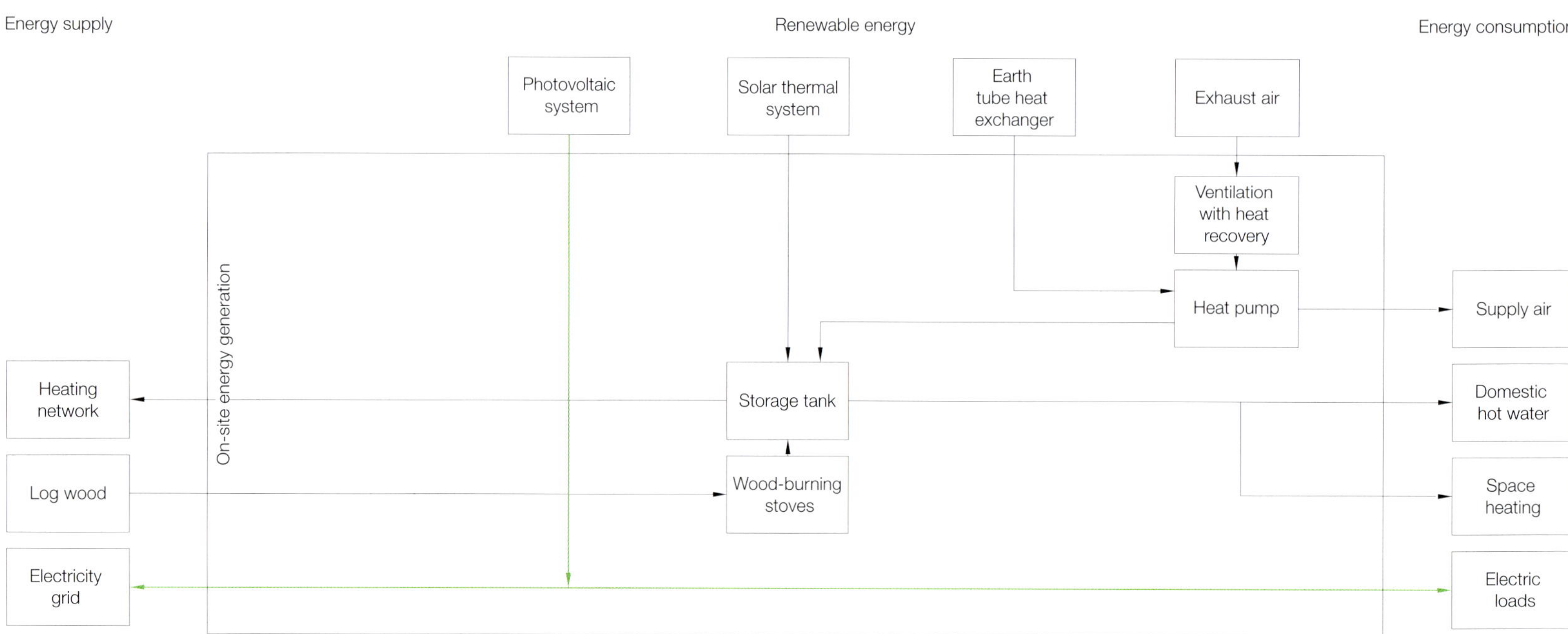

B 05.06

with water containers for heat recovery from the emissions produced by wood combustion. They supply the towel radiators in the bathrooms as well as the central 3000 l storage tank in the basement. A secondary, large seasonal 24,000 l tank is also located in the basement. This steel tank is 5 m tall with a diameter of 3 m and is completely enclosed in 50 cm of mineral wool insulation. The flow temperature of the underfloor heating is 23 to 28 °C. An air-water heat pump uses the expelled air of the ventilation system after heat recovery for pre-heating, and thus supports both the underfloor heating and the provision of hot water. This is required when the residents do not fire the small storage stoves adequately or when lower solar yields are achieved (Figs. B 05.06 and 07). The heat recovery level of the alternating current heat exchanger for the ventilation was measured 85 %. The exhaust air from the ventilation system and from the exhaust air heat pump is led into the garages and heats them to a certain degree. Even waste water is exploited for energy: by heat recovery through a wastewater heat exchanger, it heats the incoming fresh water centrally. In the first year in which measurements were made, the heat recovery of the wastewater heat exchanger costing approximately 1000 €, produced 1340 kWh and, due to the lower occupancy of the building, was below the expected figure of 2000 kWh.
Six polypropylene tubes with a diameter of 25 cm were laid parallel at 50 cm off centre with a total tube length of roughly 13 m at a depth of 2 m beneath the garden, next to the building. They form an earth tube heat exchanger for pre-warming fresh air. A downstream reverse flow heat exchanger conditions the supply air for the ventilation system in the ground with warm exhaust air before, when required, a heating coil then heats it further to approximately 20 °C. Conditioning by the heating coil functions by means of the return flow of the underfloor heating, which then flows back into the storage tank at a lower temperature, and thus increases the degree of efficiency of the solar thermal collectors. The lower temperature level of the buffer tank, which results from the cooled return flow, also allows more frequent production of heat by the solar collectors, as this becomes possible even with lower solar yields.

ENERGY BALANCE During the summer months excess and transferred heat from solar collectors heats the hot water in a neighbouring 15-family apartment building. This means that less heat is transferred to the seasonal storage tank and losses are minimised, because hot water produced is used promptly and directly. According to calculations, this is the case between June and September (Fig. B 05.10, p. 76). In an average August complete coverage can be achieved for both buildings. The transfer of heat compensates for the use of firewood for the small storage stoves in winter or the use of fossil fuels in the neighbouring building.
In sum, the solar electricity plants generate more electricity than is required to operate the building,

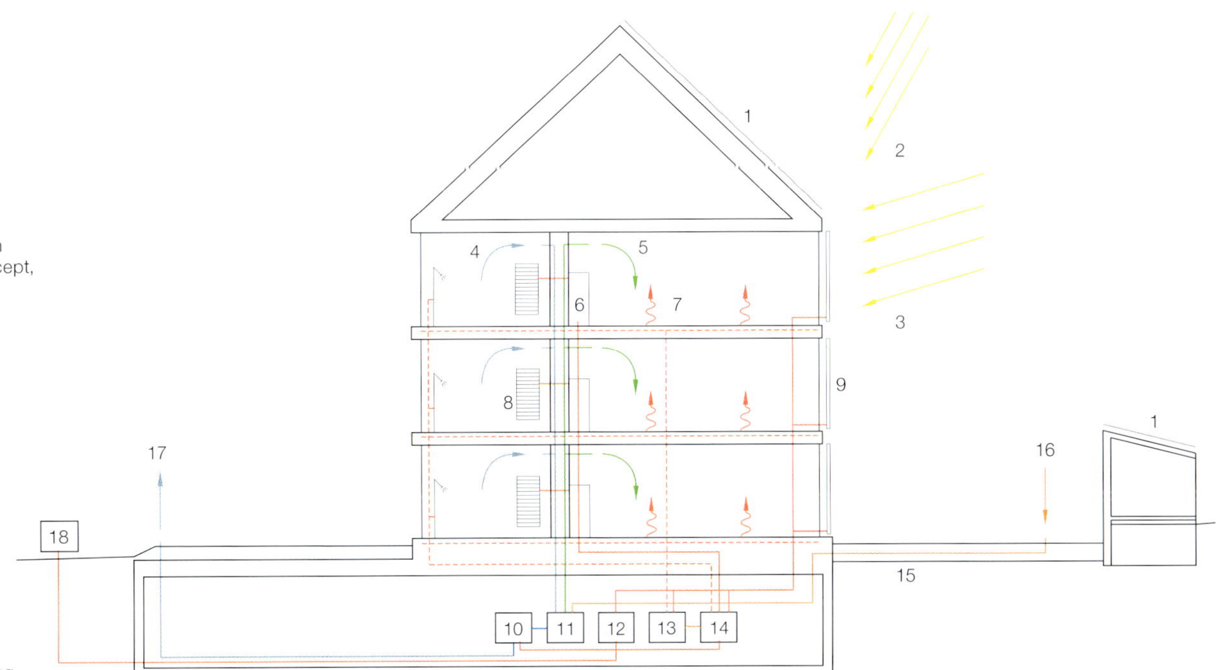

B 05.07

B 05.06 Technical schematic of energy provision
B 05.07 Section with energy and ventilation concept, scale 1:250
1 Photovoltaic array
2 Sun position in summer
3 Sun position in winter
4 Waste air
5 Supply air
6 Wood-burning stoves
7 Underfloor heating
8 Radiator
9 Solar thermal system
10 Heat recovery
11 Ventilation system
12 Buffer storage tank
13 Seasonal storage tank
14 Hot water storage tank
15 Earth tube heat exchanger
16 Fresh air
17 Exhaust air
18 Heat transfer to neighbouring building

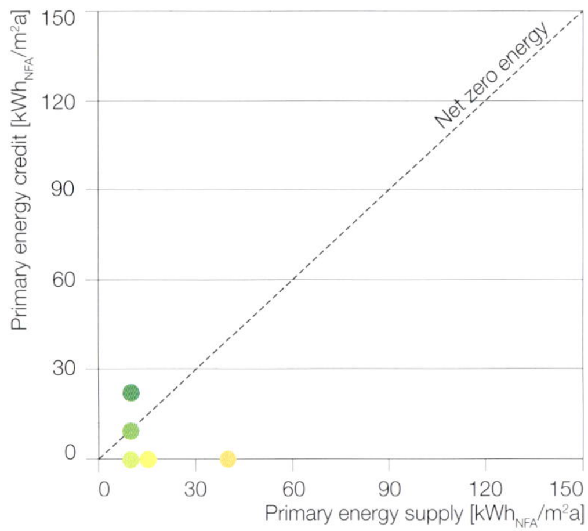

The following aspects result in the Net ZEB-11-Standard:

- Measured annual total primary energy consumption including household electricity (45 kWh/m$^2_{NFA}$a)
- Building specific primary energy consumption (16 kWh/m$^2_{NFA}$a)
- Consumption after monthly coverage of own needs by PV electricity (11 kWh/m$^2_{NFA}$a)
- Seasonal balance of remaining consumption
- Annual electricity plus and heat transfer to neighbouring building (24 kWh/m$^2_{NFA}$a)

The low consumption reveals that the building was still partly unoccupied in 2010. The energy plus achieved shows that an equalised energy balance should also be possible with full occupancy. Primary energy factors according to SIA 2031 (see Fig. A 2.07, p. 31)

including household electricity, in one year. Surplus generated electricity is fed into the public grid (Fig. B 05.08, p. 76).

LESSONS LEARNED The building was completely occupied between March 2009 and July 2010, so that the figures from the first year of operation before that do not represent the performance of the fully occupied building, and in a number of cases had to be extrapolated. Over a period of two years a program recorded the metered consumption, the interior temperatures, and the performance of the solar electricity plant, the facade collectors, and the storage system.
The first results indicate that small departures from planning data can endanger the concept of the zero or energy plus building in operation. The behaviour of the tenants differed from that of simulated "ideal tenants". The assumed room temperature of 20°C in winter was too optimistic; the reality were temperatures of 22°C. The sunshade systems were used not only during summer months to avoid overheating on days with extreme solar radiation, but also to protect against glare due to the low position of the sun in winter. This, however, reduced the passive solar heat gains, as did a grille of louvers at the level of the window parapets that was fitted later. Together with the real climatic conditions in the first winter (according to Metro Schweiz, January 2010 was the coldest month in 29 years), operative heating consumption was approximately 10% higher than initially expected.
The displays that show individual energy consumption figures had less influence on energy use than initially assumed. The balance is effective only with residents who are highly aware of the theme of sustainability or are interested in lower service costs. Despite carefully selected tenants, a bonus/malus of between 70 and 150 €/a does not seem to offer sufficient incentive for people to adapt their living habits.
On examining the performance of the plant and the interaction of the individual components, it was revealed that the heating output of the exhaust air heat pump is of critical importance for supplemental hot water provision on cold and overcast winter days. If there is no solar heat available in the thermal storage tanks, only the wood-burning storage stoves (Fig. B 05.09) and the heat pump serve as back-up. As they are subject to the influence of the users, these alternatives are difficult to control. At temperatures below -5°C the earth tube heat exchanger, relatively small due to limited space, warms the external air less strongly than had been expected and can scarcely condition the fresh air to a frost-free level. Thus, the supply air needs additional warming from the reverse flow heat exchanger or the heating coil. In a standard situation, the earth tube heat exchanger serves solely to keep the system free from frost. As result, the degree of

B 05.08
B 05.09
B 05.10

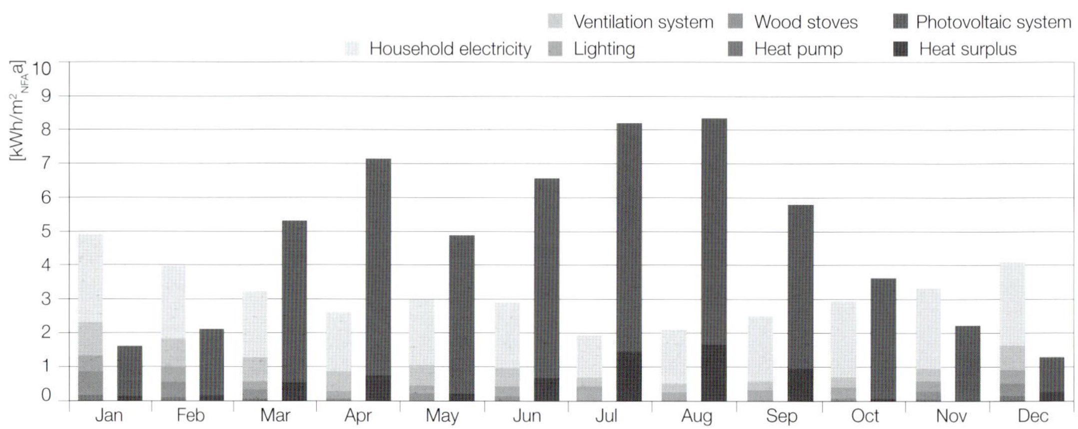

heat recovery of the heat exchanger, which is lower than planned for, appears decisive for the increased consumption of the heating coil. However, raising the air flow volume of the exhaust air heat pump would not offer any overall energy advantage on account of the increased ventilation heat losses. A volume flow regulator now allows operation at two levels: Level 1 with 50 % of the maximum air flow volume is the standard setting; level 2 allows 100 % for operation at maximum heat pump loads. For conditioning the rooms by the intake air of the ventilation system, 500 m³/h at a total interior volume of roughly 7000 m³ is sufficient. The annual performance coefficient of the exhaust air heat pump of 3.5 meets expectations. The fact that the poorly insulated heat pump cools down the utility room and the ventilation ducts fitted there, which are not insulated, is regarded as problematic.
Despite these difficulties, the annual energy balance of the first year of measurements 2009/2010 was positive. A clear electricity surplus compensates for a small shortfall in the heat balance.
The architect justifies the high personal and financial expenditure with the extensive gain in know-how that resulted from this demonstration project. Future projects will profit from this experience.
The building has already received the Swiss and the European Solar Prize, and in 2010 it was given the first Norman Foster Solar Award, thus emphasising its function as a role model.

B 05.08 Energy evaluation
B 05.09 View of interior with small wood-burning stoves and kitchen
B 05.10 Diagram of monthly balance, primary energy
B 05.11 Detail cross section roof/floor slab/exterior wall scale 1:20
1 Oak boards on heating screed 90 mm
Impact sound absorption 20 mm
Thermal insulation 10 mm
Reinforced concrete 220 mm
2 Prismatic safety glass 6 mm as cover
Facade collectors
Absorber/air cavity 42 mm
Insulation 60 mm mineral wool
Wood frame 100 × 45 mm
Rear wall 8 mm OSB
Gypsum fibre panel 15 mm
Timber joists 40 y 360 mm,
Infilled cellulose insulation
OSB-panel 15 mm
Vapour barrier
Services level 60 mm
framing, thermal insulation
Gypsum fibre panel 15 mm
Adobe render 10 mm
3 Mineral wool insulation
4 Wood-frame window with triple glazing
B 05.12 Building and energy parameters (values refer to net floor area, NFA)

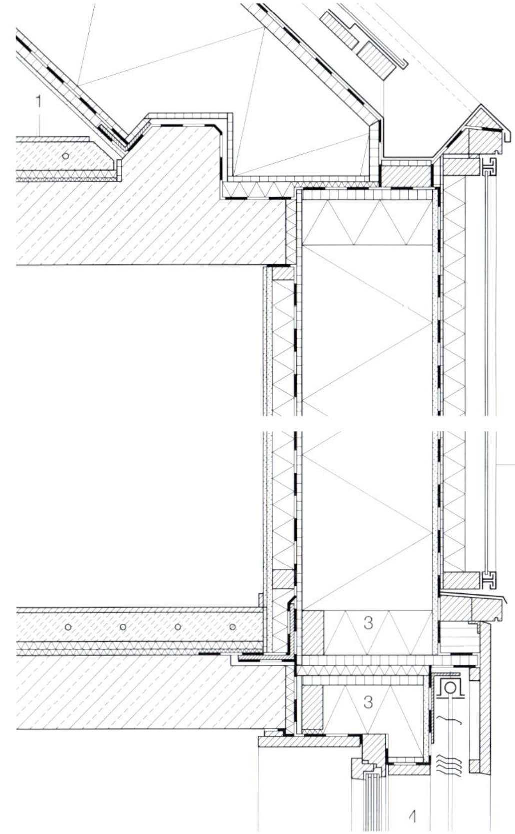

SITE	Bennau (CH)
Annual global radiation at site	ca. 1200 kWh/m²a
Annual mean temperature at site	9.5 °C
Context	rural

BUILDING ENVELOPE QUALITIES	W/m2K
U-value, exterior walls	0.11
U-value, windows (incl. frames)	0.57–0.79
U-value, roof surface	0.11
U-value, ceiling slab to basement	0.18
Mean U-value, building envelope	0.20

BUILDING EQUIPMENT PARAMETERS	
Area of solar collectors	150 m²
Area per m²	0.11 m²/m²
Thermal storage volume	27 000 l
Storage volume per m²	19.60 l/m²

Photovoltaic system area	261 m²
System area per m²	0.20 m²/m²
Photovoltaic capacity	32 kW_p
Capacity per m²	23.00 W_p/m^2

GRID INFRASTRUCTURE AND ENERGY SOURCES	
Supply infrastructure	electricity grid, deliveries
Energy source supply	log wood, electricity
Feed-in infrastructure	electricity grid, local heating network
Feed-in energy source	electricity, heat

DESIGN STRATEGIES, CONCEPTUAL FOCUS
Passive house concept, MINERGIE-P-ECO, mechanical ventilation with heat recovery, expelled air heat pump, photovoltaic arrays, solar thermal system, heat recovery from waste water, energy display for tenants, wood-burning small storage stoves, feed-in of heat

BUILDING PARAMETERS	
Net floor area, NFA	1380 m²
Gross floor area, GFA	1403 m²
Gross volume, V	3941 m³
Building envelope, A	1557 m²
Surface to volume ratio, A/V	0,39 m²/m³
Number of units	7
Total number of users	23

CONSUMPTION PARAMETERS (2010)	kWh/m²a
Space heating consumption	15
Water heating consumption	14
Site energy consumption for heat (including hot water)	11
Electricity consumption	18
Total primary energy consumption	45
Total primary energy generation	69

B 05.11
B 05.12

RENOVATION BLAUE HEIMAT

Heidelberg, D 2005

Client: GGH Gesellschaft für Grund- und Hausbesitz, Heidelberg
Architect: Gerstner Architekten, Heidelberg
Energy consultant: solares bauen, Freiburg
Building services: solares bauen, Freiburg
Monitoring: Fraunhofer-Institut für Solare Energiesysteme, Freiburg
Main stakeholders: client and energy planners

Solid construction methods were used in 1951 to erect the three-story building Blaue Heimat in Heidelberg, which forms the northern part of a block border listed housing development consisting of 155 apartments (Figs. B 06.01 and 02). Over the course of time a number of independent modernisation measures were carried out: the windows were replaced, the uppermost ceiling slab was insulated towards the unheated roof space, and a central heating system was installed. The decentralised electric or gas-operated on-demand water heaters were retained for the provision of domestic hot water. Despite these attempts at modernization, the ageing building fabric, the poor or non-existent thermal and noise insulation, as well as the outdated energy and warm water supply systems offered plenty of reasons to undertake a complete renovation in 2005.

The basis of this renovation work was initially provided by extensive construction measures that substantially reduced the building's energy requirements. The originally intended zero energy concept originally aimed at was based on primary energy credits calculated from the use of a combined heat and power plant (CHP) and a photovoltaic array on the roof of the balconies. The remaining primary energy needs were to be covered by buying regeneratively produced power (share in wind turbine). However, as no wind power shares were available

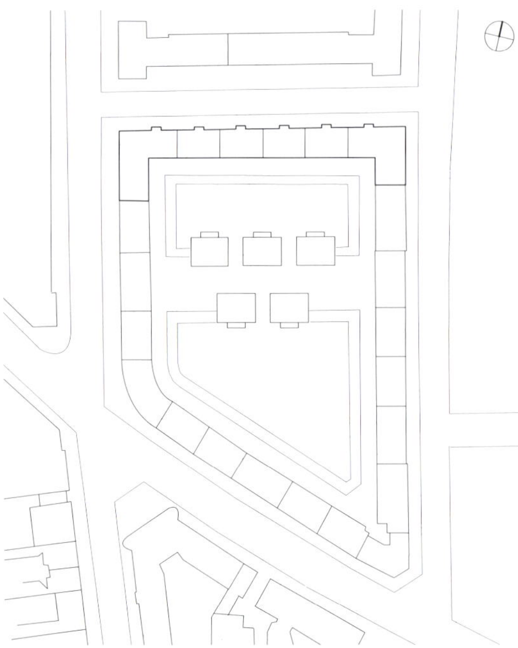

B 06.01
B 06.02

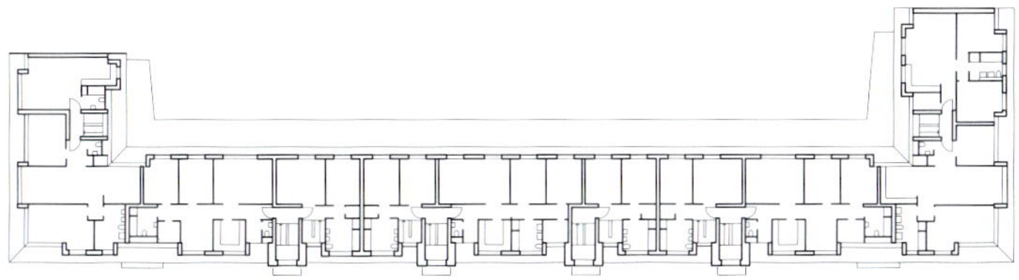

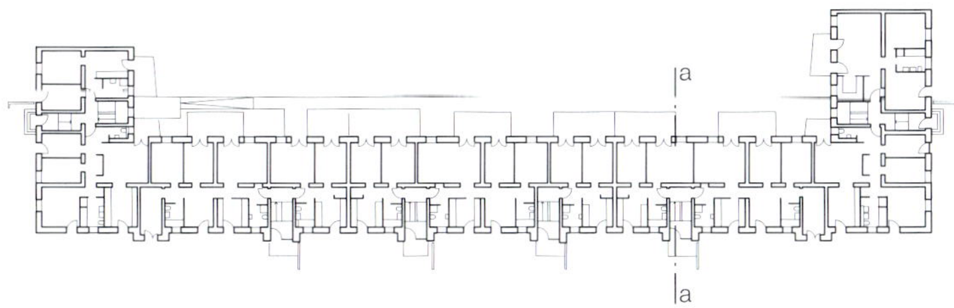

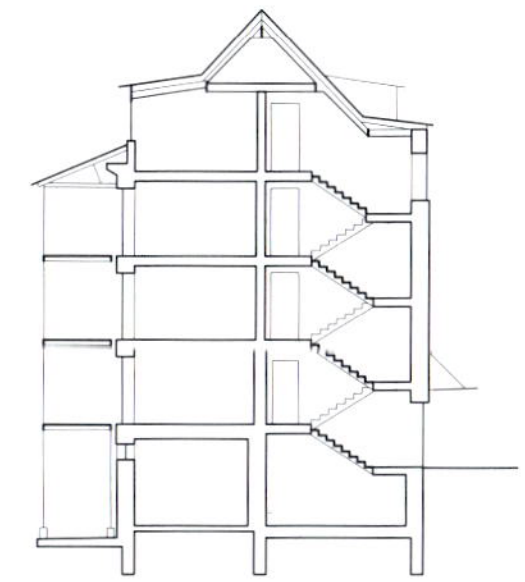

B 06.01 Site plan, scale 1:2500
B 06.02 View of the refurbished north facade
B 06.03 Attic level plan, scale 1:1000
B 06.04 Ground floor plan, scale 1:1000
B 06.05 Section, scale 1:500

B 06.03
B 06.04
B 06.05

on the market, this latter aspect was not implemented, and consequently a completely equalised primary energy balance was not achieved. Nevertheless, the concept of the Blaue Heimat represents one of the few examples of an existing apartment building that points the way to highly efficient energy supply as a step towards a zero energy building.

DEVELOPMENT, DESIGN, AND STAKEHOLDERS As well as improving the building in terms of energy consumption, the complete renovation also included giving the apartments a more attractive and contemporary design. Larger, more diverse apartment types were developed for a variety of tenant groups, offering optimal standards (balconies, attractive floor plans, efficient lighting), greater thermal comfort (central heating system, good insulation standard,) and better air quality (room air hygiene, pollen filter). To achieve this, a number of apartments were combined and additional living space was created by reusing now redundant staircases as well as the attic space. The result of these measures was that in this block border development with a total floor area of 10000 m² living space, the floor area of the northern building section was increased from 2890 m² to 3375 m². By creating larger apartments, which range in size from 2 to 4.5 rooms, the number of dwelling units was reduced from 56 to 40. The average apartment size after the renovation is 84 m², which is 30 m² more than the original average size. Due to the massive interventions required in the existing building, the renovation couldn't be carried out while the building was still inhabited, which meant that alternative accommodation had to found for the residents. The necessary increase in rents (without heating) can be balanced by reducing ancillary charges.

ARCHITECTURE The entire building has a basement with a solid concrete ceiling as ground floor slab, whereas all the other floors are constructed of timber beams covered with floorboards. On the south-facing side of the courtyard, new steel balconies were added, so that almost all apartments now have their own balcony (Figs. B 06.03 and 04). Larger areas of glazing on the south side improve the amount of daylight intake and increase passive solar gains. Other important renovation measurements included improving the sound insulation and updating the sanitary facilities according to current standards.

ENERGY EFFICIENCY To reduce the heating energy demand, the solid external walls were insulated with a 20 cm thick thermal insulation system or 16 cm thick perimeter insulation below grade. By adding 16 cm of polystyrene insulation, the U-value of the basement ceiling was improved to a figure of 0.17 W/m²K. The roof construction received 18 cm of infilled mineral wool insulation and 10 cm on top and achieves a U-value of 0.13 W/m²K. The windows are triple-glazed with wood frames, and in areas of the facade most exposed to the weather, frames are made of wood and aluminium composite. The blind stores received 7 cm of insulation. As a result, the thermal insulation mostly meets passive house standard requirements. Thermal bridges at critical points were given special attention, smaller weak spots were accepted. For instance, the bearing area between ground floor and

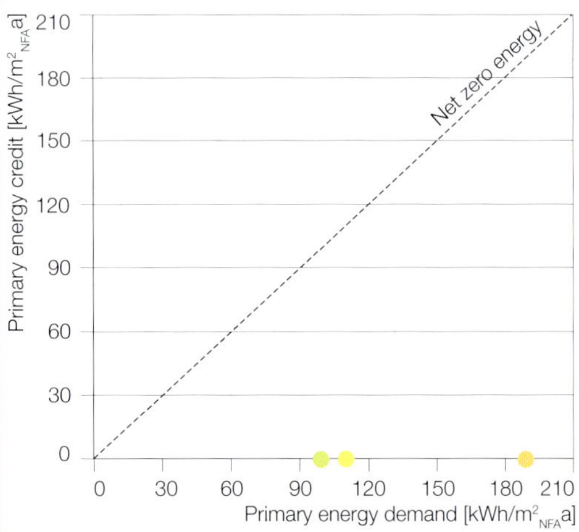

Despite the CHP and the inclusion of the yield from the photovoltaic array, the zero energy goal was not reached. During every month of an entire year, the consumption figures slightly exceed the electricity yields. No surplus remains.
The originally intended share of wind power would have to cover a deficit in electricity from renewable sources amounting to 98 $kWh_{prim}/m^2_{NFA}a$:

- Measured annual primary energy consumption including household electricity (188 kWh/m^2a)
- Building-specific consumption (112 kWh/m^2a)
- Equal monthly consumption and yields (90 kWh/m^2a)

B 06.06 Primary energy factors according to DIN 18599 (see Fig. 2.07, p. 31)

B 06.07

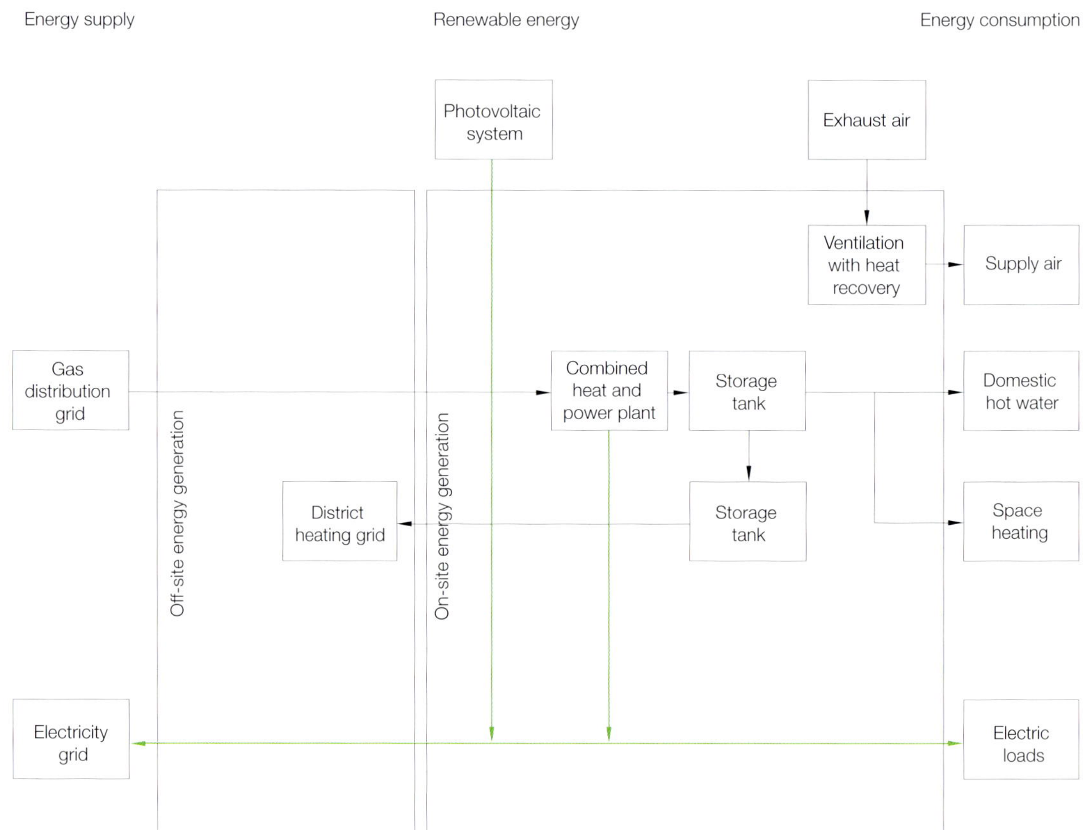

B 06.06 Energy evaluation
B 06.07 Technical schematic of energy provision
B 06.08 Photovoltaic array on the roof of the new balconies

the basement is only insulated externally. The new balconies are detached from the thermal building envelope and were connected to the building only at select points allowing thermal bridges to be reduced to a minimum.
The centralised supply system involves considerable lengths of pipework to distribute heat and hot water in this elongated building. As the pipework in the unheated basement corridors is outside the thermal envelope, it was given double the thickness of thermal insulation required by the energy directive EnEV 2002.

To reduce ventilation losses, particular attention was paid to achieving an air-tight building envelope. The air tightness according to blower door measurements is below 0.6 h^{-1}. Despite a room height of only 2.42 m, it was still possible to install a ventilation system with heat recovery (degree of heat recovery 85 %). The seven ventilation plants are semi-decentralised and located in the basement areas beside the various staircases, which means that supply lines to all the apartments are relatively short. In addition, the noise made by the ventilation appliances is confined to the basement, and the plant is easily accessible for maintenance and repair work. Maintenance costs are reduced, as only a few filters occasionally need to be exchanged at a central location. Nevertheless, the amount of air supply can be set at three different levels in each apartment by means of a volume flow regulator. Next to most of the ventilation plants, there are communal rooms for drying laundry that are also ventilated.
In order to reduce the use of electricity as well as thermal energy, the dishwashers and washing machines are connected to the warm water system. All tenants were given what is called an energy-saving package with switchable power strips and energy-saving lamps as standard fittings for bathrooms and corridors. In addition, the tenants were provided with lists of particularly economical household appliances that they were encouraged to buy, and they were educated on the subject of saving energy by means of electricity meters and targeted resident information.

ENERGY SUPPLY The energy concept is based on the supply of heating from a natural gas fired combined heat and power plant (CHP) with 80 kW thermal capacity and 50 kW electrical capacity, which covers 95 % of the heating demand. The remaining 5 % are provided on very cold days and when large amounts of hot water are required by two 92-kW peak load gas boilers. Before the renovation, a heat capacity of 485 kW was required. This joint central facility heats the renovated building (3270 m^2_{NFA}) as well as two adjoining buildings that have not yet been renovated (1950 m^2_{NFA}). Although the increased heat emission is reflected in higher gas consumption, it also means longer run times for the CHP, and thus the generation of a greater amount of electricity. To provide longer operating periods and better regulation of the heating circuit, the CHP is linked to three 1000 l storage units in the basement that are connected in succession.
The heating circuit is comprised of two decentralized storage units for hot water supply (each with a capacity of 800 l), and two further storage units, each with 325 l to supply hot water to the neighbouring buildings, served by the buffer storage (Fig. B 06.07). The distribution of heat in the buildings is by means of conventional radiators.
The generation of electricity by the CHP is supplemented by an 11.6-kW_p photovoltaic array on part of the roof. This array is financed by the customers of the plant operator, the Heidelberg municipal services, by paying a surcharge of 4 Ct/kWh on a certain percentage of their annual electricity consumption.

ENERGY BALANCE After the apartments were occupied the energy demand, and thus the efficiency of the renovation concept, was regularly examined via heat meters, and the data obtained have been analysed. With a subsidy from the German Ministry of Economics and Technology, the building underwent detailed monitoring beginning July 2009. The analysis was carried out on the basis of annual and monthly balances as well as 15 minute inter-

B 06.09
B 06.10
B 06.11

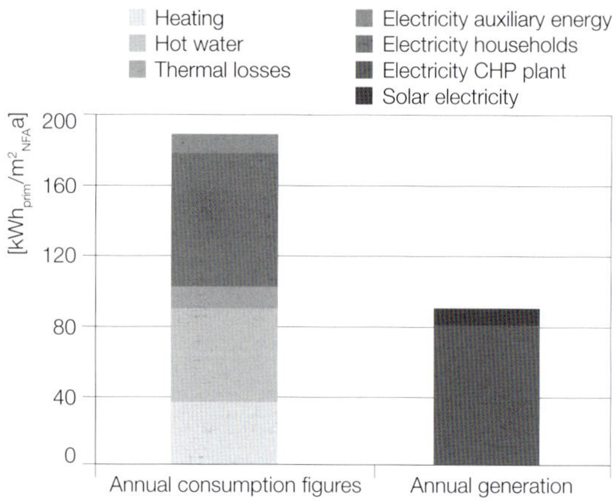

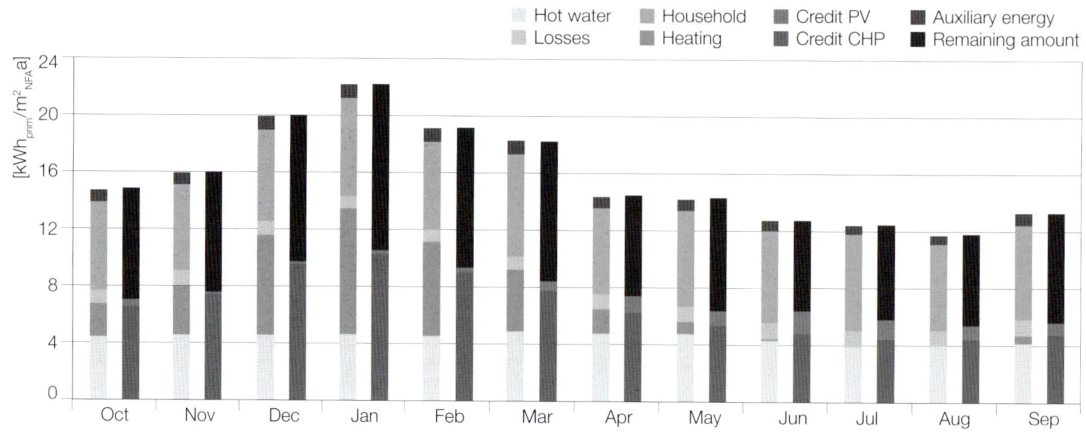

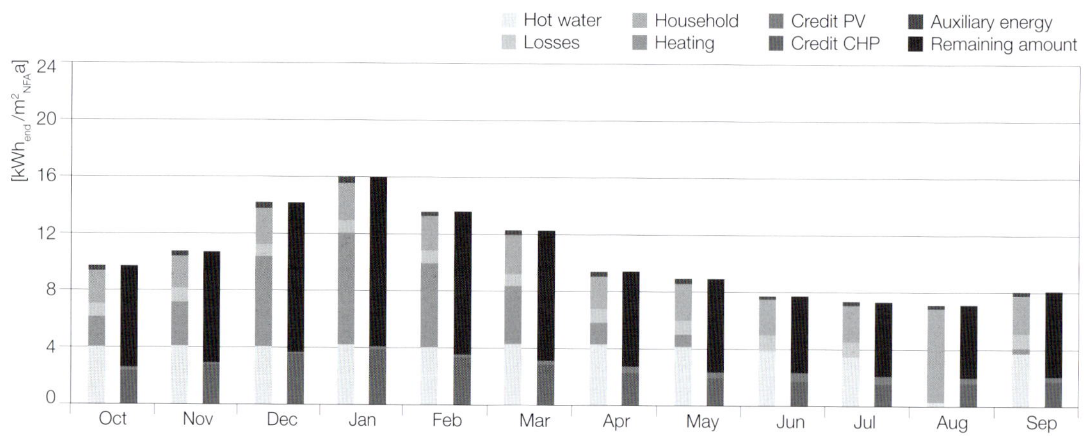

B 06.09 Primary energy balanced annual parameters
B 06.10 Monthly balance primary energy
B 06.11 Monthly balance site energy
B 06.12 Building and energy parameters (values refer to net floor area, NFA)

vals (Figs. B 06.09–11). Despite indoor temperatures in winter of over 20 °C, the annual heating energy consumption, of around 19 kWh/m²a, is very low and is almost 90 % below the level before the renovation. The hot water consumption of 11 kWh/m²a roughly matches planning figures (12.5 kWh/m²a). These low useful energy consumption figures confirm the efficiency of the construction measures employed.
An examination of the final energy balance revealed considerable losses for the provision of heat for space heating and hot water. This was in part due to the fact that the thermal efficiency of the CHP is comparatively low at 56 %, but also because the centralized supply of energy throughout a block border structure typically involves pipes and ductwork of significant lengths, in some cases more than 200 m. In combination with the seven storage units, this means that distribution and storage losses are very high. With regard to supply of hot water, the observation was made that more than half the energy used is lost through circulation and storage losses. With a figure of 12.5 kWh/m²a, the circulation losses even exceed the measurements taken for hot water consumption.
Despite the transfer of heat to neighbouring buildings and the feed-in of the electricity produced in the process, it has not been possible to achieve an equalized primary energy balance for the Blaue Heimat building. The household electricity consumption, around 30 kWh/m²a (which is 78 $kWh_{prim}/m^2_{NFA}a$), matches the average figure in Germany. After deducting the credits for input and the transfer of heat, the remaining total primary energy consumption figure is 98 $kWh/m^2_{NFA}a$. The primary energy balance shows that the credits suffice to cover the primary energy consumption for space heating, hot water, and auxiliary power.
The photovoltaic array on part of the roof provides roughly 9800 kWh of electricity annually, and thus covers about 10 % of the deficit in the balance. As the array is owned by the Heidelberg municipal services, whose electricity customers financed it as a way of compensating for their own emissions, this share cannot be credited repeatedly (Figs. B 06.06, p. 80, and B 06.08, p. 81).

LESSONS LEARNED Even though the goal of a zero energy building cannot be achieved by the local production of electricity, the project still indicates how existing buildings can achieve very efficient building and supply standards and how the option of coupling power and heat can be integrated. There is clear potential for further optimisation in the hot water supply and heat distribution. The connection to the public electricity grid allows the operation of the CPH to be harmonised in the future with time variable feed-in tariffs. Electricity is most likely to be generated if self-demands are covered or feed-in tariffs are high. This would allow the system to be used even more economically.
In addition to the improvements to comfort and the broader range of apartment types, tenants must now invest only about 0.15 € to heat one square metre living area and to provide hot water, due to high energy savings. Before the renovation, this figure was around 1.00 €/m². As a result, it was possible to halve the intended rent increase (without heating) of 1.70 €/m² following renovation to only roughly 0.85 €/m², including heating.

SITE	Heidelberg (D)
Annual global radiation at site	1050 kWh/m²a
Annual mean temperature at site	9.2 °C
Context	urban
BUILDING ENVELOPE PARAMETERS	W/m²K
U-value, exterior walls	0.15
U-value, windows (including frames)	1.20
U-value, roof surface	0.13
U-value, ceiling slab to basement	0.17
Mean U-value, building envelope	0.31
BUILDING EQUIPMENT PARAMETERS	
Photovoltaic system area	86 m²
System area per m²	0.03 m²/m²
Photovoltaic capacity	12 kW_p
Capacity per m²	3.60 W_p/m^2

CHP capacity	80 kW_{th} 50 kW_{el}
Capacity per m²	23.70 W_{th}/m^2 14.80 W_{el}/m^2
Gas boiler capacity	184 kW_{th}
Capacity per m²	55 W_{th}/m^2
Thermal storage volume	4700 l
Storage volume per m²	1.40 l/m²
GRID INFRASTRUCTURE AND ENERGY SOURCES	
Supply infrastructure	electricity grid, gas grid
Energy source supply	natural gas, electricity, green electricity
Feed-in infrastructure	electricity grid, local heating grid
Feed-in energy source	electricity, heat

BUILDING PARAMETERS	
Net floor area, NFA	3375 m²
Gross floor area, GFA	4689 m²
Gross volume, V	14,653 m³
Building envelope, A	5705 m²
Surface to volume ratio, A/V	0,39 m²/m³
Building costs (net , construction/ technical systems)	1260 €/m² (2005)
Number of units	40
Total number of users	approx. 90
CONSUMPTION PARAMETERS (2010)	kWh/m²a
Space heating consumption	19
Water heating consumption	11
Site energy consumption for heat	92
Electricity consumption	33
Total primary energy consumption	188
Total primary energy supplied	90

B 06.12

KLEEHÄUSER

Freiburg/Vauban, D 2006

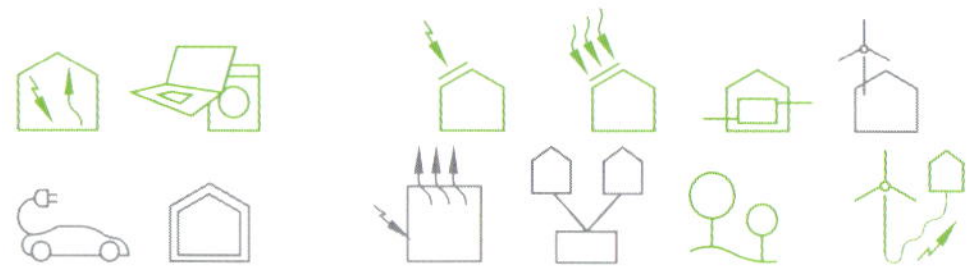

Client: Kleehäuser Ownership Association, Freiburg, Germany
Architect: Common & Gies Architekten, project architect: Michael Gies, Freiburg
Energy consultant: solares bauen, Freiburg
Building services: solares bauen, Freiburg
Monitoring: Jörg Lange, Kleehäuser, Freiburg
Main stakeholder: client

A group of twenty-four parties formed an ownership association in 2006 to build two multi-generational houses following the requirements of the 2000-watt society (see Switzerland, p. 13f.). The design goal of the two "Klee house" apartment buildings, located along Paul Klee Strasse in the Vauban district of Freiburg, Germany, was to consume an average of less than 500 watts of primary energy demand per person and for residential use. Both passive house design and consistent energy conservation enable reducing energy consumption to cover primary energy generation by a combination of small-scale combined heat and power plant, solar thermal collectors, photovoltaic systems, and acquiring shares in external wind power systems. All annual consumption-related emissions can be offset from renewable energy sources as required by 2000-watt society standards.

DEVELOPMENT, DESIGN, AND STAKEHOLDERS

The design is based on the Project Wohnen und Arbeiten (Living and Working), one of the first four-story passive houses in Germany, also planned and built in Vauban seven years earlier by the same architects and engineers. Similar to the prior project, efficiency plays a key role in all design aspects of the Kleehäuser, in regard to spatial design, construction, costs and above all, operating energy and its costs.

Two vacation apartments, thirteen privately owned apartments, and ten rental apartments with a total net floor area of 2520 m^2 were created in two buildings of differing height under the terms of a cooperative building agreement (Figs. B 07.02–04). The advantages of the cooperative include the joint use of space and equipment such as a workspace, a common recreational room with kitchen, laundry rooms with washers and dryers, and a garden. Different than purchasing completed apartments or buildings from developers, the owner association benefits financially from property taxes comprising only a fixed sum based on the property and not the completed building.

The district of Vauban, very lively and developed according to ecological principles, complements the project's attractiveness with its various infrastructure systems in walking distance, good public transport connectivity, and public green spaces. The densely populated quarter, the site of a former military base, is situated along the green fringe of Freiburg, yet close to the city centre (see Energy Plus Community, p. 94ff.). Two parking garages are located along the district perimeter, and a car sharing initiative supports the car-free community concept. Of the twenty-three Kleehäuser parties, seventeen don't own a car.

B 07.01
B 07.02

ARCHITECTURE About seventy-five individuals, from infants to eighty-year-olds, occupy the Kleehäuser since July 2006. Many were already involved in the design phase and contributed their own ideas. A key requirement is handicapped accessibility throughout the project. Generously dimensioned entrances without elevation changes, exterior walkways, and elevators guarantee that not only the living quarters, but also all common areas are accessible to individuals with disabilities. Accessibility via exterior walkways complements variations in apartment sizes and their adaptation to changing lifestyles. The construction type allows flexibly placing load-bearing partitions to create spaces with differing widths according to a modular building structure that enables joining or separating apartment units. Apartments interiors and finishes meet individual owner requirements.
Extensively glazed facades facing south optimise passive solar energy gains in winter. Cantilevered balconies oriented towards the south and extending up to two meters provide shade and prevent overheating in summer. The remaining facades feature a smaller degree of window surfaces. The layout follows the same principle: living spaces are generally oriented towards the south and other areas are on the northern side of the buildings (Figs. B 07.05 and 06).

CONSTRUCTION AND INSULATION The exterior reinforced concrete walls and the light-weight infill wood frame construction are insulated with several layers of mineral wool nearly 30 cm thick. As result, the closed exterior wall surfaces have a heat transfer coefficient of 0.17 W/m²K. Untreated wood or steel panels cover gable-end facades. The untreated materials develop a patina over the years, are recyclable, and aren't likely to require replacement for a long time. This results in both financial and material savings during operation of the two buildings in accordance with the original design principles. White fibre cement boards and triple-glazed wood frame windows are placed alternately along the vivid balcony and exterior walkway facades. 30 cm of expanded polystyrene insulation (EPS) are layered between the reinforced concrete ceiling and the extensive rooftop planting. The basement, generally used as a common area, is included in the thermal envelope and fully insulated using rigid foam panels. The supply and exhaust air system ducts feature improved thermal insulation to reduce heat loss.

BUILDING SERVICES The individually controllable supply and exhaust air ventilation system provides a comfortable interior quality. A switch mount next to each apartment entrance serves to adjust air flow as required according to three settings. In winter,

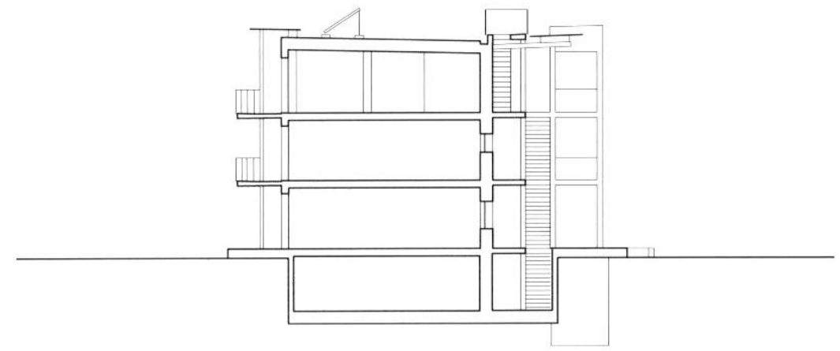

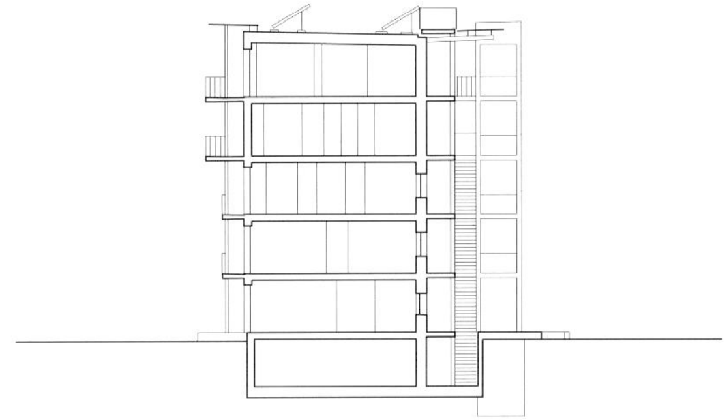

D 07.03
B 07.04

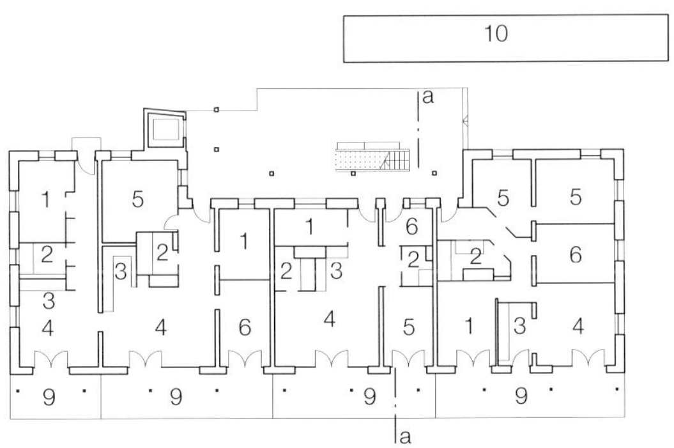

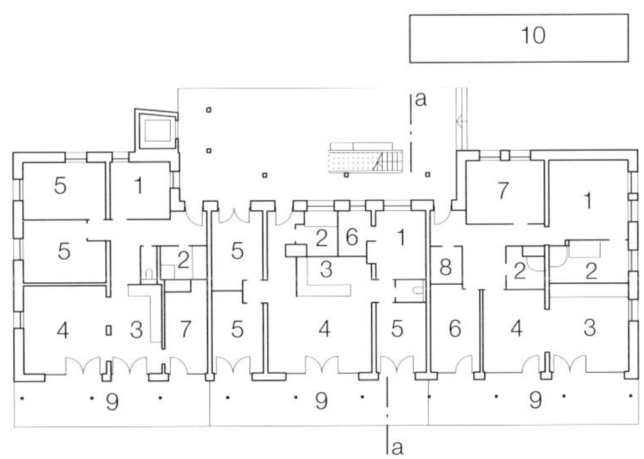

B 07.05
B 07.06

1 Bedroom
2 Bathroom
3 Kitchen
4 Living area
5 Playroom
6 Office
7 Guest room
8 Storage room
9 Patio
10 Bicycles

B 07.01 Site plan, scale 1:10,000
B 07.02 Open space between the Kleehäuser
B 07.03 Section, short Kleehaus, scale 1:600
B 07.04 Section, tall Kleehaus, scale 1:600
B 07.05 Ground floor plan, short Kleehaus, scale 1:600
B 07.06 Ground floor plan, tall Kleehaus, scale 1:600

interior air tends to be dry when it is exchanged normally, but continuously. Kleehäuser occupants can switch off the ventilation and conserve moisture when absent, so that humidity levels inside the apartment usually remain at above 40 %. In select apartments, loam render was used as a moisture buffer that contributes to maintaining comfortable humidity levels.

ENERGY EFFICIENCY The Klee houses follow the zeroHAUS standard, which was formulated by the energy consultants and strictly follows the goals of the 2000-watt society in regard to building energy requirements and is, in fact, even more rigorous: The share of median power consumption attributable to a building is in the order of 500 watts per person. zeroHAUS takes the actual energy consumed by the building, including hot water, distribution losses, and power consumption into account and contractually requires operators to verify consumption coverage from renewable sources. In addition, the CO_2 balance must be at least neutralized either by self-generated renewable energy or purchasing shares in off-site renewable energy facilities. Also, the average heat transfer coefficient of the building envelope must be at least 45 % less than required by the German energy savings directive 2002. The heat transfer coefficient of the Kleehäuser is 0.21 W/m²K. The owner association model enables all occupants to share large freezers, five washing machines with hot water connection, rarely used dryers and drying rooms in the basements, and as result, significantly reduce electricity consumption and acquisition costs. The shared common rooms reduce necessary total volume of the Kleehäuser. Energy-efficient elevators, efficient building equipment, and energy-saving or LED interior and exterior lighting also reduce electricity consumption. Further, occupants have become accustomed to avoiding unnecessary standby loads and turn off lights and ventilation systems when not required. Natural gas stoves

B 07.07
B 07.08

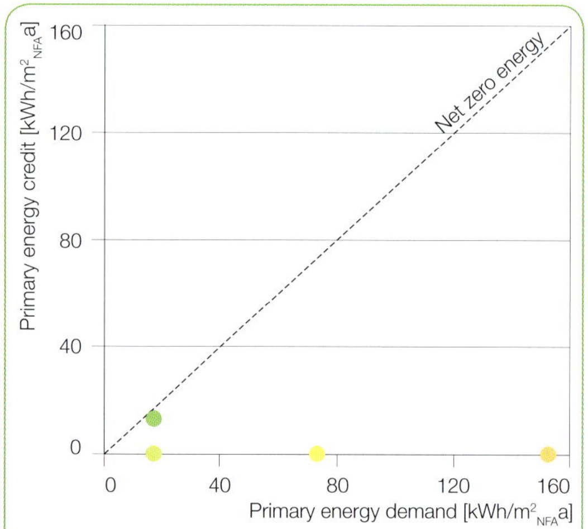

Zero energy balance cannot be achieved without applying the electricity credits from the wind power investment. Even when the renewable energy from wind power is included, consumption is not quite covered:
- Measured annual total primary energy consumption including household electricity (152 kWh/m²a)
- Residual consumption after deducting monthly energy yield (73 kWh/m²a)
- Self-demand coverage including the wind power component (135 kWh/m²a)
- Surplus renewable energy generated by wind (13 kWh/m²a)

Use of primary energy weighting factors as per DIN 18599 (see Fig. A 2.07, p. 31)

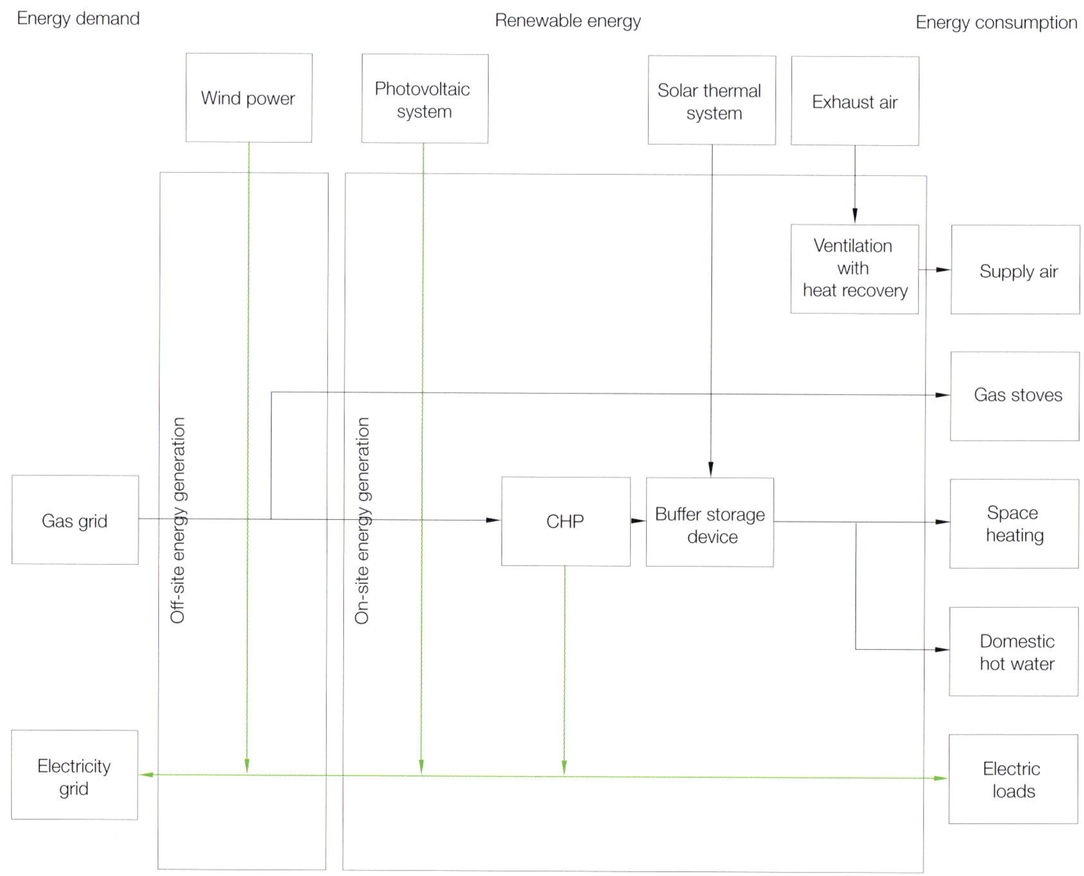

B 07.07 Energy evaluation
B 07.08 Technical schematic of energy provision
B 07.09 The Sankey diagram shows energy flows proportional to flow quantity and weighting of individual loads and generators for electricity (top) and heat (bottom); 2010 metered values.
B 07.10 Chart of monthly consumption costs per square meter for the Kleehäuser compared to the German average
B 07.11 Renewable load-match of primary energy per person

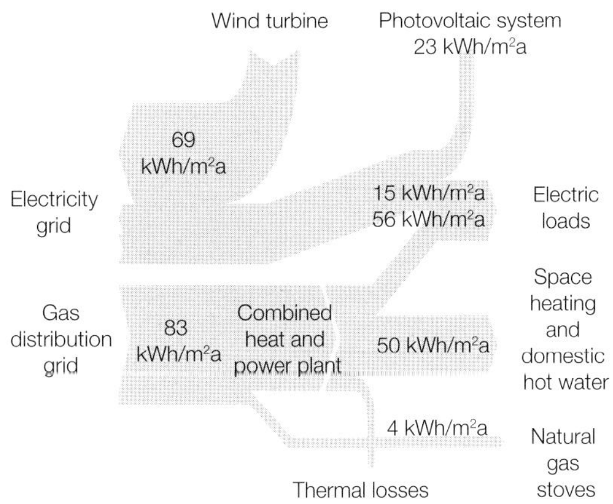

B 07.09
B 07.10
B 07.11

	German average 2008 €/m² month	Kleehäuser 2008 €/m² month
Heating	0.90	0.19
Hot water	0.28	0.10
Building services electricity	0.05	0.10
Water/wastewater	0.39	0.24
Household electricity	0.53	0.21[1]
Total consumption cost	2.15	0.84

[1] including electricity for shared washing machines

	Power/person and year
Total primary energy consumption	453 W
CHP plant feed-in	83 W
Renewable energy generated from PV and wind	354 W
Renewable load match	96 %

and energy-saving household appliances are used in all apartments.

ENERGY SUPPLY Because the apartment owners and tenants have formed an association and cooperative, and since the two buildings comprise a single legal entity, they are permitted their own local power network. A natural gas combined heat and power plant with an electrical rating of 14 kW and a thermal rating of 30 kW generates electricity for self-demand coverage, and waste heat is used to cover heating requirements. Even though additional solar thermal systems reduce runtimes, and thus, cost effectiveness of the combined heat and power plant, the decision was made to install flat collectors covering 61.2 m². They transfer heat into a 3900 l solar storage tank with external heat exchanger connected to the small-scale heating network located between the two houses. The tank is in the basement, and thus, within the building's thermal envelope. A ventilation system supplies the airtight buildings with a controlled air flow that is preheated at 20° C by the heat recovery system (85 %).
Additionally, a 23 kW_p flat roof-mount photovoltaic system provides electricity and offsets a portion of the combined heat and power plant's CO_2 emissions. Three purchased shares in a 6300 kW_{el} wind power system located in nearby St. Peter in the Black Forest cost just under € 27,000. Since the systems aren't situated in proximity to the buildings, primary energy totalling about 200,000 kWh annually flows entirely into the public electricity grid, where it raises the renewable share of the electricity mix. Thus, the wind turbines are considered part of the electricity grid (Fig. B 07.08).

ENERGY BALANCE Required maximum primary energy demands of less than 100 $kWh/m^2_{NFA}a$, including feed-in electricity as determined by the zeroHaus standard, were actually 10 % lower in the Kleehäuser. Furthermore, the various generators, including the wind power systems, can cover all of the energy consumed in the form of heating, hot water, and electricity for building services, lighting, household appliances, and entertainment. Since all consumers are considered within calculations, the vacation apartment and the guest rooms are also included on the balance sheet (Fig. B 07.07). In the annual energy balance calculation, gas required for heating and electricity generation is offset against credits for electricity feed-in from the photovoltaic system, the combined heat and power plant, and the external wind turbines. The balance sheet calculation uses primary energy weighting factors and CO_2 equivalents. In 2009 the primary energy weighting factor for electricity was changed from 2.60 from 2.70, which led to a reduction in electricity credits and, as result, negatively impacted the balance sheet calculation. Although primary energy demand (electricity component) increased only slightly, the credit component was significantly lower. The expectation is that the share of renewable energies in the electricity grid will continue to increase, which will negatively impact the overall balance sheet of the Kleehäuser in the long term. In 2010 unusually cold weather early and late in the year resulted in an increase in gas consumed by the combined heat and power plant. When heating demands are high, it operates twenty-four hours a day. However, its rating (12 W_{th}/m^2_{NFA}) is adequate. Although the primary energy demand increases accordingly, the amount of electricity generated is also higher than in earlier years. At the same time, low solar radiation reduces the amount of solar power produced. Therefore, the owners decided to generate more renewable energy on site, and a previously vacant roof became available, in addition to an excellent EEG subsidy. As result, a photovoltaic system supplying an additional 16 kW_p of power went on-line in April 2010. Only by installing the new system, achieving a positive annual energy balance became possible in 2010. In that year the Kleehäuser occupants consumed on average 453 watts per person for the "living" category, which is in accordance with 2000-watt society guidelines. This is offset by an average amount of electrical power of 437 watts generated by the combined heat and power plant, the photovoltaic system, and the wind turbine share. Thus, annual coverage via renewable energy sources reaches 96 %. The primary energy balance for 2010 was just short of 4 kWh/m²a, so that the emission balance is also almost fully met by renewable sources (Fig. B 07.07).

B 07.12
B 07.13

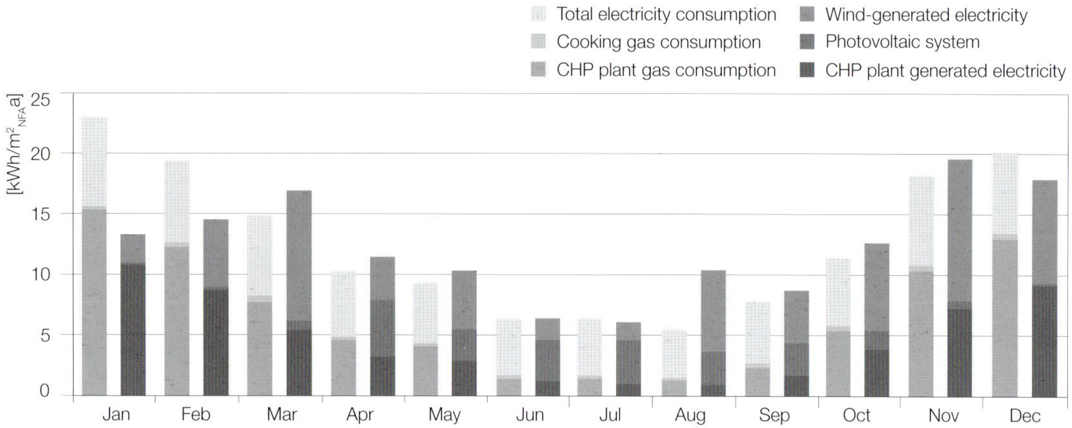

SITE	Freiburg (D)
Annual global radiation at site	1150 kWh/m²a
Annual mean temperature at site	11.6°C
Context	suburban
BUILDING ENVELOPE PARAMETERS	W/m²K
U-value, exterior walls	0.17
U-value, windows (including frames)	0.98
U-value, roof area	0.11
U-value, ceiling above basement/floor slab	0.18
Mean U-value, building envelope	0.21
BUILDING EQUIPMENT PARAMETERS	
Solar collector area	60 m²
Area per m²	0.02 m²/m²
Thermal storage tank volume	3900 l
Storage tank volume per m²	1.50 l/m²
Photovoltaic system area	202 m²
Area per m²	0.08 m²/m²
Photovoltaic capacity	23 kW_p
Capacity per m²	9.12 W_p/m^2
Combined heat & power plant capacity	30 kW_{th} 14 kW_{el}
Capacity per m²	11.90 W_{th}/m^2 5.56 W_{el}/m^2
Wind capacity	6300 kW_{el}
Capacity per m²	2500 W_{el}/m^2
GRID INFRASTRUCTURE AND ENERGY SOURCES	
Supply infrastructure	electrical grid, gas grid
Energy source supply	natural gas, electricity
Feed-in infrastructure	electrical grid
Feed-in energy source	electricity
BUILDING PARAMETERS	
Net floor area, NFA	2520 m²
Gross floor area, GFA	2965 m²
Gross volume, V	10,909 m³
Building envelope, A	4402 m²
Surface to volume ratio, A/V	0.40 m²/m³
Building costs (net, construction/ technical systems)	1154 €/m² (2006)
Number of units	25
Total number of users	75
CONSUMPTION PARAMETERS (2010)	kWh/m²a
Space heating consumption	14
Hot water consumption	10
Site energy consumption for heat (including hot water)	61
Electricity consumption	26
Total primary energy consumption	152
Total primary energy generation	148

According to the "concept for a new definition of a zero energy building" originally presented in the context of EnEV/DIN V 18599 on page 42 of the German publication "Energiebilanzierung: Praxis, Normung und Gesetzgebung" (energy balance: practice, standardization, and laws), compensation for energy consumption via off-site systems is omitted, and instead, primarily allocated to the power grid. Thus, power generated by wind turbines wouldn't be included, and net power consumed and generated wouldn't be balanced.

However, the energy consumed on a monthly basis can't be offset without the wind power component. The photovoltaic system and combined heat and power plant together provide consistent annual coverage. During cold months, when solar radiation is low, the combined heat and power plant operates longer, and thus, generates more electricity. Yet, only by adding wind energy, monthly consumption can be balanced or even enable a positive balance via seasonal surplus (Fig. B 07.12).

LESSONS LEARNED The shared ecological idea, the desired cost efficiency, and broad participation in the planning process encourage the development of a community spirit among occupants and identification with the houses. Being part of a cooperative makes it easier to achieve goals, particularly in terms of energy savings. To provide transparency in regard to savings methods, reports include utility and energy costs. The residents of the Kleehäuser maintain a highly comfortable lifestyle at a much lower cost compared to conventional apartments. This is another reason for the fact that occupants, according to their own statements, feel very much at home in the Kleehäuser (Fig. B 07.10, p. 87).

Restrained user behaviour enables savings on almost all utility and common charges, most evident in their costs for heating, hot water, and electricity. During the winter of 2009/2010 the wind turbines in St. Peter didn't generate the expected amount of power, due to freezing problems. The turbine blades were since replaced and equipped with a more efficient de-icing system. Furthermore, the third wind turbine didn't begin operating until August, and the larger on-site photovoltaic system until April 2010. Thus, the expectation is that the coverage amount from renewable energies will be significantly higher in the future.

B 07.12 Monthly primary energy balance, measurements 2010
B 07.13 Building and energy parameters (values refer to net floor area, NFA)

MULTI-FAMILY DWELLING

Dübendorf, CH 2008

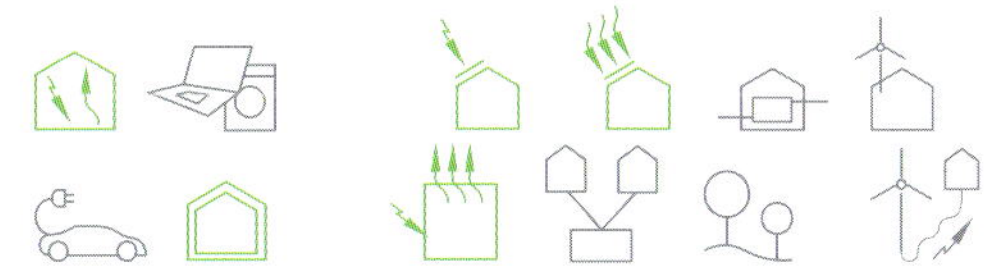

In 2008 this six-unit apartment building situated in Dübendorf was one of the first MINERGIE-P-ECO-certified buildings in the administrative district of Zurich. The new building replaced a typical single-family house from the 1920s. Its restrained design gesture and integration of solar systems into its pitched roof allow it to blend seamlessly into the surrounding residential area. The apartment building's MINERGIE-P-ECO design standard (see the MINERGIE label, p. 45) guarantees not only low energy consumption, but also ecologically sound material selection and an excellent interior climate. The photovoltaic (PV) system integrated into the pitched roof offsets the electricity consumed by the heat pump, for space and hot water heating, and the ventilation system, so that an annual net zero energy balance can be achieved. During planning, the energy balance didn't include power consumed by household appliances.

DEVELOPMENT, DESIGN, AND STAKEHOLDERS

The neighbourhood comprises small, simple houses with large gardens built in the 1920s and 1930s, with a few multi-family buildings up to four stories tall. By issuing a new building and zoning code, the municipality of Dübendorf intended to increase density near the town centre. The quality of life in this community is high and it is easily accessible, due to its proximity to Zurich. These factors prompted the owners to take advantage of the property's

Client: Bruno Hediger, Dübendorf
Architect: Beat Kämpfen, kämpfen für architektur, Zurich
Energy consultant: naef energietechnik, Zurich
Building services: naef energietechnik, Zurich
Monitoring: naef energietechnik, Zurich, and client
Main stakeholders: client and architect

B 08.01 View from south-west
B 08.02 Site plan, scale 1:1000

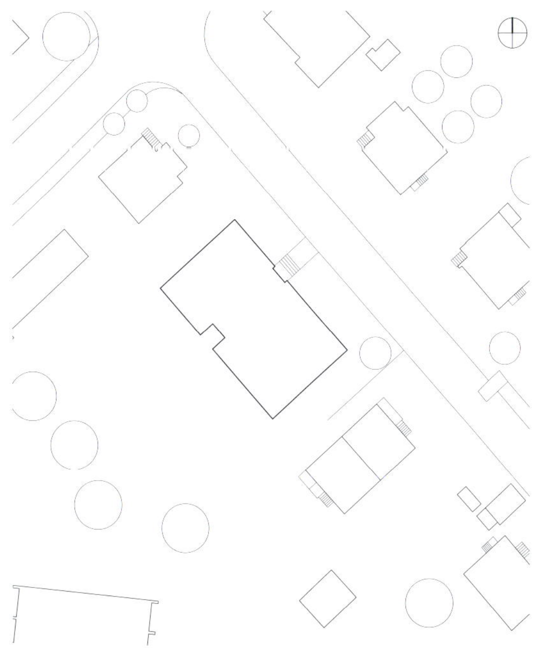

B 08.01
B 08.02

1 Entrance
2 Room
3 Bathroom
4 Kitchen/dining area
5 Living area
6 Patio
7 Loft
8 Office
9 Storage room

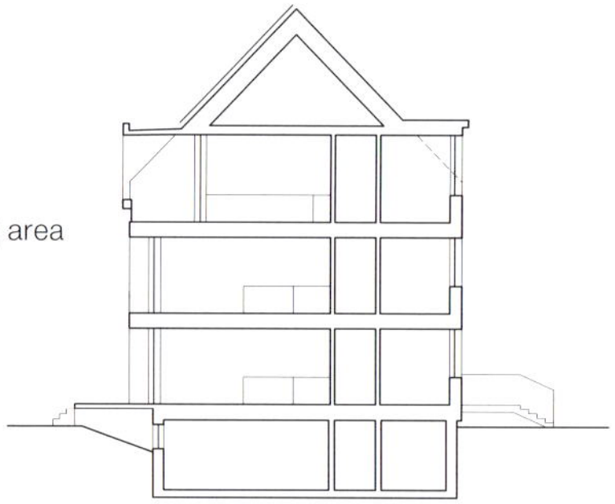

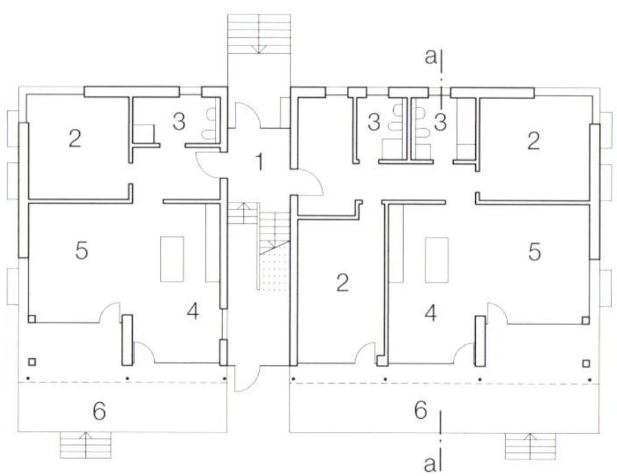

B 08.03
B 08.04
B 08.05
B 08.06

potential for an increase of building density: rental units complement their privately owned apartments. When the owners acquired the property, it included a single-family dwelling built in 1928. Because of the building's poor condition, urgently requiring renovation, the owners conducted an economic analysis and decided the house wasn't worth maintaining. In 2006 the owners and the architect collaborated in developing a concept for a building that could serve as a model house in terms of ecology and energy. The existing house was torn down in 2007 and by the end of 2008 it had been replaced by the new building featuring six apartments.

ARCHITECTURE Even though the higher density building features a volume increased by a factor of six compared to its predecessor, the new construction fits well into the small-scale housing context of the neighbourhood (Fig. B 08.02, p. 89). The completely glazed stairwell divides the building into two segments and simultaneously enables a visual connection between the street and the garden. The three residential stories are accessed via the stairwell, which serves as a weather buffer, similar to an unheated patio room. To avoid overheating in summer, vacuum tube collectors are mount on the southern oriented glazed roof and serve for space and hot water heating as well as sun protection. Over the course of the day, the collectors create varying light and shadow patterns along the exposed concrete walls of the stairwell, reminding occupants of the solar energy use (Fig. B 08.03).
The owners live in the building's slightly smaller west wing, in two stacked and interconnected 2.5-room apartments. The topmost floor is used as an office and studio (Figs. B 08.04–06). Three rental apartments are located in the building's larger east wing, each of which has a generous living and dining area with adjacent open kitchen. The apartment layouts are clearly oriented towards the south-west. One of the two bedrooms in each apartment faces south and the other north-east. The almost completely glazed main living area with balconies and patios faces south, common in solar architecture, while ancillary rooms such as bathrooms and

entrances are located on the northern side and feature only minor apertures. A key formal feature are the corner windows, which allow ample daylight intake.

CONSTRUCTION AND INSULATION The concept of three building segments is perceivable from the exterior and consequently implemented in terms of construction. The two residential cubes made of prefabricated wood elements and the glazed stairwell are structurally independent, which improves soundproofing. The walls of the lower floor are comprised of prefabricated hollow elements made of recycled concrete. All wall segments feature wood construction prefabricated in a carpenter's workshop. The high degree of prefabrication reduces on-site construction time and improves assembly precision.
The basement covers the entire building footprint. The 35 cm thick basement ceiling is made of mineral wool insulated hollow elements, while the floor slabs of the stories above consist of a wood-concrete composite construction. Thus, heated interiors are thermally separated from the unheated basement and the stairwell. Due to the insulation contained within the hollow elements of the basement ceiling, there are no thermal bridges between the basement walls and the ground floor. The basement remains cold and is separated from the surrounding earth by a 20 cm extruded polystyrene rigid foam insulating layer that prevents condensation along the perimeter walls in summer. The connections between ceilings and basement walls and stairwell also avoid thermal bridges. Horizontal components are placed on top of the wall construction's innermost layer, a load-bearing, 35 mm three-ply laminated panel. As result, the insulation along the exterior walls is continuous and homogeneous. The walls are completely insulated with 24 cm glass wool between wood studs plus an additional 10 cm with vapour barrier. The exterior consists of horizontal untreated larch siding with airspace permitting ventilation. The heat transfer coefficient of walls is 0.12 W/m²K. Wood used in construction is provided by regional Swiss forests. The roof structure features 24 cm of infilled glass wool insulation and an additional 16 cm beneath the sheet metal roofing and the integrated solar power modules, enabling a very low heat transfer coefficient of 0.09 W/m²K. The fibre-cement roof panels are sized to match the photovoltaic components, giving the roof a homogeneous appearance. The PV elements also take care of weather protection. The insulation of the entire building envelope has an exceptional heat transfer coefficient of 0.2 W/m²K. The wood frame windows are triple glazed and have a heat transfer coefficient of 0.7 W/m²K, with additional insulation on the exterior face of the frames.

ENERGY EFFICIENCY The project follows a passive solar concept and meets MINERGIE-P-ECO certificate requirements, due to very low energy demands based on excellent insulation properties and an airtight building shell. A blower door test demonstrated the required air tightness of below 0.6/per hour. The elongated, rectangular building with generously glazed southern facades utilizes passive heat gains from solar intake with a total energy transmission coefficient of 0.51. Cantilevered balconies that span almost the entire building length extend up to 2.2 m from the facades. The windows are deeply recessed due to the thickness of exterior insulation. Both features provide shade for windows and help prevent overheating of interiors in summer (Figs. B 08.08 and 09, p. 92). Fabric blinds are installed at 50 cm off the window plane along the balconies to offer further shading. The airspace between windows and blinds permits circulation of air and prevents overheating. Vacuum tube collectors shade the glazed stairwell. The wood-concrete composite ceilings and cement screed have a high thermal storage capacity. Black slate pavers are used as floor covering that also serves to maximize storage of solar heat intake.
The two wings each have their own ventilation system with 90 % heat recovery for each of the three apartments. This minimizes ventilation heat loss and improves air quality. The supply air is preheated with a water-glycol geothermal ground loop system in the infilled area on the basement level, preventing the heat exchanger from freezing. The supply air enters the living room and bedroom and is extracted in the kitchen and bathrooms and led into the heat recovery system. The building meets the MINERGIE-

B 08.03 View of stairwell with solar vacuum tube collectors providing shade in front of south-facing glazed wall
B 08.04 Section, scale 1:400
B 08.05 Top floor plan, scale 1:400
B 08.06 Ground floor plan, scale 1:400
B 08.07 Energy evaluation

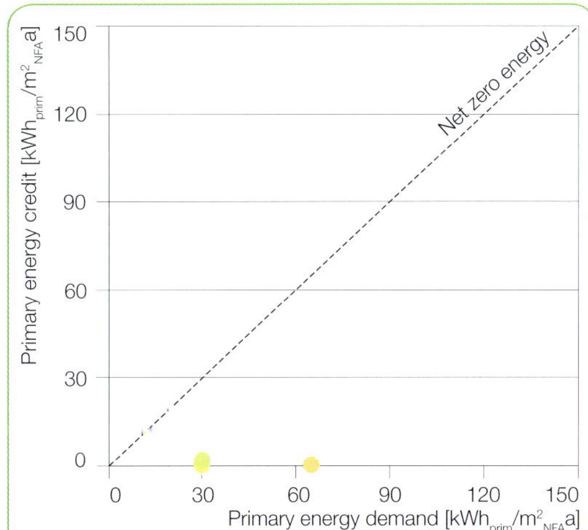

By including household electricity consumption, which actually isn't part of the balance calculation, total energy balance isn't achieved. The photovoltaic system is dimensioned to only cover building equipment consumption, including the heat pump.

- metered annual total primary energy consumption for building and household power consumption as per SIA 2031 (66 kWh/m²a)
- self-demand covered by monthly energy yield (35 kWh/m²a)
- residual electricity feed-in (1 kWh/m²)

The high monthly energy consumption coverage is the result of the PV system, although it is too small for this balance. The amount of energy consumed is usually greater than the amount generated, even during summer (eleven months of the year). Primary energy factors as per SIA 2031 (see Fig. A 2.07, p. 31)

B 08.07

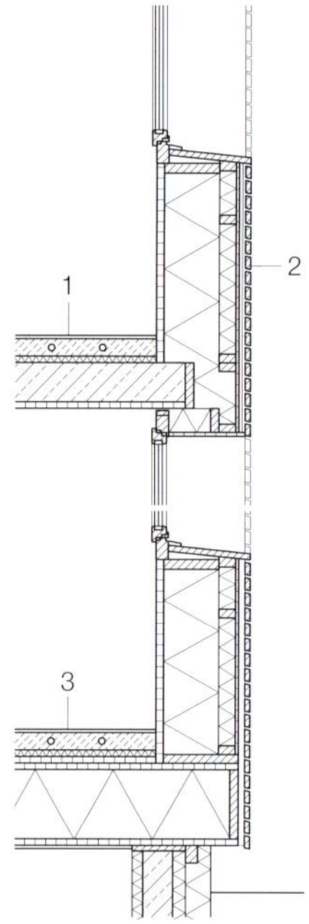

1 Floor construction:
flooring 15 mm
cement screed 80 mm
underfloor heating 80 mm
polyethylene membrane
impact sound insulation 30 mm
concrete 180 mm
solid wood panel 30 mm
2 Wall construction:
three-ply laminated wood panel 35 mm
studs, infilled insulated 260 mm
framing, infilled insulated 80 mm
MDF panels, 15 mm
building wrap, black
ventilation air space 30 mm
larch cladding 25 mm
3 Floor construction:
flooring 15 mm
cement screed 80 mm
underfloor heating 80 mm
polyethylene membrane
impact sound insulation 30 mm
gravel 30 mm
laminated wood panel 30 mm
fibreglass insulation 320 mm
solid wood panel 30 mm, fibre reinforced gypsum board 15 mm

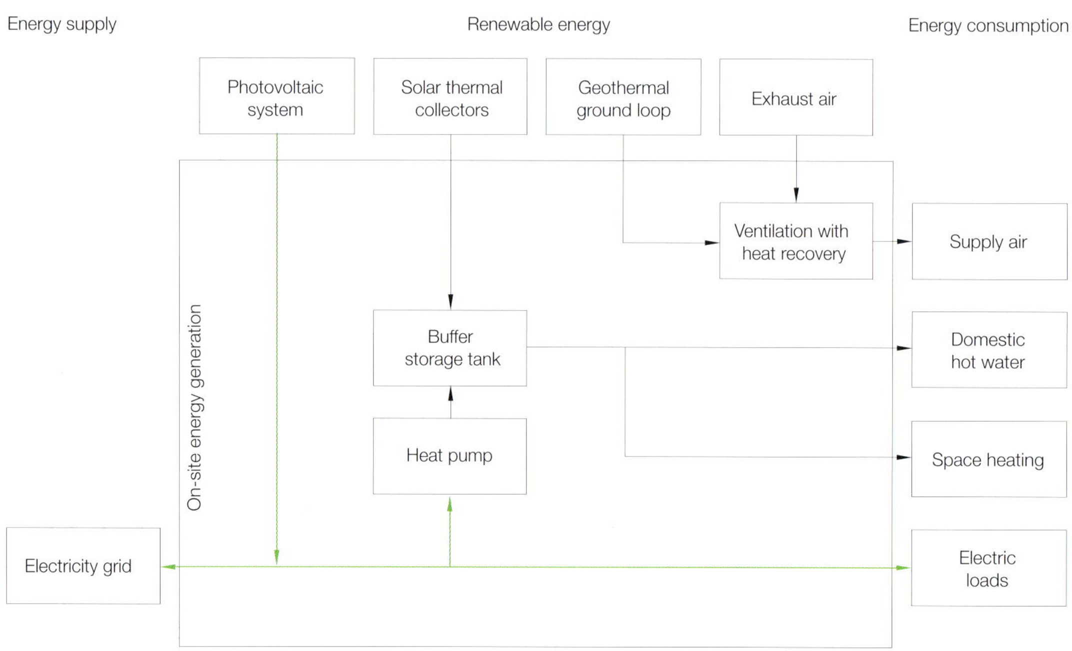

B 08.08
B 08.09
B 08.10

P standard's air exchange specification of 0.33 m^3/m^2h.

Tenants are encouraged to use energy efficient lamps to save electricity. Movement sensors, daylight sensors, and LED lighting are installed in the stairwell. All washing machines have a hot water connection. The dryers have an inbuilt heat pump, and remaining household appliances are Label A or above to meet the specifications required for MINERGIE-P-ECO certification.

ENERGY SUPPLY An initial concept for heating supply included use of a heat pump with a geothermal heat exchanger located as deeply as possible in the earth plus a photovoltaic system. However, this concept was considered too expensive and was rejected.

The energy for space and hot water heating is now provided by 14 m^2 of vacuum tube solar collectors and an air-to-water heat pump. The heat pump has a thermal capacity of 11.7 kW, consumes 3.3 kW of electricity, and is located in a free-standing wooden shed adjacent to the building. Heat from the two systems is transferred to a 1800 l combination hot water storage tank with integrated solar circuit heat exchanger. The heat is distributed at a low supply temperature of 30 °C through an underfloor heating system and small radiators in the bathrooms according to heat pump efficiency rates.

A 14-kWp photovoltaic solar power system generates electricity to offset the electricity consumed by the heat pumps and the ventilation systems. It is integrated into the southern halves of the two pitched roof surfaces angled at 45°. The system is somewhat larger than originally planned in order to compensate for the extra power required by the air-water heat pump as compared to the geothermal heat pump that was planned, but not installed (Fig. B 08.10).

ENERGY BALANCE The solar power system generated 14 kWh/m^2_{NFA} of electricity during its first year of operation, entirely fed into the public grid. The generated electricity didn't fully offset the total demand of nearly 17 kWh/m^2_{NFA} a required for space heating, ventilation, and hot water heat-

ing. As result, net zero energy balance for technical building equipment wasn't achieved during the monitoring period between October 2009 and September 2010. In 2009 and 2010 the annual solar intake on-site was about 10 % below average, which led to reduced solar electricity and heating supply and an increase in the power consumed by the heat pump (heat pump annual performance coefficient is 3.01.).
Household electricity consumption is not fully balanced. Based on an average consumption of 23 $kWh/m^2_{NFA}a$ (see Fig. A 2.10, p. 33), an additional 140 m^2 of PV modules would be necessary to achieve a balance. Not all of the rooftop area is available for solar power generation, and the remaining alternative is to reduce electricity demands. A life cycle analysis conducted within the MINERGIE-P-ECO certification process shows that the wood construction with its low embodied energy substantially reduces negative impact on the environment. In order to minimize the grey energy required, the owners were able to avoid building an underground garage typical for Switzerland and often required by authorities. From the basement ceiling upward, the building is a prefabricated wood construction. Use of pollutant-free construction materials received particular attention. The highly insulated wood construction stiffened by reinforced concrete walls made of recycled concrete requires approximately one-quarter of the grey energy of a comparable solid structure. Based on a projected life cycle of sixty years, the total environmental impact of the apartment building's energy consumption (see company headquarters in Kemptthal, p. 120ff) and of its CO_2 equivalents (see Fig. A 2.06, p. 31) are about 60 % less than for comparably sized multifamily dwellings that don't meet MINERGIE-P-ECO label requirements. This calculation includes the energy required for producing building materials, construction, maintenance and repair, and disposal; the energy consumed for its operation; and also the solar power generated on site.

OVERALL EXPERIENCE With building heights and the sloped roof determined by local construction codes, the A/V ratio is fundamentally advantageous. It permits active and passive use of solar energy, although the steep 45° roof slope isn't optimal for electricity generation. However, the many projecting elements and recesses along the south-western facade, also a result of code requirements, increase the building envelope, complicate the design, and lead to an increase of thermal bridges. The roof windows required to illuminate interiors and the balcony recesses reduce the roof area and minimise use of roof surfaces for active solar energy.
Tenants in the rental apartments often keep their blinds closed in winter to avoid glare and provide privacy, despite explanations regarding their intended purpose. This reduces solar heat gain and increases space heating requirements. A decision on automated control of blinds is pending. The stairwell receives no shading other than by the fixed vacuum tube solar collectors along the south-western glazed roof surface. As result, solar intake on the north-eastern side generates considerable heat on summer mornings. Lack of exterior sun protection also leads to higher temperatures on upper floors and in the corner rooms (north-east and south-east). Here as well, adjustable sun protection would be preferable, despite in part 388 mm deep recesses. The blinds installed flush with the balcony face supplement sun protection of south-western facades, also shaded by projecting balconies. However, this obstructs circulation of warm air on balconies. Blinds consisting of a reflective material or a second blind layer installed directly in front of the windows could improve this situation.

B 08.08 Deeply recessed windows to provide shade for high sun positions in summer
B 08.09 Exterior wall section, scale 1:50
B 08.10 Technical schematic of energy provision
B 08.11 Building and energy parameters (values refer to net floor area, NFA)

SITE	Dübendorf (CH)
Annual global radiation at site	1090 kWh/m^2a
Annual mean temperature at site	8.5 °C
Context	suburban
BUILDING ENVELOPE PARAMETERS	W/m^2K
U-value, exterior walls	0.12
U-value, windows (including frames)	0.70
U-value, roof	0.09
U-value, ceiling above basement/floor slab	0.11
Mean U-value, building envelope	0.20
BUILDING EQUIPMENT PARAMETERS	
Solar collector area	14 m^2
Area per m^2	0.02 m^2/m^2
Thermal storage tank volume	1800 l
Storage tank volume per m^2	2.50 l/m^2
Photovoltaic system area	94 m^2
Area per m^2	0.13 m^2/m^2
Photovoltaic capacity	14 kW_p
Capacity per m^2	19.30 W_p/m^2
GRID INFRASTRUCTURE AND ENERGY SOURCES	
Supply infrastructure	electricity grid
Energy source supply	electricity
Feed-in infrastructure	electricity grid
Feed-in energy source	electricity
DESIGN STRATEGIES, CONCEPTUAL FOCUS	
MINERGIE-P-ECO concept, mechanical ventilation with heat recovery, air-water heat pump, photovoltaics, solar thermal collectors, ecologically sound materials	
BUILDING PARAMETERS	
Net floor area, NFA	727 m^2
Gross floor area, GFA	986 m^2
Gross volume, V	3266 m^3
Building envelope, A	1300 m^2
Surface to volume ratio, A/V	0.40 m^2/m^3
Building costs (net, construction/technical systems)	2400 €/m^2 (2009)
Number of units	6
Total number of users	10 persons
CONSUMPTION PARAMETERS (2010)	kWh/m^2a
Space heating consumption	11
Water heating consumption	21
Site energy consumption for heat (including hot water)	9
Electricity consumption	26
Total primary energy demand	66
Total primary energy generation	36

B 08.11

SOLAR COMMUNITY

Freiburg, D 2006

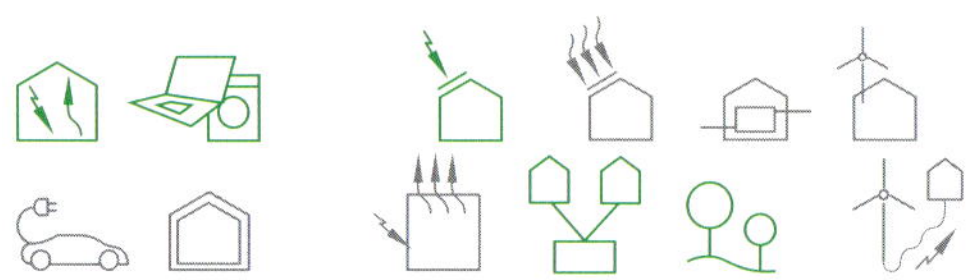

Client: Solarsiedlung, Freiburg
Architect: Rolf Disch SolarArchitektur, Freiburg
Energy consultants: Phoenix Solar, Sulzemoos
Building services: Phoenix Solar, Sulzemoos
Monitoring: Bergische Universität Wuppertal
Main stakeholder: architect

Planning for the solar community (Solarsiedlung) situated along the Schlierberg in Freiburg began in 1996. The project was completed between 2000 and 2006 as one of the world's first energy plus projects on the scale of a residential development. The goal of energy plus status was successfully achieved by building passive houses, installing large-scale photovoltaic systems, and participating in a wood-fired district heating grid located in the city's nearby ecological showpiece district of Vauban.

DEVELOPMENT, DESIGN, AND STAKEHOLDERS

In 1996 the architect and project initiator, Rolf Disch, designed an ecologically friendly community that was intended to go beyond the typical passive houses built at that time. The design goal of the development wasn't only to not impact residents' health, provide energy savings, sustainability, and rely on a renewable energy supply and, as result, achieving the smallest possible energy-related ecological and environmental footprint. It was also supposed to even produce more energy than it consumed, which prompted Rolf Disch to coin the term "energy plus house (Plusenergiehaus)". A master plan addressed quality of life, social community aspects, and high property values (approximately 500 €/m² for undeveloped land). As result, by maximising the floor space index, as well as conducting shading studies and evaluating the potential to

B 09.01
B 09.02

B 09.03
B 09.04
B 09.05

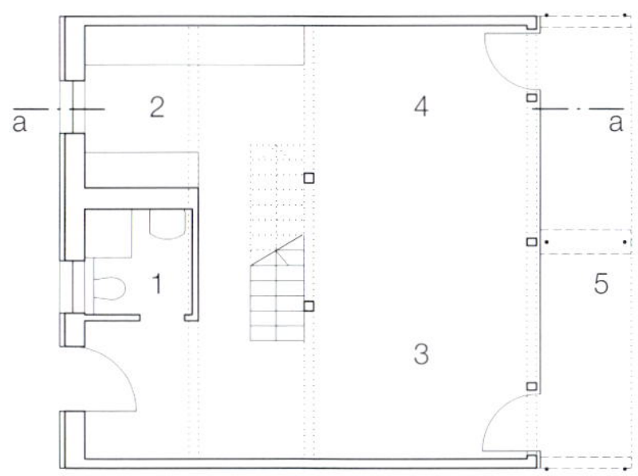

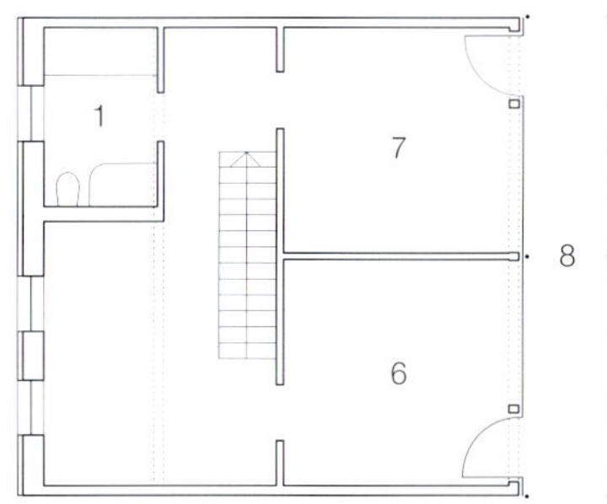

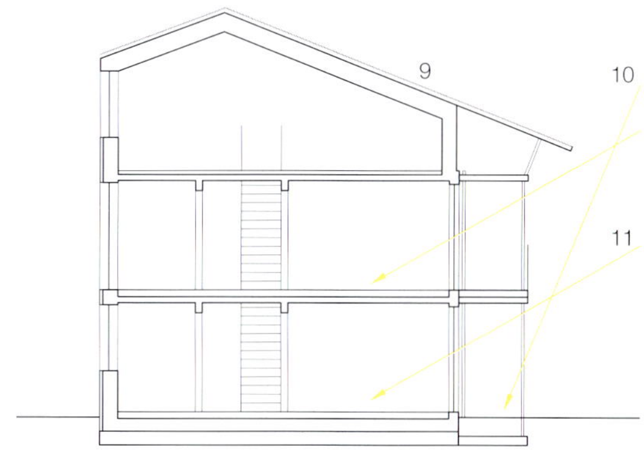

1 Bathroom
2 Kitchen
3 Living area
4 Dining area
5 Patio
6 Bedroom
7 Rooms
8 Balcony

B 09.01 Site plan, energy plus community, scale 1:1500
B 09.02 View from south-west with bordering Sun Ship
B 09.03 Ground floor plan, typical mid-row house, scale 1:200
B 09.04 Upper floor plan, typical mid-row house, scale 1:200
B 09.05 Section, scale 1:200
9 photovoltaics
10 sun position, summer
11 sun position, winter

utilize solar energy, it defined the orientation of houses arranged in rows, their roof geometry, and settlement density. The goal was to offer photovoltaic roof surfaces free of shading throughout the year and southern elevations exposed to sunlight in winter. Urban centres often can't offer these qualities, which is why this development is located towards the urban periphery.
Freiburg's location in a very sunny region of Germany that also experiences mild winters is a key factor in achieving a positive energy balance in the community.
Due to financing difficulties and high costs of both real estate and photovoltaic systems, the architect was required to reduce the size of the development and planned recreational spaces. Several "solar funds" were also established. The purpose of this innovative financing concept in residential construction was to facilitate roof-mount solar units throughout the development, even if the buyers of some houses couldn't cover the cost. Closed-end real estate funds cover funding gaps for high construction costs. Agreements were met that address the fact that roof-mount PV areas of units are sometimes owned not by the owner of the house, but a third party. The contracts stipulate that owners permit roof-mount photovoltaic systems, allow third parties to operate them, and also waive their own operation rights, as well as guarantee optional repair of any damages that may be caused to their roof. The legal framework including solar funds forms part of the sustainable and ethical concept of the community. The investment in renewable energies is transparent, because projections on surplus are relatively exactly defined, both locally and by Germany's Renewable Energies Directive (EEG). Within this governance model, third-party private investors have a share in the completion of fifteen of the fifty-nine houses.

RESIDENTIAL COMMUNITY CONCEPT The key priority for all planning aspects related to spatial design, construction methods, materials selection, and above all, operating energy is resource efficiency. The fundamental principle of the planning concept is to maximize active and passive utilization of solar energy. Deliberation on the open spaces between the rows of houses, their height, and roof shapes contributed to creating shade-free southern elevations in winter, while aiming at maximizing the floor space index of the very expensive properties as part of economic concerns. The result was the characteristic image of the community with its densely arranged row houses and asymmetrical pitched roof structures with clear southern orientation (Figs. B 09.01 and 02).

The energy plus community offers its 170 residents both the social advantages of an integrated residential project and the benefits of city life close to nature. The city centre is easily accessible by bicycle or streetcar, and car sharing is a fixture of community life. The neighbouring quarter of Vauban features kindergartens and schools. This former military base served as an experimental area for ecological construction methods since the 1990s and is a model for Freiburg's policies aimed at sustainable growth. The residents of the energy plus community also have access to the diverse green spaces in the Vauban

B 09.06
B 09.07

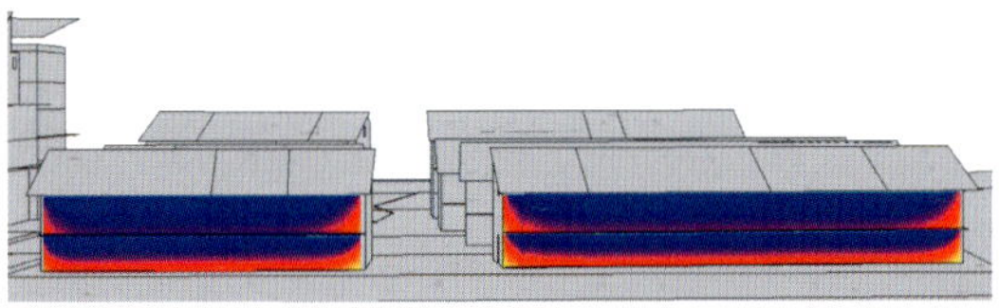

a

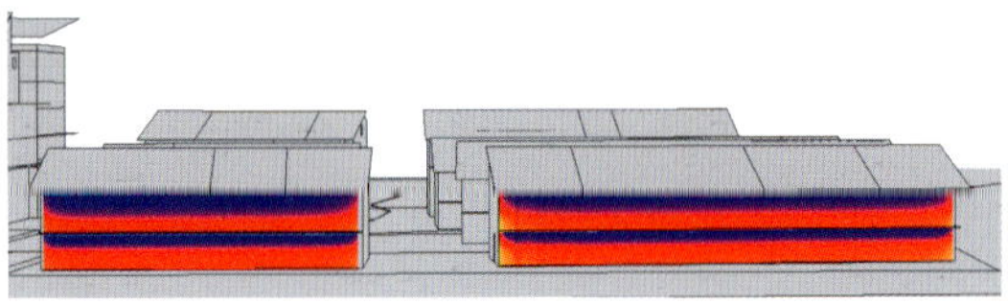

b

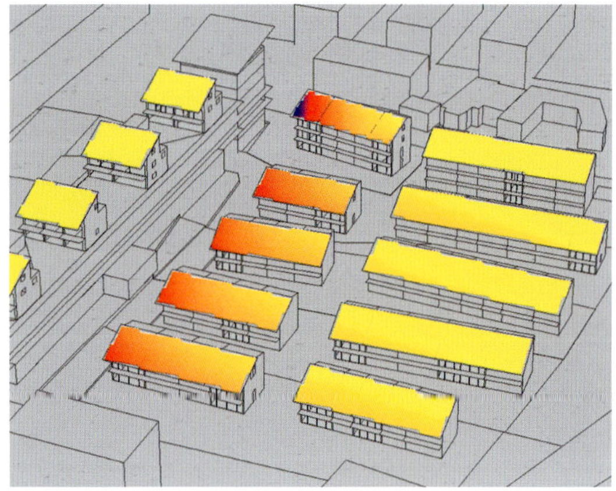

to B 09.06	to B 09.07
100 %	114 %
92 %	109 %
84 %	105 %
76 %	100 %
68 %	95 %
60 %	91 %
52 %	86 %
44 %	81 %
36 %	76 %
28 %	72 %
20 %	67 %

district and its excellent infrastructure featuring a district heating grid.
Fifty row houses facing south and arranged in ten separate blocks of varying sizes are situated on the 11,000 m² site, separated by a central access road that serves only for loading and unloading of cars. Parking is available in the underground garage of the "Sun Ship (Sonnenschiff)", a three-story office and service centre also designed as an energy plus building, or in the nearby "solar garage". Access to the row houses is via narrow walkways between every second row, alternately from the north or the south. The remaining outdoor area comprises individual private gardens; there is no communal area.

The Sun Ship's roof offers space for nine three-story penthouses, also designed as energy plus buildings. The longitudinal building volume separates the residential development from the main road and shields it from noise, providing a calm interior atmosphere. Access to the penthouses is from the main street via stairwells in the Sun Ship. A ground floor row of shops gives residents the opportunity to purchase groceries close to home.

ARCHITECTURE The two-story row houses and the northern three-story row houses have varying widths. As result, unit floor spaces range from 75 to 200 m². The layouts follow classic solar house design concepts (Figs. B 09.03–05, p. 95): living areas are on the southern side, hallways in the centre, and service areas, including kitchen, bathroom, and utilities are in the north. An entrance with vestibule intersects the living area only in houses accessed from the southern side. The clear southward orientation and high building density produce a very homogeneous overall appearance. Different building colours that conform to a master colour scheme and the slight rotation of certain rows add degrees of complexity. Most exterior walls are comprised of prefabricated, slender wood I-joists as studs and insulated with 30 cm of mineral insulation. Prefabrication reduces construction time, and the wood joist system offers flexibility for required construction spans, while small cross-sections reduce thermal bridging. The curtain-wall facades are mostly clad in wood originating in regional, sustainable forestry. Due to their row house design, building volumes are very compact. Since there are no basements, exterior sheds offer storage space along the walkways and structure the small private outdoor areas. Each building has two separate water distribution grids. The second grid makes use of rainwater that doesn't seep into the community's drainage ditch.

ENERGY EFFICIENCY The energy plus buildings are based on the passive house concept and to date demonstrate low energy consumption. This results from certain design decisions: They are very compact, thermal bridging is minimized, and they are very well insulated with 30 cm of mineral insulation in the outer walls, 36 cm beneath the roof, and triple-glazed thermal windows. They also feature decentralised, compact ventilation systems with integrated heat recovery along their northern perimeter. Although no air ducts are required due to the decentralized design, compromises have been made in terms of energy efficiency. The installed ventilation systems both vent exhaust air and provide fresh air by reversing the fan direction. Air passes through a heat sink, which is less efficient than using a central system with a cross-flow heat exchanger. In row house groups, the ventilation systems are interconnected and operate in push-pull mode to avoid excess or insufficient pressure.

Passive solar yield from the southern elevations with their very large windows reduces heating load. At the same time, this permits significant daylight intake, while roof eaves and balconies shading the southern-oriented windows prevent overheating in summer. Limited numbers of windows on the north, east and west walls reduce heat loss and construction costs. Plumbing valves that support water conservation are also in use. Energy-saving appliances and lighting, in addition to energy-conscious user behaviour, also help reduce household energy use.

ENERGY SUPPLY A photovoltaic system and a district heating grid actively deliver energy to the energy plus community (Figs. B 09.09 and 11, p. 98).

B 09.06 Shading of the southern walls in summer to mitigate summer heating (a) and in winter for passive heating via solar intake (b)
B 09.07 Solar potential of roof surfaces for active solar power utilization on-site, simulated for an entire year (100% is equivalent to maximum yield on a horizontal surface on-site)
B 01.08 Bird's eye view, solar community
B 09.09 Technical data for energy supply from 2007 to 2009. The district heating grid's combined heat and power plant has since been replaced by a wood-fired boiler.

ELECTRICAL POWER SUPPLY The central feature of the energy plus community is roof surfaces sloping southward at 22° and completely covered with photovoltaic modules. The PV elements are mounted on steel frames set on top of a waterproofing layer, while an air space in between enables ventilation and improves solar system yield. Projecting PV surfaces expand the size of southern-oriented roof surfaces. Due to the considerably shorter northern elevations, asymmetrical, pitched roof shapes are formed. On their southern side, 400 kW_p of peak power are generated. Altogether, the PV systems generate more than the total amount of energy consumed by the community.

HEATING The district heating grid supplies heat for space heating and hot water. This grid is fed by a combined heat and power plant fired by wood chips and natural gas, located in the Vauban district. Transfer stations in each building supply heating circuits and decentralized hot water storage tanks. Thus, the need for continuously operating the district heating grid in summer is avoided and corresponding losses are reduced. There are no solar thermal collectors.

ENERGY BALANCE The balance method for primary energy in the case of an energy-plus building is applied. The site sets the spatial limits for the balance. All types of consumption enter the consumption calculation. In addition to consumption specified

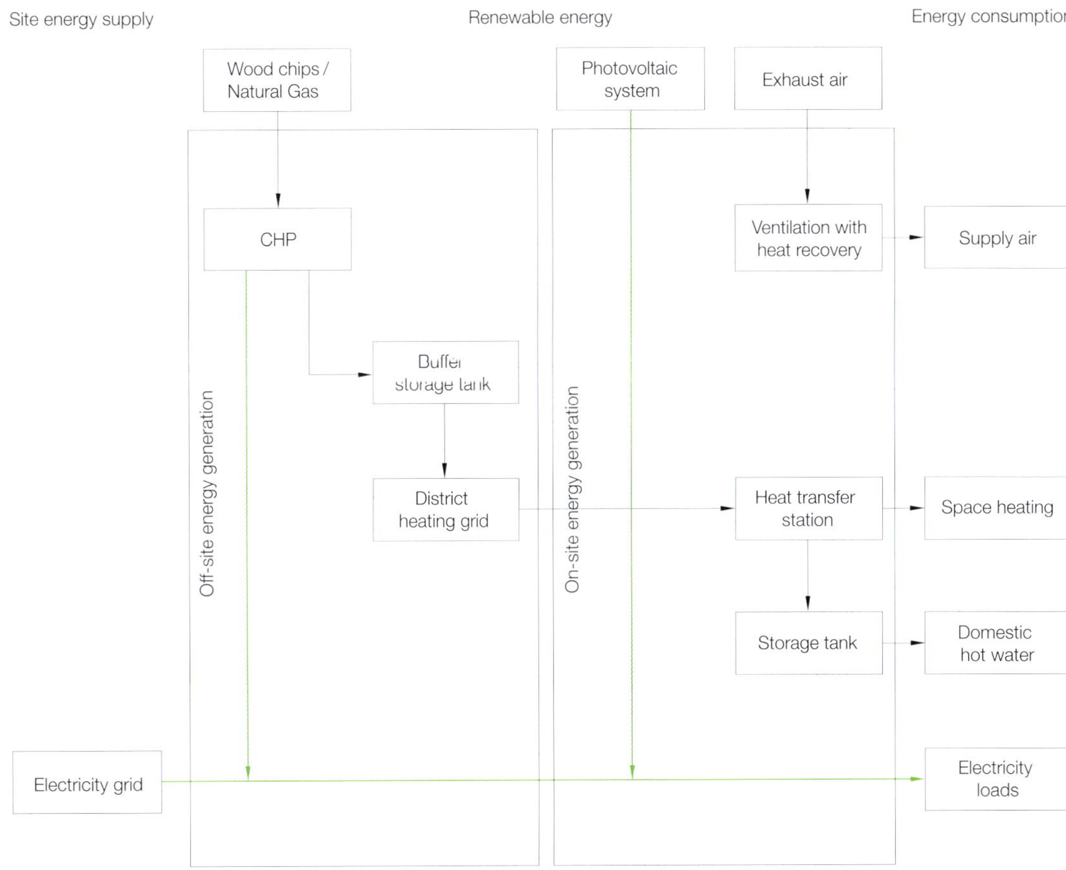

B 09.08
B 09.09

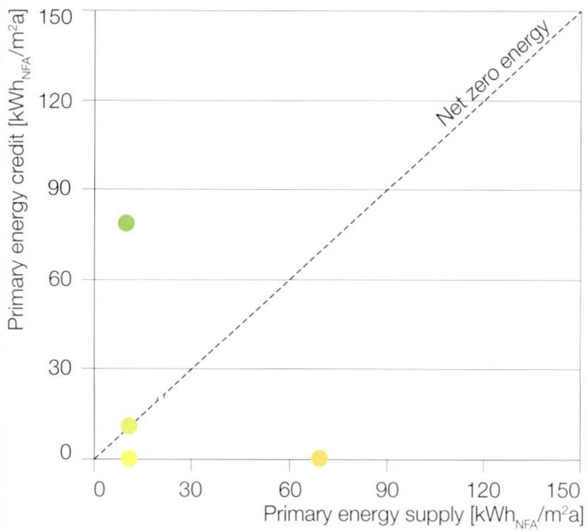

The following aspects qualify the Net ZEB-10 standard:
- Metered annual total primary energy consumption including household electricity (70 kWh/m²a)
- Remaining consumption after deducting monthly creditable energy yield (60 kWh/m²a)
- Seasonal compensation of remaining consumption
- Annual energy surplus, public grid feed-in (82 kWh/m²)

The seasonally very high self-demand coverage is driven by photovoltaic systems that are capable of covering low consumption levels, even in spring and fall. Primary energy factors according to DIN 18599 (see Fig. A 2.07, p. 31)

by the German energy code for space heating and hot water, calculations also include electricity for all other building services, lighting, household appliances, and entertainment (Fig. B 09.10). The fuel mix consisting of regionally produced wood chips and natural gas is utilized with high efficiency together with power-heat coupling in the combined heat and power plant that supplies the district heating grid. The primary energy factor for heating is at a low 0.60, which can be offset in the balance by completely feeding the large photovoltaic system's summer electricity yield into the public grid. It generates significantly more electricity throughout the year than is consumed overall. Except for the cold and less sunny months from November to January, enough solar power yield generated by the photovoltaic system can be credited to cover all energy demands for any month, even based on monthly balances. Only 10 kWh_{prim}/m²a are required to offset the higher winter demand in the seasonal energy balance. The photovoltaic systems feed a corresponding amount into the public grid as surplus during the summer months. The fed-in electricity offsets heat generated from wood/gas. The remaining energy surplus fed into the public grid raises its renewable energy component.

Until 2010 solar electricity in Germany was typically entirely fed into grids. The preferred concept of compensation for photovoltaic power generation according to the Renewable Energies directive (EEG) supported feeding all electricity generated into grids, as opposed to generating it primarily to cover building self-demands. Property owners of the solar community were guaranteed 0.49 €/kWh of electricity fed into the grid for a duration of twenty years from the date of system start-up in 2006. Beginning in mid-2010, the EEG revised these regulations. Now they encourage covering self-consumption of electricity generated from the photovoltaic systems on-site. Credits for using on-site generated electricity (0.23 €/kWh) plus savings achieved by avoiding consumption of electricity from the public grid are more than the compensation for feeding electricity into the grid (0.33 €/kWh to 30 kW). Using this accounting method requires the installation of corresponding meters.

LESSONS LEARNED Between 2007 and 2009 Wuppertal's Bergische University evaluated the energy balance of the fifty row houses. The analysis was based on annual simultaneous consumption data for heat, domestic hot water, and total electricity use on the debit side, and generated electricity on the credit side. Other than initial problems with inverters, there were no technical problems. The analysis showed that the average house generated a surplus of about 36 kWh/m²a of primary energy per annum. A few buildings didn't achieve this balance (Fig. B 09.13). In some cases, the living area to

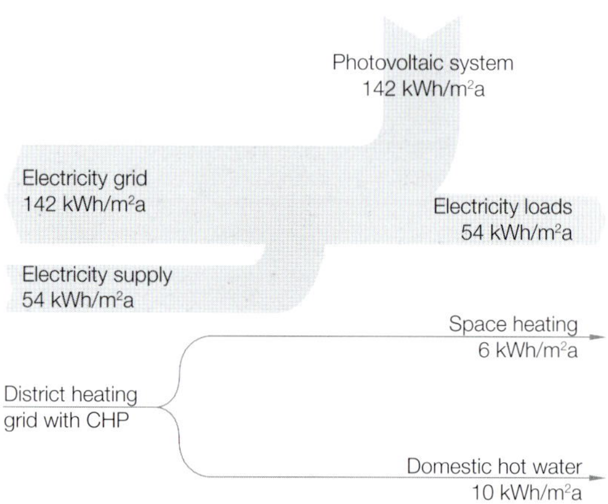

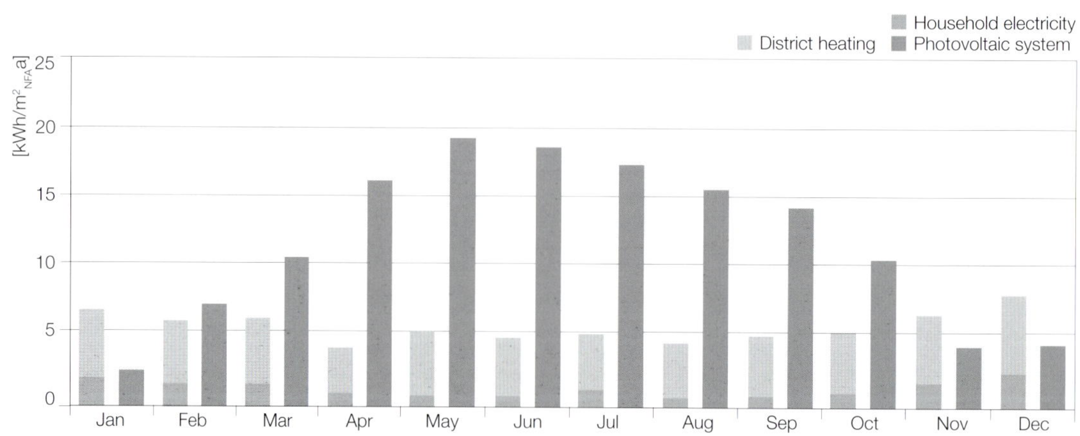

B 09.10
B 09.11
B 09.12

PV surface area ratio wasn't adequate, as in the case of the northern row of three-story houses that feature less roof surface area for the photovoltaic system (0.22 m² instead of the average 0.36 m² PV per m² heated living area). In addition, a few of the row house end units required more heating energy because of their larger building envelope. Some units were used as offices, and thus, consumed more electricity. There were also differences in user behaviour and needs.

Approximately one-quarter of primary energy is equally distributed among space heating and hot water; three-quarters were used for electricity. The total electrical power consumed by the community is only slightly below the German average, which supports the conclusion that there is further potential for cutting back on electricity consumption.

The efficiency of the buildings, together with the generation of renewable energy, is key to achieving a positive balance. The average energy plus community house would have primary energy demands of 185 kWh/m²a according to the 2004 Energy Saving Directive, while its electrical power savings variant assigns 165 kWh/m²a. A target value of 98 kWh/m²a is possible through further improvement of efficiency via the passive house standard. The use of heat energy from the district heating grid decisively reduces primary energy consumption, due to its advantageous primary energy factor. As result, credits for solar electricity can generate a significant surplus that would, however, be lower if the energy source was natural gas alone. Yet, the difference is minor, due to the houses' very low heating demands (Fig. B 09.14).

Wood is planned as near-exclusive fuel for the district heating grid, with gas to be used only for peak loads. However, during operation, 30 percent of fuel used was comprised of natural gas, since the combined heat and power plant displayed numerous problems. As result, it was shut down, and the district heating grid switched to 100 percent wood as fuel.

B 09.10 Energetic characterisation, typical for a metered row house
B 09.11 Sankey diagram for electricity (top) and heat (bottom), metered row house as example
B 09.12 Monthly balance chart, primary energy for an example house
B 09.13 Energy balances for various metered houses compared to the entire community. The balance sheet for the entire community and for the average house is positive. The community average figure compensates varying performance of individual houses.
B 09.14 Comparison of primary energy balance in the case of different energy sources for space heating and hot water consumption (community median value)
B 09.15 Building and energy parameters (values refer to net floor area, NFA)

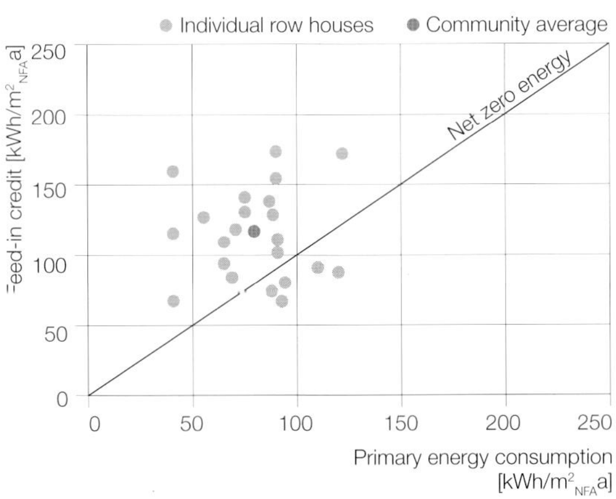

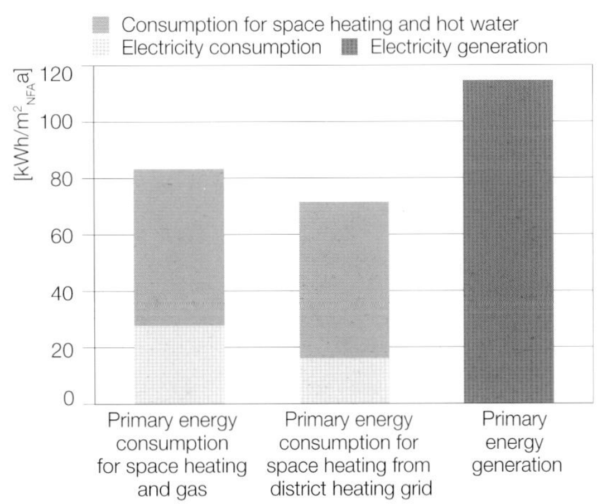

SITE	Freiburg (D)
Annual global radiation at site	1100 kWh/m²a
Annual mean temperature at site	11.6°C
Context	urban
BUILDING ENVELOPE PARAMETERS	W/m²K
U-value, exterior walls	0.12
U-value, windows (including frames)	0.80
U-value, roof surface	0.12
U-value, ceiling above basement/ floor slab	0.16
Mean U-value, building envelope	0.28
BUILDING EQUIPMENT PARAMETERS	
Photovoltaic system area	3205 m²
System area per m²	0.40 m²/m²
Photovoltaic capacity	400 kW$_p$
Capacity per m²	50.70 W$_p$/m²
GRID INFRASTRUCTURE AND ENERGY SOURCES	
Supply infrastructure	electricity grid, district heating grid
Energy source supply	district heating grid, electricity
Feed-in infrastructure	electricity grid
Feed-in energy source	electricity

DESIGN STRATEGIES, CONCEPTUAL FOCUS
Passive house concept, mechanical ventilation with heat recovery, daylight intake optimization, district heating grid (wood chip-fuelled), photovoltaics, fixed sun protection, ecologically sound construction materials

BUILDING PARAMETERS	
Net floor area, NFA	7890 m²
Gross floor area, GFA	8112 m²
Gross volume, V	24,416 m³
Building envelope, A	13,722 m²
Surface to volume ratio, A/V	0.56 m²/m³
Building costs (net, construction/ technical systems)	1940 €/m² (2006)
Number of units	59
Total number of users	170
CONSUMPTION PARAMETERS (2007)	kWh/m²a
Space heating consumption	10
Water heating consumption	16
Site energy consumption for heat (including hot water)	26
Electricity consumption	21
Total primary energy consumption	70
Total primary energy generation	142

B 09.13
B 09.14
B 09.15

ENERGY PLUS COMMUNITY

Weiz, A 2006 (phase 1), 2008 (phase 2)

Client: Gemeinnützige Siedlungsgenossenschaft Elin (Elin non-profit community association), Weiz
Architect: Arch ° Buero Kaltenegger, Passail
Energy consultant: Arch ° Buero Kaltenegger, Passail
Building services: TB Bierbauer, Passail
Monitoring: AEE – Institute for Sustainable Technologies
Main stakeholder: Architect

This energy plus residential development (Plusenergiewohnen) is located in the Weiz Gleisdorf "energy region" of East Styria, Austria. Its regional energy association formulated a vision for a regional zero energy balance by 2020. On average, this passive house community with its twenty-two row houses generates surplus energy by photovoltaic systems that produce more electricity than is required to operate the decentralized heat pumps and other electric loads in the all-electric buildings. The modular residential structures were built in two construction phases from 2006 to 2008 and represent the culmination of a project for the development of a mass produced, prefabricated house that meets energy and cost efficiency specifications.

ARCHITECTURE Each of the six house rows consists of three, four, or five connected row houses with two different floor plan types each (93 m² or 105 m²). The rows aren't strictly oriented towards the south. Instead, the architecture is integrated into the landscape in order to create a desired state of harmony between both. The southern orientation of the individual rows varies by 12° to 34° (Fig. B 10.01). As demonstrated, the impact of this variation on the photovoltaic systems yield is minor, and the resulting metered yield difference for a row of houses is only about 3 %. The alternating open and closed exterior spaces have a significant impact. Fir used for the prefabricated light wood frame originates entirely in the heavily forested Styria region (Fig. B 10.02). The post-and-beam exterior wall construction is covered with vapour permeable wood fibre sheathing towards the exterior and gypsum board on the interior. The space in between features 35 cm of blow-filled cellulose insulation. The resulting heat transfer coefficient is 0.09 W/m²K. An air space between the spruce-clad, weather-resistant facade and the wall construction enables ventilation. Solid, cross-laminated spruce wood plates comprise load bearing elements of ceiling and roof construction. The buildings have no basement, and instead, uninsulated metal containers covering 4.50 m² serve as replacement and are located on the entrance side of each residential unit.

B 10.01
B 10.02

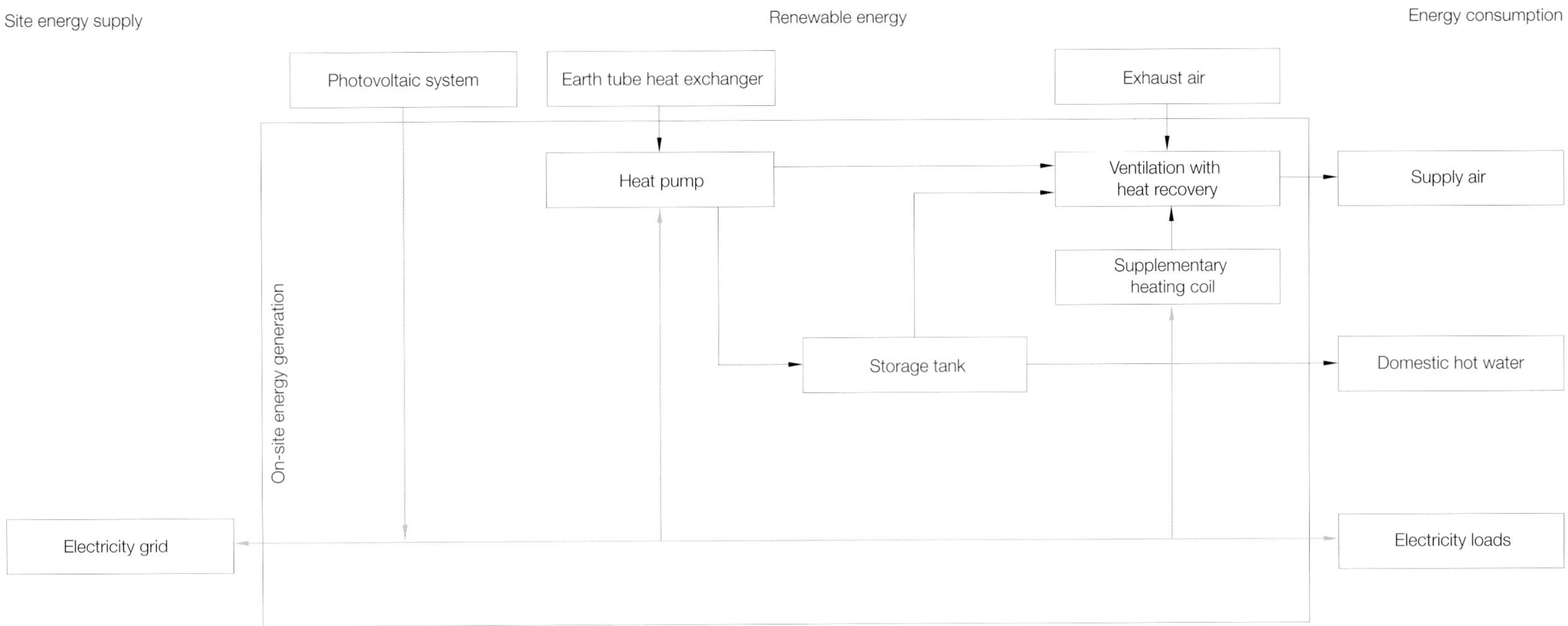

B 10.03

B 10.01 Site plan, scale 1:2500
B 10.02 View from north-east
B 10.03 Technical schematic of energy provision

ENERGY EFFICIENCY The houses are extremely well insulated with an airtight building envelope (0.37–0.50/h at 50 Pa measured) and an average heat transfer coefficient of 0.20 W/m²K. They feature a decentralized ventilation system with heat recovery and offer living spaces with large glazed surfaces towards the south. These aspects enabled the project to meet passive house standards already during planning. The design, together with the location and orientation of the building rows, and their extremely compact size (supported by the metal storage boxes) are the reasons for very low residual energy demands as a fundamental precondition for "surplus energy living."

ENERGY SUPPLY The low space heating demand of 15 kWh/m²a is covered by the ventilation system and an air-to-air heat pump. Fresh air is supplied and preheated via the earth tube heat exchangers installed beneath the buildings. The ground loop pipes are 25 cm in diameter and between 45 and 60 m long. Each building includes a 1 kW_{th} heat pump that combines a domestic hot water heat pump and a ventilation system as heat source. A 185 l hot water tank is integrated into the ventilation system to facilitate hot water supply. A supplementary heating coil with a maximum electrical capacity of 1.05 kW is integrated into the compact ventilation and heating unit to cover peak loads. Solar thermal collectors aren't in use. A double-row photovoltaic system installed on the flat roof of buildings generates all electric power consumed over the course of a year. The front row provides shade for large windows facing south and simultaneously acts as a roof for the balconies included in select units. A separate system consisting of multicrystalline modules with a surface area of 40 m² and an installed capacity of 4.95 kW_p is located on the rooftop of each unit,. The total area and capacity for buildings of both construction phases are 520 m² and 64 kW_p. Similar to the decentralized heat pumps, a photovoltaic system with its own inverter and meter is assigned to each row house. They are currently operated as a single system by the client as responsible entity for its construction, investment, approval, operation, and maintenance (Fig. B 10.03).

ENERGY BALANCE The energy demand for space heating varies, depending on particular location and related building envelope of individual units, ranging between 13 and 15 $kWh/m^2_{NFA}a$. After three years of operation between 2006 and 2008, the consumption and yield data for the first building phase, and thus, the first nine row houses was assessed by the Institute for Sustainable Technologies (AEE Intec). The balance is based on primary energy and includes the yield of the assigned pho-

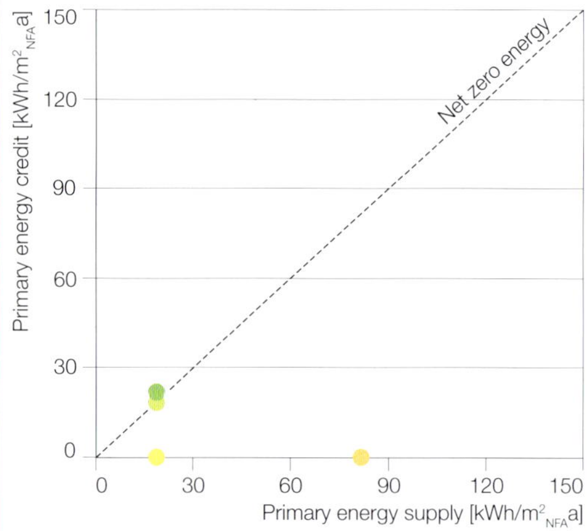

Net-ZEB-18 Standard is qualified by the following averaged aspects:

- Measured annual total primary energy consumption including household electricity (79 kWh/m²a)
- Self-demand coverage by allowable monthly energy yield credits (18 kWh/m²a)
- Offset of residual consumption by monthly surplus
- Annual energy surplus (3 kWh/m²)

Primary energy factors as per GEMIS 4.5 (see Fig. A 2.07, P. 31)

tovoltaic system for each unit. Consumption includes total unit-based energy consumption for space heating, domestic hot water, ventilation, lighting, household appliances, and entertainment. The difference between the yield for the individual photovoltaic systems and total unit power consumption was mostly positive, with few exceptions. As result, the energy balance for the entire community is positive. The power generated by the photovoltaic system is entirely fed into the public grid (Figs. B 10.04 and 05).

LESSONS LEARNED An evaluation of the photovoltaic systems yield of the first three house rows indicated that the varying angles of the buildings resulted in only minor differences. In comparison, there were large variations in the amount of energy consumed for heating, hot water, and ventilation of individual units, even though they are almost identical. Higher and lower consumption can be explained by the less advantageous locations of some units in terms of energy and by user specific consumption behaviour. However, considerably higher consumption in some units can't be directly attributed to a greater number of occupants. This shows that the users have a major impact on operation. Nevertheless, projected energy component parameters were achieved according to the community average, and the living comfort level was assessed as good.

SITE	Weiz (A)
Annual global radiation at site	1110 kWh/m²a
Annual mean temperature at site	9.0 °C
Context	suburban
BUILDING ENVELOPE PARAMETERS	W/m²K
U-value, exterior walls	0.09
U-value, windows including frames)	0.800
U-value, roof	0.11
U-value, floor slab	0.10
Mean U-value, building envelope	0.20
BUILDING EQUIPMENT PARAMETERS (per row house)	
Photovoltaic system area	40 m²
System area per m²	0.43 m²/m²
Photovoltaic capacity	5 kW_p
Capacity per m²	53.80 W_p/m^2
Thermal storage tank volume	185 l
Storage tank volume per m²	2.00 l/m²
GRID INFRASTRUCTURE AND ENERGY SOURCES	
Supply infrastructure	electricity grid
Energy source supply	electricity
Feed-in infrastructure	electricity grid
Feed-in energy source	source electricity

DESIGN STRATEGIES, EMPHASIS
Passive house concept, photovoltaics, air-air heat exchanger, ventilation system with heat recovery, fresh air intake via earth tube heat exchanger, ecologically sound and locally available construction materials

BUILDING PARAMETERS (for a typical row house)	
Net floor area, NFA	93 m²
Gross floor area, GFA	130 m²
Gross volume, V	403 m³
Building envelope, A	291 m²
Surface to volume ratio, A/V	0.72 m²/m³
Building costs (net, construction/ technical systems)	1420 €/m² (2008)
Number of units (community in total)	22
Total number of users (per unit)	3
CONSUMPTION PARAMETERS (average 2006–2008)	kWh/m²a
Space heating consumption	15
Water heating consumption	20
Site energy consumption for heat (including hot water)	24
Electricity consumption	61
Total primary energy consumption	79
Total primary energy generation	82

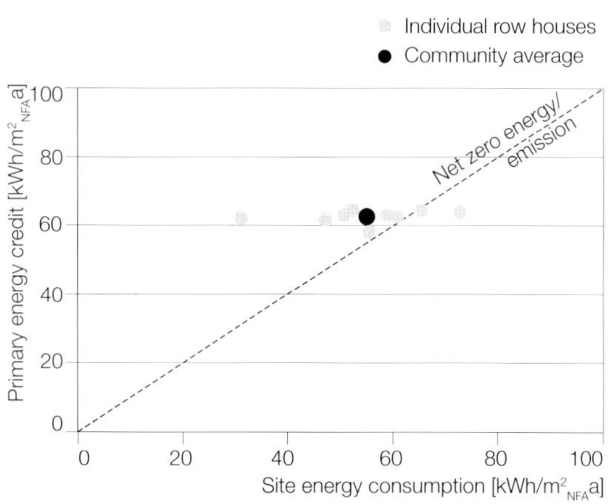

B 10.04
B 10.05
B 10.06

B 10.04 Energy evaluation
B 10.05 Graph of primary energy balance of the individual row houses of the first construction phase and the community average. The consumption and yields represent a three year average for 2006 to 2008.
B 10.06 Building and energy parameters (values refer to net floor area, NFA)

BEDZED COMMUNITY

Sutton near London, GB 2002

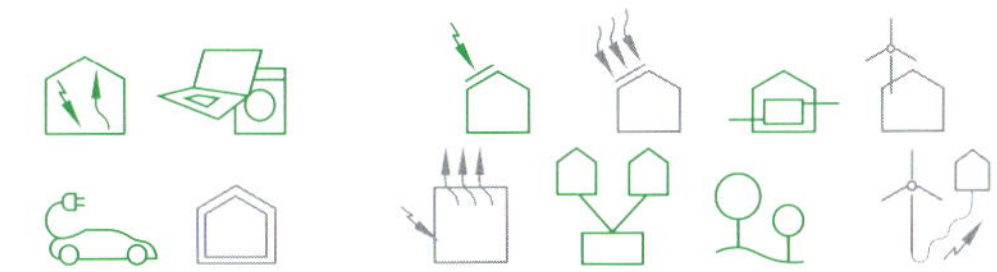

Client: Peabody Trust, London
Architect: Bill Dunster, ZEDfactory, Surrey
Energy consultant: Arup & Partners, London
Building services: Arup & Partners, London
Monitoring: BioRegional, London
Main stakeholders: architect and client

Great Britain's first net zero energy community was built in 2002 in the village of Beddington in London's Sutton District. Beddington's Zero Energy Development (BedZED) was built on the vacant site of a former water treatment facility. The design is based on an energy concept similar to a passive house and features solar power systems and a wood-fired combined heat and power plant. The aim is to achieve a zero energy balance and use 90 % less heat and 33 % less electricity than specified by British standards for new construction, while also taking into account the total household electricity consumed.

DEVELOPMENT, DESIGN, AND STAKEHOLDERS

In the mid-1990s British architect Bill Dunster, an advocate of ecologically sound construction, designed and built an active solar energy-powered, low-energy prototype house for his family. Called the Hope house, it served as a model for the simultaneously developed Hope town design with multiple use, high density terraced units. The design subsequently came to fruition as the Beddington Zero Energy Development. This was primarily made possible by the collaboration of the entrepreneurial, ecologically focused charitable organisation BioRegional and Peabody Trust, one of London's biggest housing construction companies.

B 11.01 View from south-west
B 11.02 Site plan; ground floor plan, scale 1:1500

1 Kindergarten
2 Fitness centre
3 Café
4 Office

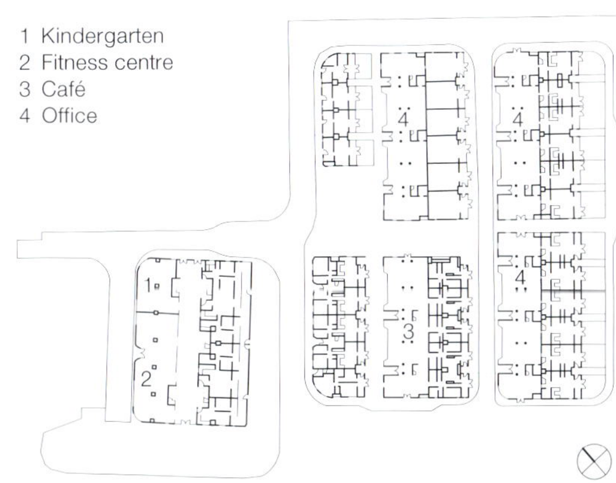

B 11.01
B 11.02

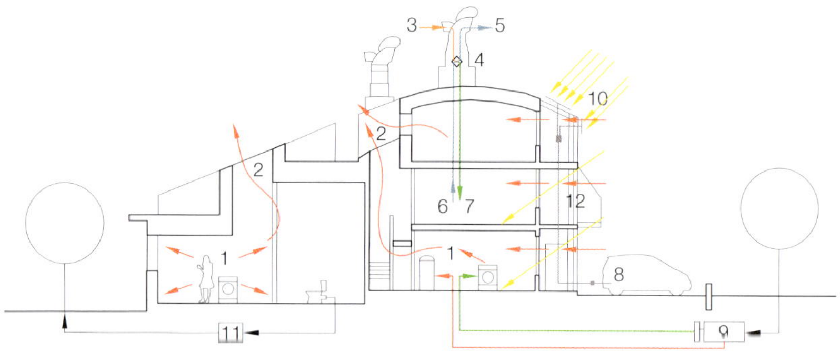

	Living [m²]	Parking [m²]	Streets [m²]	Infrastructure [m²]	Offices [m²]	Green spaces [m²]
Conventional row house design	1938	1568	2454			5105
Parking space reduction via ecologically sound traffic concept		968				
Concept with car-free secondary access			540			
Addition of office and workspaces				770	1216	
Roof terraces and green roofs	2378	968	540	770	1216	4621 and 1182
Completed net area, BedZED community	7446			1096	1404	5803

B 11.03
B 11.04
B 11.05

BioRegional secured funding to promote the project via the World Wide Fund for Nature (WWF), while Peabody Trust assumed the role of developer. In 1998, the three partners bought a property in the town of Beddington with the intent to develop a arbon-neutral residential community on-site. BioRegional worked with local key individuals to gain support for the project and organised two exhibits to include Beddington's residents in the decisionmaking process.

COMMUNITY CONCEPT The development features eight buildings, most of them three stories tall, the orientation, location, shape, and function of which are aligned with the energy concept and density of the community (Figs. B 11.01 and 02, p. 103). Together with the mix of uses for living and working, these aspects constitute the character of the project. The design takes public transportation into consideration, leading to a significant reduction in the amount of space required for parking and traffic. In addition, particularly high densities were achieved on site by optimising and stacking usable areas and green spaces in the remaining areas (Fig. B 11.04). Each unit has its own private green space or outdoor area. Roof gardens are incorporated into the terraced houses, which supports reducing setbacks between units and boosting solar yield. The fifty residential units, 120 workspaces per hectare, and a social mix of residents characterise community life. Of the eighty-two units, twenty-three are partially government funded, ten receive rental subsidies, and fifteen are rented as public housing. The cost of the condominiums is comparable to that of nearby conventional houses in London's south. Social interaction in the community is promoted by the many common areas and the differentiated design of private, semi-private, and public outdoor spaces. Small bridges cross the public access routes between the house rows and link residential units, adjoining office spaces, and roof terraces, creating thoroughfares on multiple levels.

CAR FREE LIVING Because the development features infrastructure such as kindergarten, community rooms, fitness centre, cafés, offices, and commercial spaces for small businesses, services are in walking distance, which reduces traffic and allows residents to avoid CO_2 emissions and travel expenses. Due to quick access to public buses and trains, in addition to 115 m² allocated for bicycle parking, residents don't have to rely on cars to commute. The number of kilometres travelled by car per person reportedly declined by 65 %. Addressing energy, waste, food, and water complements the integrated ecological approach. A car sharing program replaced an originally planned pool of electric cars, in part because electric cars aren't being developed as fast as anticipated. In addition, less electricity from renewable energies is being generated on-site than originally projected. Thus, the electric charging station for forty electric cars is currently used only by a small number of residents who own electric vehicles.

ARCHITECTURE On a site covering 12,000 m², the BedZED community integrates four different house types featuring various townhouses and maisonette apartments, as well as one and two room apartments. Southern oriented, terraced apartments with their own workspaces and front gardens are situated along quiet auxiliary streets. The houses alternate with a total of 2500 m² of office space located along car-free pathways and include gable-end access to publicly used areas such as cafés, shops, and communal buildings that offer child care facilities, studios, and fitness rooms.

LAYOUT A basic module consists of a three-story row house terraced on its north-western side. Flexible layouts permit splitting into apartments of varying size as required. Interior arrangement and orientation includes three zones: double-glazed, full-story patio rooms that serve as thermal buffer are placed along the south-eastern perimeter. They are partially shaded by horizontal glazing with integrated solar cells. Living rooms, office spaces, and special use areas are oriented along a north-western axis and receive ambient daylight from the north, while living rooms are heated by sunlight from the south. The southern facade is almost completely open and features triple glazed, argon-filled thermal wood frame windows. Despite high densities, roof shapes and setbacks between house rows are designed to minimise shading of neighbouring buildings. These conditions are especially advantageous for passive and active solar energy use. Because of soil contamination due to the former on-site waste treatment facility, buildings don't have basements.

CONSTRUCTION AND INSULATION Aside from operating energy, the buildings' embodied energy is also taken into account via a material registry established as part of planning, documenting the ecological quality of usable materials. The materials used for the buildings' solid construction primarily originate within a radius of fifty kilometres from the site. In addition to low-emission or emission-free natural materials such as certified wood and natural stone, recycled concrete and other recyclable, durable, and easy-to-repair materials were used. Composite materials were largely avoided. Traditional brick facing and untreated oak cladding were used to cover the solid concrete construction along the mainly closed east and west facades. The space between the load-bearing concrete block walls and the brick facing is filled with 30 cm of mineral wool insulation. 30 cm of rigid foam insulation were applied to the prefabricated concrete roof components and beneath the reinforced concrete floor slabs to provide the building envelope with an average heat transfer coefficient of 0.21 W/m²K.

ENERGY EFFICIENCY The southern-oriented, terraced structure optimises solar gains and allows daylight to enter living spaces. Heat is stored in the unfinished concrete walls, ceilings, and ceramic tile floors, all of which have a high thermal storage capacity. The airtight building envelope with low degree of thermal bridges minimises heat loss via controlled ventilation. Air is continuously exchanged by a passive ventilation system with a heat recovery ratio of 70 %. The colourful wind cowls follow the wind and are a characteristic feature of the residential community (Fig. B 11.06). The rooftop-mount, revolving wind cowls serve as the ventilation system's air intake. Without ventilators, they generate sufficiently high pressure to draw

B 11.03 Energy concept, scale 1:250
1 Internal heat sources
2 Natural ventilation/passive cooling
3 Outside/intake air
4 Heat recovery
5 Exit air
6 Exhaust air
7 Supply air
8 Electric cars
9 Combined heat and power plant
10 Photovoltaics
11 Fermentation/biomass utilisation
12 Solar yield

B 11.04 Green spaces optimised in varying ways on-site and resulting net areas for living, infrastructure facilities, offices, plus entire green space

B 11.05 View of terraces and inner courtyard

B 11.06 The community's characteristic cowls (back) with photovoltaic systems (front)

B 11.06

fresh air through a heat exchanger, lead preheated air into the living areas, and vent the exhaust air from the kitchens and bathrooms. During summer nights, the building is chilled by the cool air downflow via the wind cowls, which also enable venting thermal energy stored in the building mass (Fig. B 11.03, p. 104). Water-saving plumbing fixtures are used to reduce hot water demands. Energy efficient lighting, computers, and household appliances that were purchased in bulk reduce acquisition costs and save electricity.

ENERGY SUPPLY The power generated by 777 m² of photovoltaic modules covers about 15% of the community's total power consumption. A 250-kW$_{th}$ gas-fired combined heat and power plant (CHPP) was originally planned to cover remaining energy demands for hot water, space heating, electricity, and electric vehicles. The biogas originally intended as fuel was supposed to be created on-site by a wood gas generator. Due to major problems with the wood gas generation, the combined heat and power plant is fired by natural gas since 2005. A new woodchip-fired combined heat and power system is scheduled to produce the necessary heat and electricity starting in 2011. A district heating grid distributes the heat, and decentralised hot water tanks in each residential unit and office are filled several times a day to minimise grid losses. Solar thermal collectors aren't used (Fig. B 11.07).

ENERGY BALANCE After seven years of operation, readings taken at seventeen representative residential units indicate that the amount of electricity per household is 45% less than in other parts of Sutton District. However, this is slightly higher than projected. The metered 38 kWh/m²a or 1,900 kWh per single residential unit contrast the predicted consumption of 34 kWh/m²a or 1,700 kWh. The excess average consumption per town house actually is 12 kWh/m²a. Overall, the average electricity consumption for the entire community in 2007 was measured at 34 kWh/m²a. Households use about 48 kWh/m²a of energy for space and hot water heating, 80% less than comparable households in the area. Because of the significantly worse primary energy factor of gas compared to regionally generated biomass, the net zero energy objective has not been achieved. Only 20% of the energy consumed on-site is offset. The total energy demand is reduced by 60% compared to residential buildings in England that aren't optimised in terms of energy consumption. The balance may improve significantly in 2011 with the installation of the new combined heat and power facility. In order to calculate all advantages of the integrated ecological approach, all cost expenditures, energy consumption, and emissions were accounted for and evaluated. The developers focused on comparing additional construction costs to higher revenue by optimising building volumes as well as savings from reduced material costs without changing area for rent. Compared to designs typical for the region, their initial additional

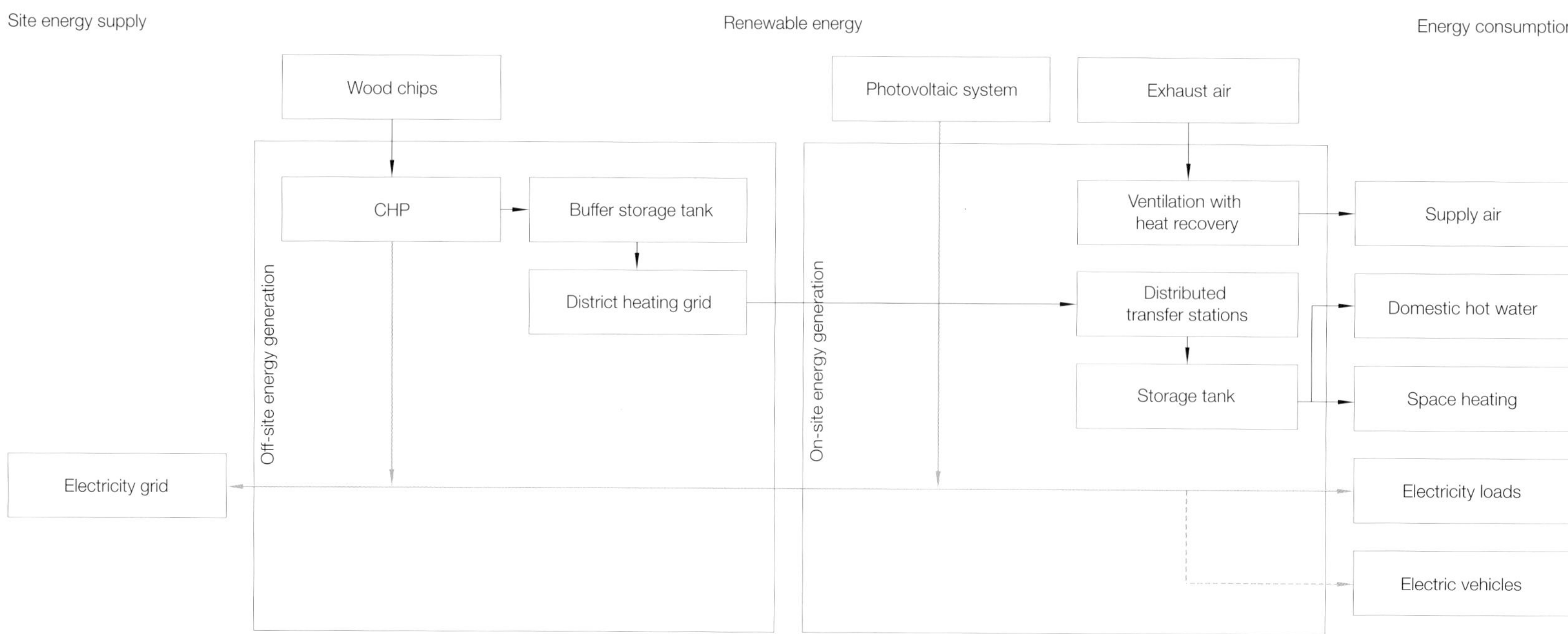

B 11.07

cost is € 800,000 for a terraced unit for six families. This is offset by € 820,000 of additional rental income. For environmental protection reasons, CO_2 emissions and water saved are included in the balance. Savings of 150 tonnes of CO_2 emissions annually can be generated by using the wood gas combined heat and power facility.

OVERALL EXPERIENCE Despite considerable success in reducing heat energy consumption, the residential community didn't attain the goal of achieving a net zero energy balance, because it was unable to use renewable resources for the combined heat and power plant, as well as higher-than-projected electricity consumption. The wood gas combined heat and power plant, newly developed in 2002, was installed without prior testing. An uninterrupted supply of wood gas couldn't be provided, because of problems with the automatic wood chip delivery mechanism, among other things. As result, some residential units initially didn't receive continuous and automatic supply with heat and hot water. As soon as temperatures dropped below 18 °C, thermostats activated the electric heating rods that were intended only for emergency heating in bathroom hot water tanks, as well as the ventilating system's heating registers. Many occupants purchased electric space heaters as a temporary solution, which led to a considerable increase in electricity consumption. Readable metres in the kitchens and user education provided by the community now supplement efficiency measures and lead to significantly lower consumption rates. The community's residents take the integrated ecological approach seriously and try very hard to succeed, in order to live up to the community's goal of serving as a model. The reduced complexity of the technical systems results in lower operative, maintenance, and energy costs. The combination of living and working reduces the consumption of fossil fuels and associated costs and improves BedZED's attractiveness.

B 11.07 Technical overview of energy provision
B 11.08 Interior view
B 11.09 Building and energy parameters (values refer to net floor area, NFA; primary energy parameters are exemplary and based on operating the biomass-fired combined heat and power plant)

SITE	Sutton near London, (GB)
Annual global radiation at site	800 kWh/m²a
Annual mean temperature at site	9.7 °C
Context	suburban
BUILDING ENVELOPE PARAMETERS	W/m²K
U-value, exterior walls	0.11
U-value, windows (incl. frames)	1.20
U-value, roof	0.10
U-value, skylights (incl. frames)	1.20
U-value, floor slab	0.10
Mean U-value, building envelope	0.21
BUILDING EQUIPMENT PARAMETERS	
Photovoltaic system area	777 m^2
System area per m^2	0.09 m^2/m^2
Photovoltaic capacity	108 kW_p
Capacity per m^2	12.20 W_p/m^2
Combined heat and power plant capacity	250 kW_{th}/120 kW_{el}
Capacity per m^2	28.20 W_{th}/m^2/ 13.60 W_{el}/m^2
NETWORK INFRASTRUCTURE AND ENERGY SOURCES	
Supply infrastructure	electricity grid, gas grid, wood chip delivery
Energy source supply	natural gas, electricity, wood chips
Feed-in infrastructure	electricity grid
Feed-in energy source	electricity

DESIGN STRATEGIES, CONCEPTUAL FOCUS
Passive house components and concept, wind-pressure driven ventilation with heat recovery, combined heat and power, photovoltaics, small-scale district heating grid, community, user education, ecologically sound traffic concept

BUILDING PARAMETERS	
Net floor area, NFA	8850 m^2
Gross floor area, GFA	10,388 m^2
Gross volume, V	24,465 m^3
Building envelope, A	12,346 m^2
Surface to volume ratio, A/V	0.48 m^2/m^3
Building costs (net, construction/ technical services)	1580 €/m² (2002)
Number of units	82 residential units
Commercial and administration areas	2500 m^2
Total number of users	220
CONSUMPTION PARAMETERS (2007)	kWh/m²a
Space heating consumption	16
Water heating consumption	23
Site energy consumption for heat (incl. hot water)	48
Electricity consumption	34
Total primary energy demand	172
Total primary energy generation	184

B 11.08
B 11.09

MASDAR URBAN DEVELOPMENT PROJECT

Masdar, UAE 2008–2025

Client: Mubadala Development Company , Abu Dhabi
Architect, master plan: Foster + Partners, London; Masdar Institute, Masdar
Energy consultant: Transsolar Klimaengineering, Stuttgart
Building services: Transsolar Klimaengineering, Stuttgart
Monitoring: Masdar Institute, Masdar
Main stakeholder: client

Masdar is a newly designed city planned for 40,000 future residents and 50,000 commuters in the peri-coastal desert belt of the United Arab Emirates. It is situated about 25 km from Abu Dhabi, a city of 900,000, and borders its international airport on its north-eastern perimeter (Fig. B 12.01). The worldwide design consortium led by the architects responsible for the city's master plan, Foster + Partners, and the Mubadala Development Company's Masdar Initiative, aims at creating a completely carbon neutral city including all associated construction and infrastructure processes (Fig. B 12.02). This presented a challenge not only in terms of construction, but also in the development of a mobility concept.

The goal is to provide an incentive for 1,500 companies to open offices in Masdar. A special business zone serves to draw foreign investors by offering low taxes and favourable conditions for research, development, and locating in the city. Various international corporations that embrace the topic of sustainable urban development and renewable energy intend to establish offices and a research centre in Masdar. A significant number of their employees is expected to live in the city, so that it doesn't merely become a satellite city of Abu Dhabi.
Solar intake in Masdar is twice as high as in central Europe. On the one hand, this is an advantage for generating solar electricity, but on the other hand,

B 12.01
B 12.02

B 12.03
B 12.04
B 12.05

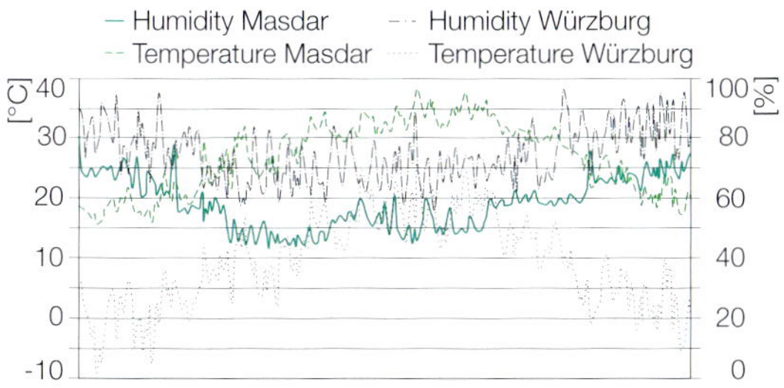

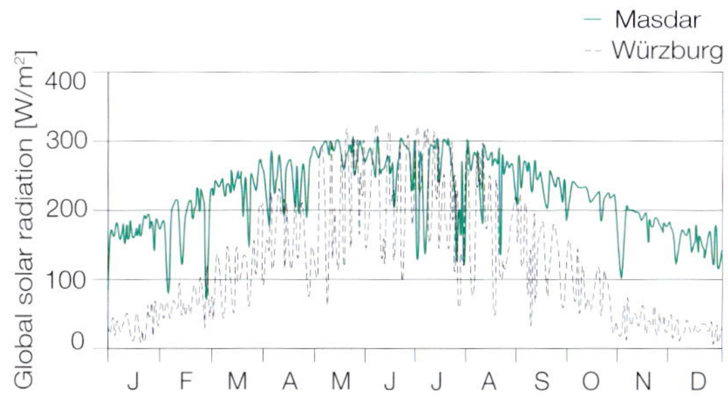

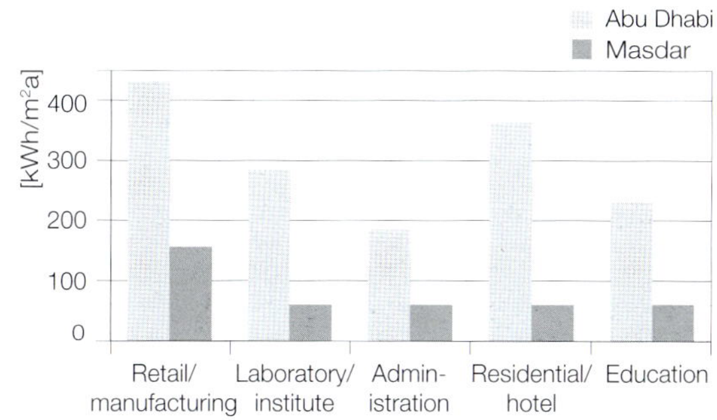

B 12.01 Masdar, geographic location
B 12.02 Master plan

B 12.03 Average daytime temperature in Masdar compared to Würzburg, Germany
B 12.04 Average daily global radiation in Masdar compared to Würzburg, Germany

B 12.05 Projected savings for electricity demand in kWh/m²a for Masdar compared to nearby Abu Dhabi, for various building types

it results in high annual temperatures of up to 48 °C (Figs. B 12.03 and 04). Seasonal differences are moderate in the desert climate, which helps solve the problem of storing seasonal excess energy. Electricity (0.03 €/kWh for local population) and water are subsidized by the government, and as result, energy-efficient architecture becomes difficult. Corresponding user habits cause consumption of enormous amounts of energy due to air-conditioning in summer, while grids are overloaded and generation capacities stretched to the limit. The region's annual energy consumption growth is in the double digits. At first glance, due to the extremely high energy consumption of typically oversized buildings and prevailing construction and use habits, the energy-reduction targets appear utopian.
The paramount design goal is to consume 70 % less total primary energy than in the nearby city of Abu Dhabi (Fig. B 12.05). Primary energy consumption per person in the United Arab Emirates is 169,000 kWh per annum, which is supposed to be reduced to 50,000 kWh per annum in Masdar, roughly equivalent to current consumption in Western Europe (see Fig. A 1.08, p. 14). The residual demand is to be covered by energy from grid infrastructure largely based on renewable energies.

DEVELOPMENT, DESIGN, AND STAKEHOLDER

After an initial two-year design phase, construction of the planned city started in 2008 within an ongoing design process. The total projected budget was $ 22 billion U.S. with a scheduled completion date of 2015. The master plan was revised as construction proceeded and modified to adjust to the current situation and ongoing technological advances. The energy concept was developed together with the German engineering company Transsolar. At times, over 500 engineers were on site to work on and implement planning.
"Masdar" means "source" and "origin" in Arabic. The city is both a prototype and test site; a place where sustainable design processes and environmentally sound urban development are rethought and realised. It is becoming an international centre for integrated, cooperative research on innovative technologies for climate friendly and energy efficient urban development concepts. Actual urban planning is part of the Mubadala Development Company's Masdar Initiative, which serves to manage all activities related to renewable energies in the United Arab Emirates. This includes Masdar Power, which builds solar power stations in the emirate, Masdar Carbon, which promotes carbon capture and storage, and the Masdar Institute university and research centre.
The initiative is led by Abu Dhabi Future Energy Company (ADFEC) and Sheikh Muhammad Zayid Al Nahyan.
Following the financial crisis in 2008/09 the real estate market in Dubai and the Emirates was also impacted, leading to far-reaching financial consequences that also affected the Masdar project.
In addition, the Arab Emirates' oil industry profits declined below accustomed levels. In 2010 the Masdar management team had to revise all design and construction projects. Engineering-based alterations also took place, since certain concepts proved difficult, and in some cases also extremely expensive. Beyond that, the construction budget was cut, the transportation concept changed, and the completion schedule extended to 2025. The planning process became more flexible in order to improve capacity for incorporating lessons learned. The first phase of construction covering 60 ha is now scheduled for completion by 2015. The first building complex to be completed, the Masdar Institute, was also designed by Foster + Partners. Construction started in 2008 and it was handed over to the clients in mid-2010.

ENERGY EFFICIENT URBAN DEVELOPMENT

The project developers conceptually oriented their master plan on Arab building traditions and attempted to design new urban structures based on traditional building typologies. As result, Masdar's square urban plan, oriented toward the north-west and south-east, is reminiscent of historic Arab cities such as Medina and Fes. The shape, volume, and orientation of buildings, infrastructure, and open spaces supports the creation of a pleasant microclimate. Two offset squares of differing size are situated on a site covering about 700 ha. The intended population

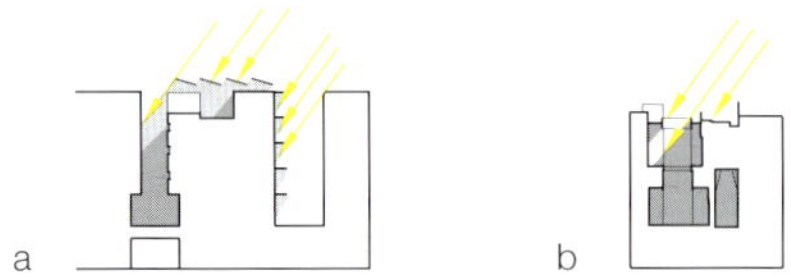

B 12.06 Shading comparison
a Masdar
b Fes
B 12.07 Wind patterns for a street longer/shorter than 75 m
B 12.08 Wind tower principle
B 12.09 The first completed wind tower

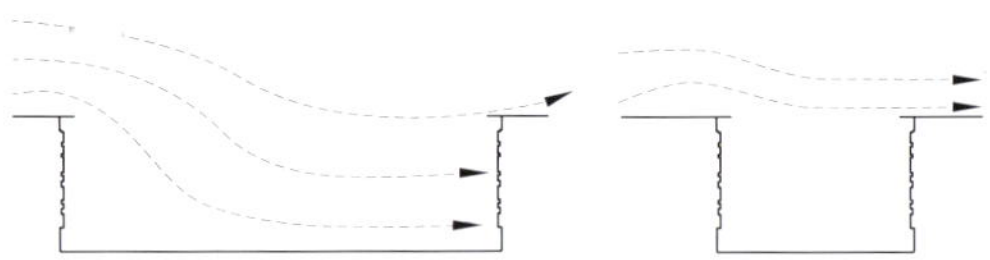

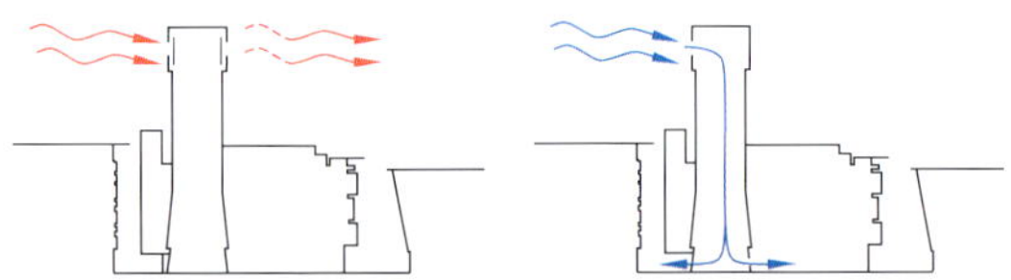

B 12.06
B 12.07
B 12.08
B 12.09

density on-site is 140 inhabitants per hectare. During peak periods, the density is expected to reach 240 persons per hectare including commuters, compared to a rate of 386 per hectare of persons living and working in Berlin. The goal of this relatively dense development in Masdar is to create a compact city that is welcoming and accessible due to short walking distances.

SHADING AND VENTILATION Rotating the squares and the urban structure by 45° enables improving shading and providing good daylight utilisation. Limiting widths of walkways enables the dense structure comprised of various building types to shade facades of adjacent buildings, and thus, prevent them from overheating (Fig. B 12.06). In addition, roof-mount photovoltaic systems extending beyond facades shield the public walkways and streets from direct sunlight (see Fig. A 1.03, p. 11). Because of the sun's near-perpendicular position in June, completely shading walkways at noon isn't possible. In open areas, folded umbrella structures provide sun protection.
It is decisive for the city's microclimate that very humid, north-western daytime winds that heat up above the sea up to 47 °C don't enter the city's small-scale urban structure. Ancillary streets that don't exceed 75 m in length and run from north-west to south-east serve precisely this purpose. Because the streets are relatively short, the wind doesn't enter the network of streets, but remains above roof levels (Fig. B 12.07). The streets are very narrow and branch out. In certain places, they open towards small courtyards filled with greenery and serving as recreational areas. These as well don't exceed a maximum diagonal width of 75 m, yet allow daylight to enter the buildings.
Two parks following an axis from north-west to south-east serve as the city's fresh air corridors and support air circulation along the narrow walkways. This helps cool the skin of inhabitants and makes temperatures seem more bearable. During the day, the hot north-western winds flowing through these corridors can be cooled down by evaporation. The heat required for this is drawn from the environment, which cools off as result. This effect is reinforced by numerous fountains and open pools located in the area. Evaporative cooling is also provided by spraying plants using sprinkler systems fed by the water distribution system. At night, streets and adjacent buildings are cooled by breezy winds from the eastern desert. The outdoor temperature within the city then declines to 34 °C.
Wind towers that utilize natural air circulation to support ventilation of streets and squares are positioned at the ends of the branch roads. The towers are conceptually rooted in the traditional Persian Bādgire (wind catcher) that have been in use to ventilate buildings since centuries (Figs. B 12.08 and 09). Here, in their updated form, they are used to cool public spaces. Incoming winds are detected by sensors and are directed vertically by dampers inside the wind tower through a duct made of fabric and sprinkled with water. The chilled air vented at the base of the tower flows into the pedestrian areas and inner courtyards and cools them. The design also calls for using air from the cool pedestal level at 7 m above grade to chill the pedestrian areas. The pedestal level was originally supposed to house the entire transportation system and infrastructure. However, only the part belonging to the first construction phase has been realised, due to cost considerations. Underground ducts also serve to cool building supply air.
Arcades that follow Arab examples and are permanently shielded from direct sunlight intake are situated on the ground floor level, where mostly retail uses are located and people pass by during daytime. At night, the solid floor and wall construction emits the heat captured at daytime, due to their thermal storage properties, thus providing cooling during the day. The sum total of the various passive measures is intended to reduce exterior temperatures in Masdar's outdoor spaces by 20 °C compared to nearby Abu Dhabi, permitting occupancy even in summer.

ENERGY EFFICIENT BUILDINGS Beyond optimising urban planning and building placement, when operating residential, administrative, and especially laboratory buildings on site, building envelope quality in relation to air tightness, insulation, thermal storage, degree of glazing, and shading is of key importance to prevent buildings

from overheating due to heat intake. The Masdar Energy Design Guidelines (MEDG) specify the appropriate design parameters.
For example, the student dormitory comprises a multilayer facade. The exterior is reminiscent of window grilles typically found in Arab architecture (see Fig. A 1.03, p. 11). This exterior wall is made of glass fibre reinforced concrete with desert sand aggregate and represents a playful interpretation of traditional regional architectural forms that also protects interiors from direct and diffused sunlight. The second layer consists of a shaded airspace that also includes small balconies. The next layer features a copper-coloured anodised aluminium element with integrated windows that reflects heat and seals the facade. It is followed by a 25 cm strong polystyrene insulation layer and a solid concrete wall that encloses the interior space. Manufacturing an airtight facade constituted a major technical problem for local manufacturers.
MEDG specifications mandate energy-efficient appliances and lighting (highest efficiency class) inside units to minimise heat loads and reduce electricity consumption. Only household appliances that meet European Standard A++ are permitted inside the apartments. In office buildings, laptops are used instead of PCs, since laptops consume less energy. Furthermore, the buildings are supplied via smart grids and include smart metering and control systems to achieve more balanced grid loading and cap peak loads.

ENERGY SUPPLY Thanks to an average eleven hours of sunlight daily and throughout the year as well as an average solar radiation of 225 W/m², the electricity generated by the photovoltaic systems covers up to 90 % of demands. The photovoltaic systems are installed on suitable roof surfaces. Some project visibly beyond the roof edges towards the streets, while providing additional shade and demonstrating renewable energy generation to Masdar's inhabitants. Roof-mount lightweight frame constructions optimise the orientation of the photovoltaic systems towards the sun, improving their efficiency. A 10 MW open-field PV system was also installed.

ELECTRIC POWER SUPPLIED BY SHAMS The largest solar thermal power station in the world is under construction on a 200,000 m² site in the less dust-prone interior of the country: the so-called "Shams", which is the Arab word for sun. A system of mirrors collects solar radiation within an absorber, and steam turbines and a generator transforms heat into electricity. Molten salt heat storage tanks store excess solar energy after sunset, enabling its nighttime use to generate electricity. Rating is phased, and the first phase provides 100 MW, to be increased incrementally to 200 MW. Thus, Masdar is intended to cover peak loads of surrounding districts connected via a common power grid, in addition to its own energy demands. Additional solar power stations with ratings of 100 MW are subject to planning. An initial wind turbine was also erected for testing purposes. However, the observation was made that wind conditions in the Emirates don't permit cost-effective operation of wind turbines.

HEATING Deep geothermal probes and vacuum tube collectors provide heat and cover the city's domestic hot water demands. The goal is to also use the 2,500 m deep probes for cooling. An initial geothermal plant to support the central air-conditioning system is in operation as a pilot project. All three of Masdar's grids, electricity, air conditioning, and water, are tied into the neighbouring municipal systems and are obliged to compete with the highly subsidised energy and water prices of the adjacent city. This complicates employing new technologies. The required energy to achieve a zero CO_2 balance is to be supplied by further solar power stations located at other, more suitable sites (Fig. B 12.10, p. 112).

COOLING Because of high outdoor temperatures, air conditioning systems of buildings consume the majority of loads in summer, when electricity consumption in the Emirates is twice as high as in winter, and air-conditioning systems consume 70 % of electricity demands.
Two primary strategies are used to cool buildings. The first is latent cooling by use of air handling systems with energy recovery or liquid desiccant systems that extract moisture from air. The second is sensitive cooling with energy recovery in combination with chilled beams (passive or active cooling convectors). This significantly increases efficiency of air conditioners. In the cooling convectors, water chilled by a central system flows through a heat exchanger. In the active version, the air is channelled through ducts to the condenser, which improves heating and cooling performance. Chilled ceilings and thermo-active building systems are also planned.
During the first construction phase, cooling is provided via Abu Dhabi's district cooling grid. However, Masdar is intended to have its own system in the near future. Solar absorption chillers are planned for this central chilling system. Geothermal sources are supposed to provide required heating energy. Waste heat generated from waste incineration will serve as supplemental heat source. The first pilot plants began operation in 2010 and are currently subject to testing. Remaining energy demands for the operation of decentralised compression systems are also covered by renewable energies.

WASTE CYLE Already during construction, attention is paid to producing a minimum of waste and to utilising whatever waste is produced.
The intention is to recycle or compost about 60 % of the waste generated by systematically separating waste materials. The remainder, nearly 40 %, is supposed to be incinerated in a waste management facility. The intention is to use generated heat to operate an absorption chiller and to drive turbines for producing electricity. Since the volume of waste isn't sufficient to guarantee cost-efficient operation in Masdar, a waste incineration concept with energy recovery is to be developed jointly with the city of Abu Dhabi.

WATER MANAGEMENT Even though Masdar is located in proximity to the coast, water is a rare commodity. This is mainly because significant amounts of energy are required to desalinate sea water. Water is to be prepared in solar desalination plants and stored as potable water for users. The residual salt will be sold to the industry. Solar desalination plants of the size required for Masdar are currently not available. Thus, Masdar is connected to Abu Dhabi's municipal fresh water distribution system. Grey water and the marginal amount of rainwater that falls during the region's six rainy days per year are used to irrigate green spaces. The biomass from processing waste water in the waste treatment plants is also utilised.

TRANSPORTATION CONCEPT Private cars are completely banned from the city. The residents park their cars at the city limits in covered car parks and walk towards the city centre along a fine-meshed street grid that exclusively serves for pedestrians, cyclists, and local public transport. The street grid consists of shaded, individual pathways for pedestrians and cyclists, the light rail line, and features a metro connection to the airport and Abu Dhabi. The central component of the transportation design was a personal rapid transit system (PRT) that was to be located within the pedestal level. These automatic electric vehicles with seating for four are controlled by a central computer that selects the quickest route and prevents collisions. The nearest stop would be less than 200 m from any location in the city. The system was supposed to handle about 150,000 trips per day. However, the observation was made that a system of this size and complexity would exceed reasonable expenditure. As result, the PRT system is only partially installed within the first construction phase. This necessitated a revision of the transportation concept, which now focuses on electric buses at street level. In addition, the pedestal level will be omitted in future project phases. It would have required significant amounts of massive construction and concrete, thus violating other criteria of carbon-neutral construction. Supply and disposal services have now been installed conventionally beneath the streets. Therefore, the city can be built more flexibly and respond faster to new developments.

BUILDING MATERIAL ECOLOGY Both the operation and construction of the city are supposed to be as carbon-neutral as possible. Inspectors supervise construction and educate participating companies and manufacturers about measures required to achieve carbon-neutrality. Computer programs track energy consumption, waste generation, and materials used, as well as prepare a monthly balance sheet. A large share of the building materials are sourced in the region or are manufactured there, such as the desert sand used in concrete on-site.

DESIGN AND CONSTRUCTION PROCESS
Masdar is a planned city. However, it incorporates experiences and developments of existing cities in this climate. In this case, the plan called for a very compact, mixed-use city, rather unusual for the region and only extant in this form in historic cities. This requires a significant amount of reflection among regional users and planning authorities. Equally unusual are respect for pedestrians and the obligation to park cars and use other means of transportation. The goal of the carbon-neutral city demands a maximum of creativity from all

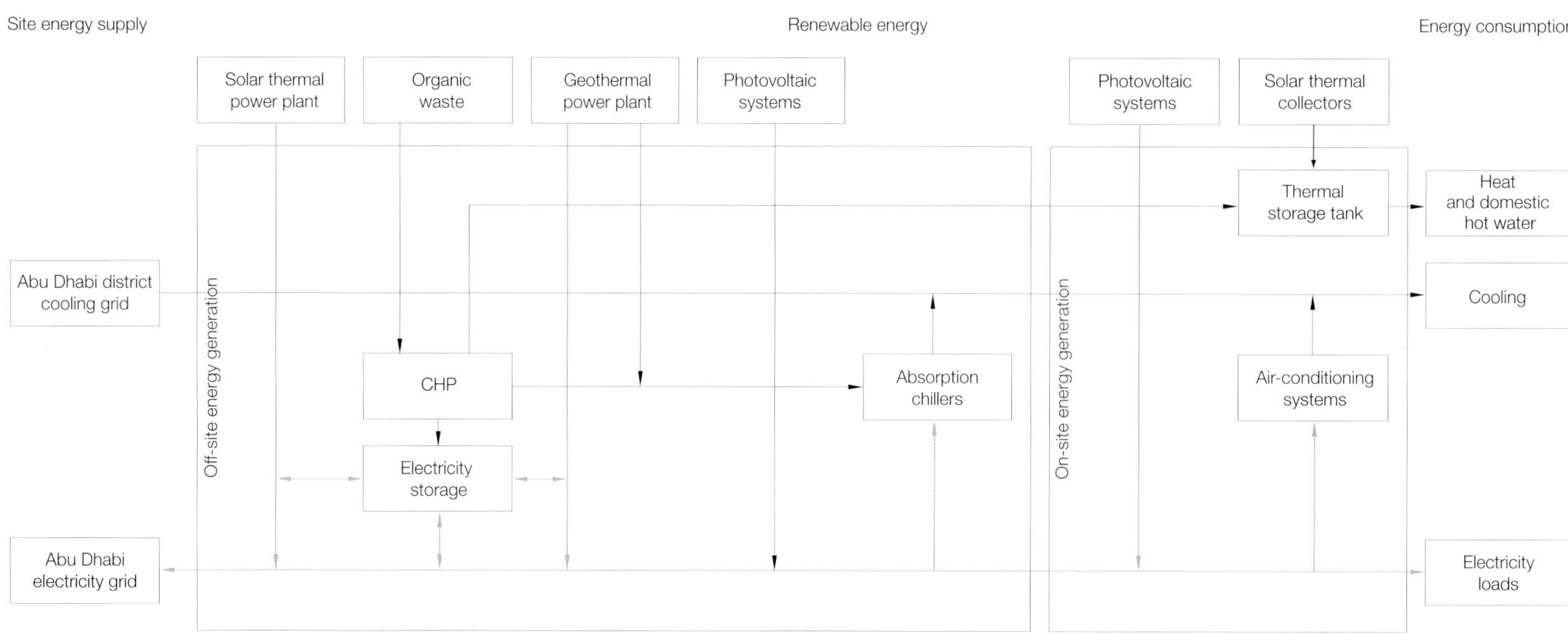

B 12.10

stakeholders, leading to successfully learning new skills. This impacts the entire region in a positive way and also influences legislation. For example, in the Emirate of Abu Dhabi, a mandatory rating system for buildings called "Estidama" was introduced along with sustainability guidelines for urban planning. The MEDG specifies benchmarks for all new construction and requires use of energy efficient systems, appliances, and lighting. In order to optimise load and energy profiles, user behaviour will not only be supported electronically, but also incentivised by variable tariffs. Feed-in legislation is also subject to discussion, and energy and water subsidies are being reduced step by step. Now that a general framework has been established and inhabitants begin living in the city, future user demands also become an influencing factor. Masdar has now gained the initial skills in terms of technical and economic feasibility and is systematically striving towards new approaches: for example, by improving the integration of globally networked engineering partners. The master plan is flexible enough to adapt to new user needs in correspondence to ongoing developments. Rather than an ideal plan that can supposedly solve all problems, Masdar comprises a process-driven plan that can be implemented in various stages. Masdar will certainly not work exactly as planned, but the city's role as a testing ground leaves enough room for various opportunities and possibilities. How well this model city will deal with changes to come remains to be seen. In any case, many important lessons are learned to advance future sustainable urban development in similar climates.

B 12.10 Technical overview of the energy supply
B 12.11 First construction phase of the Masdar Institute, Foster + Partners, 2010. Current degree of completion: additional new buildings will produce a network of shaded walkways.
B 12.12 City and energy parameters

SITE	Masdar, United Arab Emirates
Annual global radiation at site	2000 kWh/m²a
Annual mean temperature at site	27.2 °C
Context	urban
GRID INFRASTRUCTURE AND ENERGY SOURCES	
Supply infrastructure	electricity grid, district cooling grid
Energy source supply	cold, electricity, green electricity
Feed-in infrastructure	electricity grid, district heating grid, district cooling grid
Feed-in energy source	electricity, heat, cold
DESIGN STRATEGIES, EMPHASIS	
Photovoltaics, solar thermal energy, CHP, solar thermal power plant, geothermal power plant, waste incineration, closed circuit supply chains, eco-friendly construction materials, district cooling grid, district heating grid, integration of public and private transportation	
PARAMETERS	
City area	700 ha
Projected total number of inhabitants	90,000

B 12.11
B 12.12

OVERVIEW OF PROJECTS AND THEIR CHARACTERISTICS – PART 2

CHARACTERISTICS OF THE NON-RESIDENTIAL BUILDINGS With different functions such as education (19 %), administration (41 %), production (3 %), and a mix of other typologies (37 %) among the net zero energy non-residential buildings, the weighting between building and user-specific consumption also varies, and with it, the possibilities of increasing energy efficiency as well as the demands made on energy technology to achieve an equalised balance. In comparison to residential buildings, the proportion of projects that include user-specific consumption types in their balance is smaller.

ENERGY EFFICIENCY Similar to the discussed residential buildings, the analysed non-residential buildings (▷ project list p. 178), are also mostly located north of the equator. This fact results in a need to lower the heating demand, even though in office or factory buildings other consumption sectors are of equal importance (see Fig A 1.16, p. 18).

Whereas the number of certified MINERGIE- or passive houses in the non-residential sector is small, the concept and the individual components find frequent use (Fig. B 2.01). The fact that the average U-value is higher than in residential buildings is above all due to the large areas of glazing required for an even distribution of daylight at the workplaces or in the classrooms (0.32 W/m²K). The ratio of window to net floor area in the office and education buildings analysed is around 42 %.
Essentially these buildings, which are larger than the residential buildings, utilise the potential of a lower surface to volume ratio (A/V). The school renovation project in Schwanenstadt (Fig. B 2.01) demonstrates the importance attached to this aspect. Here, a comprehensive thermal renovation of the building envelope with the aim of achieving passive house level and a new programming of spaces including focussed demolition and extension measures made it possible to improve the compactness of the building to a low figure of 0.28 m⁻¹. Conse-

B 2.01
B 2.02

quently, the new A/V ratio is clearly below the average figure of 0.41 m^{-1} for known net zero energy school buildings; the office buildings analysed have an average figure of 0.35 m^{-1}.
Whereas in residential buildings solar gains from large south-facing areas of windows are used, in non-residential buildings the avoidance of heat loads in summer plays a major role. Only through this approach can the need for energy-intensive cooling systems be avoided. Low-e glazing and internal sunshade elements are generally insufficient. In almost all the office and educational buildings discussed here, either fixed sunshade systems such as (roof) overhangs (▷ B 13, p. 120ff.) or horizontal elements (▷ B 14, p. 125ff.; Fig. B 2.04) or external blinds and shutters shade the windows. The latter frequently consist of two parts and provide more natural lighting by means of deflecting daylight (▷ B 18, p. 144ff.). If they are connected to the building automation system, the flexible systems react to the amount of daylight or the solar radiation (▷ B 17, p. 138ff.). As well as providing sunshade, special windows allow further functions for the better exploitation of daylight and are deliberately employed as design elements (▷ B 13, p. 120ff.; B 21, p. 158ff.; B 22, p. 163ff.).
The principles of passive cooling are employed throughout. In addition to effective sunshade and low internal heat loads, the activation of the thermal storage mass in the building and its targeted dissipation are expedient. Mechanical ventilation systems remove heat from the thermal storage mass by increased night-time ventilation in combination with night-time flow openings in the facade (windows that can be set at an angle, controllable supply air elements) and discharge it (▷ B 19, S. 150ff.). Downstream heat pumps can lead the heat taken from the waste air into buffer storage tanks. (▷ B 18, p. 144ff.). In the Solarfabrik in Freiburg (▷ project list p. 178; Fig. B 2.04) a glass hall in combination with an earth tube heat exchanger and openable roof lights causes uplift ventilation that cools the building.
To reduce ventilation heat losses and to improve the quality of the air in non-residential buildings, which are more densely occupied than residential buildings, ventilation systems with heat recovery are used almost everywhere. The efficiency of the heat exchangers is, on average, above 80 %. Office buildings generally utilise centrally located plants that are coupled with ground registers for pre-warming and protection from frost (▷ B 13, p. 120ff.; B 15, p. 129ff.). In educational buildings, no uniform approach regarding the use of central ventilation systems can be identified. De-centrally positioned systems offer more flexible use by individual user groups (▷ B 19, p. 150ff.) or use in sections, according to need (▷ B 22, p. 163ff.). However, central plants are also used (▷ B 21, p. 158ff.).

SAVING ELECTRICITY The necessity to save electricity results in non-residential buildings with a net zero energy approach based on the motivation to lower both energy consumption and internal heat

B 2.01 South facade of school building in Schwanenstadt: entrance situation with the renovated building on the right, the solar electricity plant integrated in the facades in the middle, and a new building on the left
B 2.02 Green Lighthouse (950 m^2_{NFA}), Copenhagen (DK) 2009, Christensen Arkitekter
B 2.03 The mean U-value of the building envelope of non-residential buildings as a function of their compactness. The figures above a mean U-value of 0.6 W/m²K represent, among others, the Pixel Building and Solar XXI as two buildings in considerably warmer climates.
B 2.04 575 m² of solar electricity modules provide shade for the glass hall at the office wing and together with a plant oil CHP (65 W_{th}, 45 W_{el}) as well as a 220-kW wood boiler cover the electricity and heating needs of the 3500 m² Solarfabrik. Freiburg (D) 1998, Rolf + Hotz Architekten
B 2.05 The first passive house supermarket in Austria is heated solely through the waste heat from the cooling plant and the bakery ovens (heating heat load 22 kW). The photovoltaic arrays on the facade and roof with an area of 395 m² offset a major part of the electricity demand. Thening (A) 2003, Poppe-Prehal-Architekten

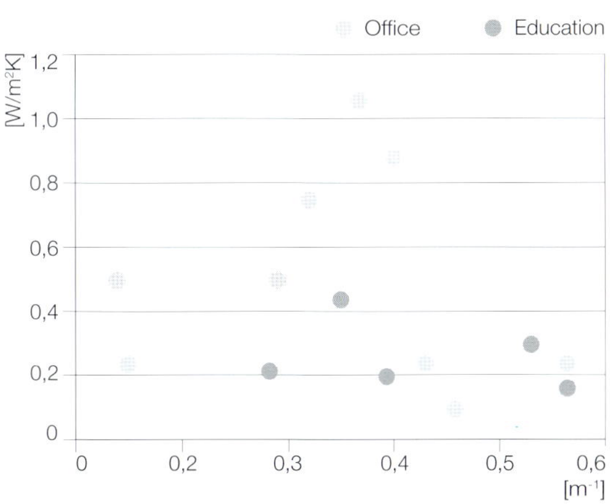

B 2.03
B 2.04
B 2.05

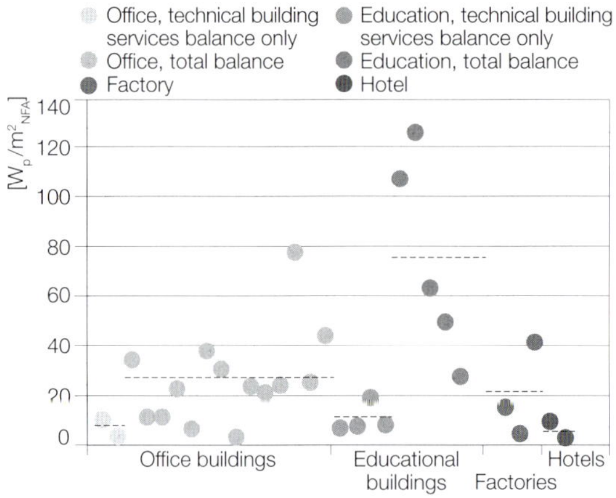

B 2.06
B 2.07
B 2.08

loads. The measures chosen range from central IT technologies with the focussed discharge of heat to economic lighting and combined daylight and artificial light concepts. (▷ B 18, p. 144ff.; B 16, p. 134ff.). The latter employ motion sensors or incorporate sunshade and light deflection features by means of building control technology (▷ B 17, p. 138ff.).
In comparison to residential buildings, the average user-specific electricity consumption in office buildings at almost 25 kWh/m^2a is somewhat higher, but varies considerably. In schools and kindergartens, about only one-fourth of the electricity used in residential buildings is required. As the floor areas are greater than those of residential buildings, in sum the total energy consumption is clearly higher. As regards to the roof areas available for the generation of solar electricity, this raises specific problems (see Figs. B 1.06, p. 52, and 07, p. 53).
The electricity consumption by the technical services in office buildings, an average figure of 7 $kWh/m^2_{NFA}a$, amounts to only 22 % of the total consumption. The figures in the reviewed literature are clearly below this. In educational buildings, the consumption by technical services is dominant and amounts to 12 $kWh/m^2_{NFA}a$ (see Figs. B 1.06, p. 52, and 08, p. 53).

ENERGY SUPPLY All office, school and factory buildings that aim at a net zero energy balance use solar electricity plants. For the primary energy balance of the energy used by building services, office buildings require around 7 W_p/m^2_{NFA}. If the demand is expanded to include user-specific consumption, this figure rises to 27 W_p/m^2_{NFA}. The figure for factories is 21 W_p/m^2_{NFA}. In educational buildings, the figures vary greatly, depending on the type of use (school, academy, kindergarten) (Fig. B 2.06). In the case of the Solar Academy, the concept of an "insular operation" increased the photovoltaic capacity (Fig. B 2.07). All in all, 152 kW_p (32 kW_p facade, 59 kW_p roof, 61 kW_p tracking) and a 70 kW_{el} biogas CPH with a 230 kWh battery pack should make it possible to operate the building at times without grid electricity.
Augmenting the yields from solar electricity plants by employing combined heat and power plants (▷ B 18, p. 144ff.), stations (▷ B 16, p. 134ff.), or by buying green electricity (▷ inverter factory, project list p. 178, Fig. B 2.08) is usual in large, non-residential buildings and is generally also necessary. However, the potential for using wind energy at the building is low, so that only few projects that make use of this approach are known (▷ B 16, p. 134ff.; Hotels in Freiburg and Vienna, project list p. 178).

COVERING HEATING AND COOLING DEMAND
To cover the heating needs of non-residential buildings, different systems are used and combined with each other. The only slightly higher thermal performance data of the systems (in comparison to the residential sector) indicate that the passive house standard is the efficiency goal (Fig. B 2.09). The use of biomass or combined heat and power allows the primary energy value to be reduced (see Fig. A 3.01, p. 40; Halle 8, company headquarters in Wörrstadt, project list p. 178). As larger CHP plants can now be run on rapeseed oil, these two are now combined with each other, in a number of cases leading to very low primary energy factors (▷ B 18, p. 144ff.; Solarfabrik, project list p. 178 and Fig. B 2.04, p. 115). However, the restricted long-term availability of rapeseed oil from sustainably cultivated land is a critical factor.
In comparison to the buildings with heat pumps, those with other heat supply systems have clearly higher installed thermal outputs. (Biomass) boilers and combined heat and power units are less effective in covering short peak outputs and buffer the produced heat in storage tanks. Around 40 % of the non-residential buildings are "all-electric-buildings". These are generally small buildings that come close to a domestic passive house and have a lower heat demand (▷ B 13, p. 120ff.; IBA Dock, Naturalia Bau, elementary school, project list p. 178). The heat sources vary from waste air and ground water to geothermal probes. In the "WWF-Headquarters in Zeist" (▷ B 14, p. 125ff.), the latter also serve as a heat sink for the reversible heat pump that can, thus, be used for active cooling. This is an economic and energy-efficient measure, as the heat stored in the ground during the summer can be drawn in winter, and thus the storage potential of the ground is not reduced in the long-term. Combined heat, power, and cooling, or the use of absorption cooling

machines are exceptions in zero energy buildings (▷ B 17, p. 138ff.).
Solar thermal plants are used in only half of the educational and office buildings, generally integrated with the heating circuit. Here, the collector area is less than 0.1 m² per m^2_{NFA}. in educational buildings, various kinds of integration are employed including feeding heat into a district heating system (Fig. B 2.10).

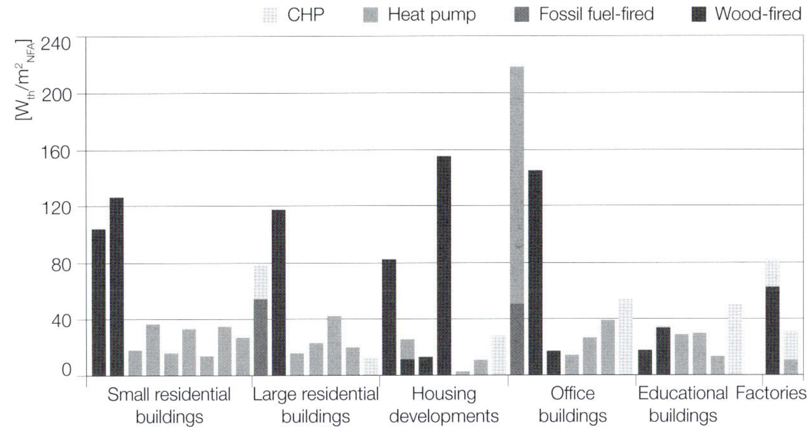

B 2.09

B 2.10

B 2.11

B 2.06 Installed capacity of the solar electricity plants per m^2_{NFA} in the non-residential sector, separated into buildings that aim at a completely equalised balance and those that only balance the primary energy demand of the building services

B 2.07 In the Solar Academy, users of solar energy systems receive their training on a floor area of 1400 m². Niestetal (D) 2010, HHS Planer + Architekten

B 2.08 Invertor factory (26 000m^2_{NFA}), Niestetal (D) 2009, HHS Planer + Architekten

B 2.09 Installed heating output per m^2_{NFA} in the non-residential buildings examined. Connections to district heating systems are not taken into account here. The omitted column in educational buildings represents the university building in Saint-Pierre as a comparison for warm climates.

B 2.10 PV modules on the roof generate 65 kWp electricty. 1500 m² of water coils beneath them absorb heat and transfer it to a sports hall and a residential district. A central heat pump outside the school building is connected to the heating grid. Christiaan Huygens College (7250 m^2_{NFA}), Eindhoven (NL) 2011, Rau Architects

B 2.11 In the balance, a 14,155 m² solar electricity plant covers all consumption by the stadium infrastructure in the period of one year. As sports events are held in the evening, while solar radiation occurs during the day, up to 1.1 MWh/a are fed into the public electricity grid of the surrounding housing estates. Sport stadium, Kaohsiung, Taiwan (RC) 2009, Toyo Ito & Associates, Architects

B 2.12

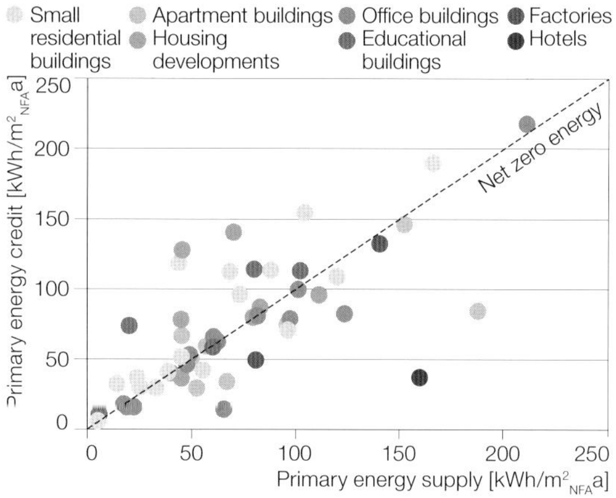

B 2.12 Comparison of floor area-related (net floor area) primary energy use and credits for the respective recorded consumption of the technical building services and use-specific electricity consumption. Data not adjusted for climate and the use of green electricity.

B 2.13 Comparison of typical strategies towards achieving zero -energy buildings after the assessment of 80 reviewed concepts. the technical solutions show a quantitative number and the qualitative performance of selected systems. The -typology groups factory, hotel and supermarket, as well as plants for the production of cooling are missing, as few applications are known.

COMPARISON OF ENERGY EVALUATIONS AND CONCEPTS The evaluation of 80 projects covering many different building types (see B01 to B22 and Project List p. 176ff.), all of which aim at a net zero energy balance, reveals that most of the projects achieve the targets set. A comparison of the use of primary energy for energy consumption and the primary energy credit for locally generated energy clearly shows that efficiency is a basic prerequisite for reaching this goal. The lower the consumption figures that have to be offset, the greater the likelihood that this will succeed (Fig. B 2.12).

ENERGY BALANCE As regards the housing sector, the statement can be made that, with the current possibilities for the generation of renewable energy at the building, an average total primary energy consumption of 73 $kWh_{prim}/m^2_{NFA}a$ is the target figure for residential buildings in central Europe. These buildings have an average primary energy credit of 80 $kWh_{prim}/m^2_{NFA}a$. With a credit of 111 $kWh_{prim}/m^2_{NFA}a$ educational buildings require about 77 $kWh_{prim}/m^2_{NFA}a$ of primary energy. In the case of office buildings the difference is smallest; a surplus of 4 $kWh_{prim}/m^2_{NFA}a$ remains after a total supply of 80 $kWh_{prim}/m^2_{NFA}a$ (Fig. B 2.12).
The evaluation of the annual primary energy balances shows that a number of the case studies planned as net zero energy buildings failed to achieve their goal. Without yield reserves, they are very susceptible to failure as a result of even small increases in consumption. In the projects that were monitored, increases in the number of staff, in the volume of production, or in the size of the family are cited as the reasons for increased consumption (▷ B 04, p. 68ff.; B 13, p. 120ff.; B 18, p. 144ff.). Climatic conditions such as colder winters (▷ B 05, p. 72ff.; B 07, p. 84ff.) or lower amounts of solar radiation and technical problems with systems that are, in part, new were also responsible (▷ B 04, p. 68ff.). The users remain a decisive dimension. Their behaviour cannot be determined in simulations or calculations, even if they are given "technical assistance" to operate the building such as building automation systems (▷ B 04, p. 68ff.; B 05, p. 72ff.). User-specific consumption parameters amount to a major part of the total consumption in very efficient buildings as well (see A 2.08, p. 31; Fig. B 1.06, p. 52).

MONTHLY COVERAGE RATE The evaluation of the energy characteristics of the central European projects presented in Part B (see p. 56ff.) in accordance with the system of drawing up a balance already described (see Infobox, p. 41; Fig. A 3.03, p. 42) shows that in a majority of projects large proportions of the monthly energy consumption are covered by yields. Therefore, the electricity grid is not completely utilised as seasonal storage. In residential buildings this proportion amounts to between 82 % (housing development buildings) and 60 % (small residential houses). With non-residential buildings, too, the monthly chargeable yields are generally more than 60 % of the consumption parameters. The figure for buildings that use biomass are particularly high as, due to the low primary energy factors, the amount used for the provision of space heating is reduced and even lower solar electricity yields can offset a greater proportion of consumption (▷ B 03, p. 64ff.). In addition, buildings with very large solar electricity plants or combined heat and power plants (winter electricity generation) achieve high coverage rates (▷ B 09, p. 94ff.; B 22, p. 163ff.).

PATHS TO AN EQUALISED ENERGY BALANCE
Elements of the MINERGIE or passive house concept are regarded as the basic means for reducing heating demands in all typologies. Solar thermal energy plants take over the heating of hot water in residential buildings, support heating systems and – connected to local heating grids – supply neighbouring buildings. An efficient, generally external sun-shade system reduces the danger of overheating presented by the large areas of south-facing windows that are used (at least in residential buildings) for the passive exploitation of solar energy. Optimisation of the supply of daylight is incorporated. Passive cooling concepts remove the need for air conditioning or cooling systems. The utilisation of energy-saving equipment and appliances in housing, offices and classrooms reduces user-specific consumption of energy.
To cover energy consumptions, photovoltaic plants are required. The potential of the sun is utilised, both integrated in buildings and in proximity to the location. Use of wind, which is independent of the time of day, is rarely encountered in or on buildings (▷ B 16, p. 134ff.). The first wind power turbines integrated in buildings reveal that they cannot completely offset the energy demands of buildings (▷ B 16, p. 134ff.; Hotels in Freiburg and Vienna, Project List p. 178). Where sufficient heat is generated, combined heat and power plants offer an alternative and a supplement with the potential to both use and generate energy at the same time (▷ B 22, p. 163ff.). Small-scale units and the burning of biomass will be developed further in the future. Due to the increased use of heat pumps, electricity is the main energy source for net zero energy buildings. Biomass, for example in the form of wood pellets, wood chips, or rapeseed oil is gradually replacing fossil fuels. Local or district heating grids are being used increasingly in housing developments (Fig. B 2.13).

Typology	Agreed strategies for energy efficiency of built examples, solar collectors with details of frequency of use	
SMALL RESIDENTIAL BUILDINGS	Mean U-value	0.23 W/m²K
	Surface to volume ratio	$0.67\ m^{-1}$
	Passive house concept	
	Solar collectors (78 %)	$0.07\ m^2/m^2_{NFA}$
LARGE RESIDENTIAL BUILDINGS	Mean U-value	0.24 W/m²K
	Surface to volume ratio	$0.39\ m^{-1}$
	Passive house concept	
	Solar collectors (88 %)	$0.04\ m^2/m^2_{NFA}$
HOUSING DEVELOPMENTS	Mean U-value	0.22 W/m²K
	Surface to volume ratio	$0.56\ m^{-1}$
	Passive house concept	
	Solar collectors (45 %)	$0.13\ m^2/m^2_{NFA}$
OFFICE BUILDINGS	Mean U-value	0.38 W/m²K
	Surface to volume ratio	$0.28\ m^{-1}$
	Passive house components and passive cooling measures	
	Solar collectors (44 %)	$0.05\ m^2/m^2_{NFA}$
EDUCATIONAL BUILDINGS	Mean U-value	0.26 W/m²K
	Surface to volume ratio	$0.45\ m^{-1}$
	Passive house concept	
	Solar collectors (56 %)	$0.06\ m^2/m^2_{NFA}$

Frequencies of use for heat and electricity generation including average capacity dimensions

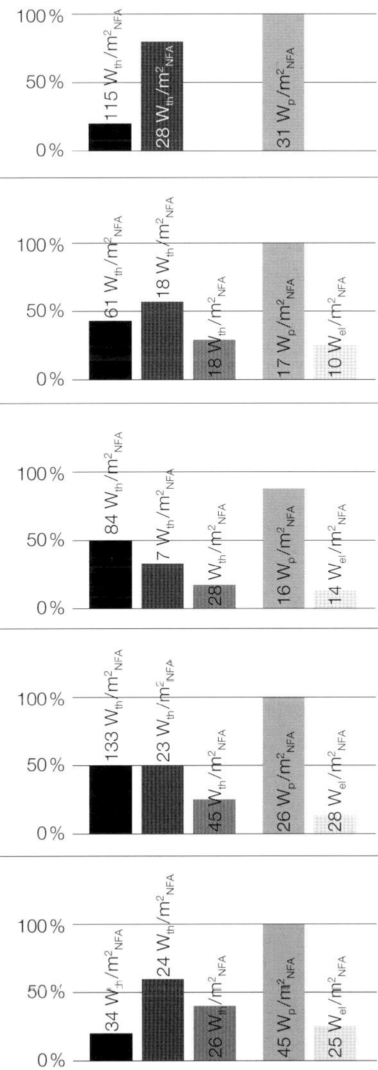

- Boiler
- Heat pump
- CHP (heat)
- Photovoltaics
- CHP (electricity)

Distribution of the energy carriers used for building heating

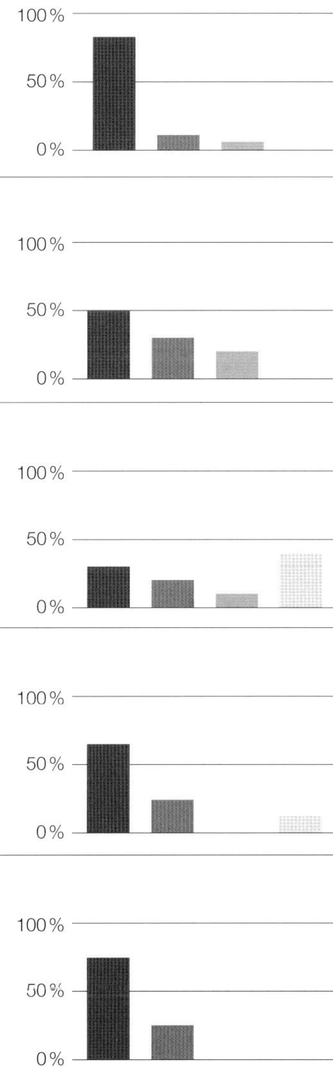

- Electricity
- Biomass
- Fossil fuels
- Local/district heating grid

B 2.13

OVERVIEW OF PROJECTS AND THEIR CHARACTERISTICS – PART 2

CORPORATE HEADQUARTERS

Kemptthal, CH 2007

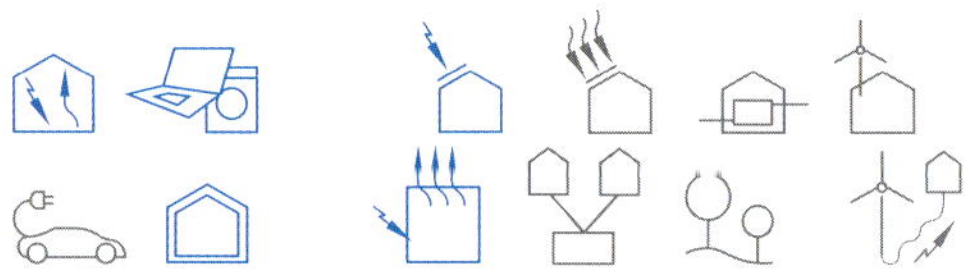

Client: Marché Restaurants Switzerland, Kemptthal
Architect: Beat Kämpfen, kämpfen für architektur, Zurich
Energy consultant: naef energietechnik, Zurich
Building services: naef energietechnik, Zurich
Main stakeholders: architect and client

The three-story corporate headquarters of the Marché Restaurant group, a division of Mövenpick, is located about five kilometres south of Winterthur near the Kemptthal autobahn service area. Sustainability and ecologically sound practices are part of the company philosophy, and the goal of providing high quality workspaces for 50 employees was central to the design for the Kemptthal office building. The MINERGIE-P-ECO-certified building (see Switzerland, p. 43ff.) combines formally restrained architecture with active and passive solar energy utilisation. It is a typical all-electric building featuring a photovoltaic system and a geothermal heat pump that produce electricity on-site to match all consumption over the course of the year.

DEVELOPMENT, DESIGN, AND STAKEHOLDERS

Twelve months lie between begin of planning and handing over the building, and no additional costs arose compared to a typical commercial building according to SIA, the Swiss Association of Engineers and Architects. Construction, technical installations, finishes, and form as central conceptual aspects were kept simple, in accordance with corporate image, tight budget, and limited construction time. These aspects are complemented by energy-related features such as special glazing and the photovoltaic system. Overall, parameters such as functionality, workplace quality, and design, as well as criteria such as sustainability and energy consumption were treated equally.

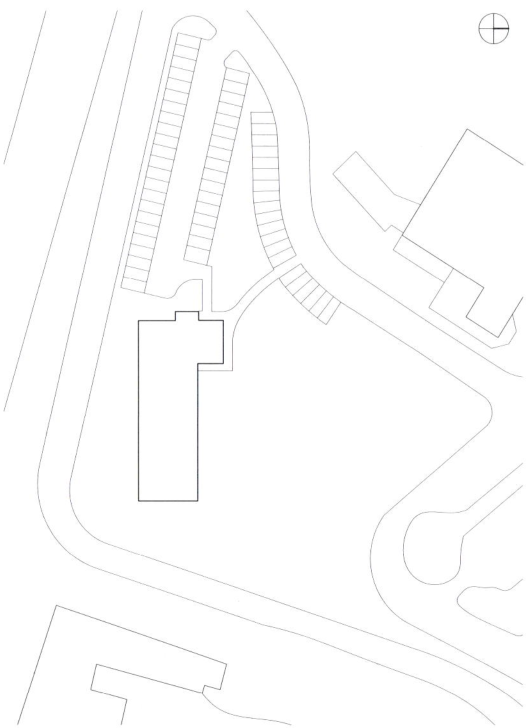

B 13.01
B 13.02

B 13.03
B 13.04

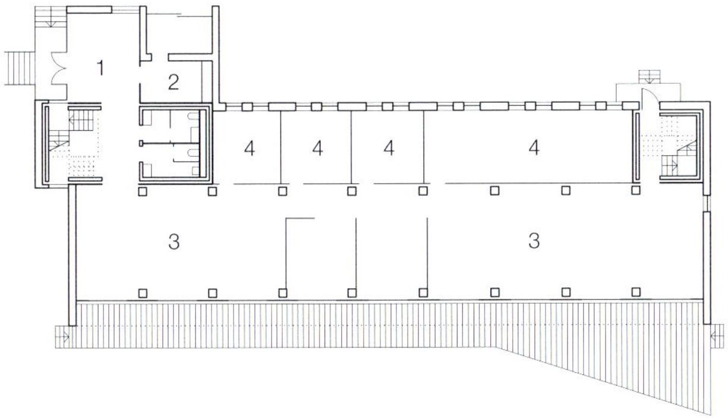

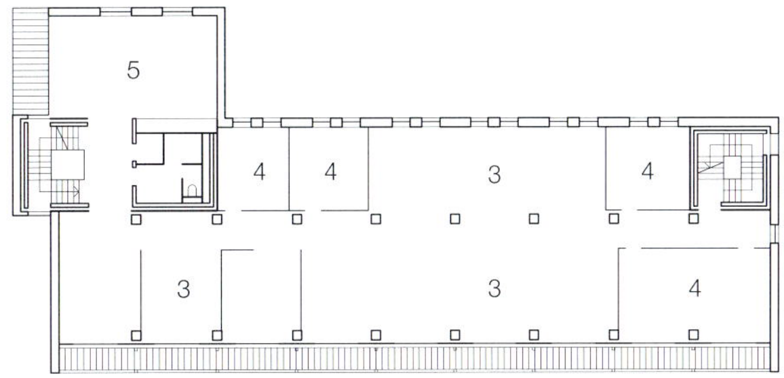

1 Entrance
2 Vestibule elevator
3 Open plan office
4 Individual office/ meeting room
5 Employee area

B 13.01 Site plan, scale 1:1500
B 13.02 View from south
B 13.03 Ground floor plan, scale 1:500
B 13.04 Upper floor plan, scale 1:500

ARCHITECTURE Neither the building's restrained design gesture nor its position on the narrow site refer to the context of the surrounding buildings or the street. Rather, it was exclusively designed to meet energy-related requirements (Fig. B 13.01 and 02). The strict southern orientation of the building's main facade allows both optimum passive solar energy use through its large window surfaces and active solar energy use via the photovoltaic system integrated into the roof. Except for the two stairwells facing north, the building consists of wood panel construction. The pre-fabricated facade elements of up to 4 m × 12 m in size reduced construction time to six months and improved assembly precision, essential to creating the airtight envelope with a measured air tightness was 0.57 h^{-1}.

CONSTRUCTION AND INSULATION The wood structure is self-supporting. The two solid stairwell cores have no load-bearing function, yet consist of concrete poured on-site due to fireproofing purposes. A 30 mm gap separates the stairwells from the remaining structure to insulate it from impact sound (Figs. B 13.03 and 04). Concrete pavers cover the wood hollow element ceilings to increase weight of floors, and thus, improve impact sound absorption (Fig. B 13.11, p. 124). They compensate the missing weight of the wood construction and contain pipes for underfloor heating. Wood cement composite panels as flooring ensure good heat transfer into the thermal storage mass below.

BUILDING SERVICES The utility room for heating, hot water, and ventilation is located in the attic above the stairwells. To reduce costs, hot water pipes are installed only in the front stairwell and the rooms immediately adjacent to the vestibule. The air, heat, and media ducts are horizontally installed along the floor of the unheated area beneath the shed roof and are covered with 30 cm of cellulose insulation. This allows quick access for maintenance and repairs. The unusually wide column grid of 4 m is designed to provide space for two workstations and ensures laying out the space flexibly.

FACADE Continuous balconies extending 1.30 m beyond the building facade and outdoor fabric blinds with an F_c-value of 0.11 shade the completely glazed southern facade. Triple-glazed wood frame windows with a heat transfer coefficient of 0.73 W/m²K comprise 50 % of glazed surfaces. The remainder consists of a solar facade made of special glass. Within the triple glazed thermal windows, the space between panes towards the interior features a heat transfer coefficient of 0.48 W/m²K and is filled with a salt hydrate phase change material (PCM, see p. 19). It stores heat from solar intake, while heat release is delayed depending on interior temperatures. As result, temperature fluctuations are equalized and temperature pikes are avoided. The PCM compensates missing thermal storage mass of the wood construction, especially in summer. A 15 mm layer has a thermal storage capacity of 1185 Wh/m² for a phase change temperature range of 26 to 28 °C. Increased interior surface temperatures lead to an increase in spatial comfort in winter.

The PCM usually melts in winter and becomes almost transparent. In summer crystallised PCM leads to diffusion of sunlight, and as result, the degree of ambient light in the office is high. Solar intake doesn't reach floors, and indoor temperatures rise less quickly. The light transmission factor for crystallised PCM is 8 to 28 %, for liquid PCM 12 to 44 %. An additional prism glass in the gap between panes facing the exterior only lets sunlight pass when its angle is shallow, as in winter. When the sun's position is high in summer and the angle of incoming sunlight is greater than 40° the prism glass reflects

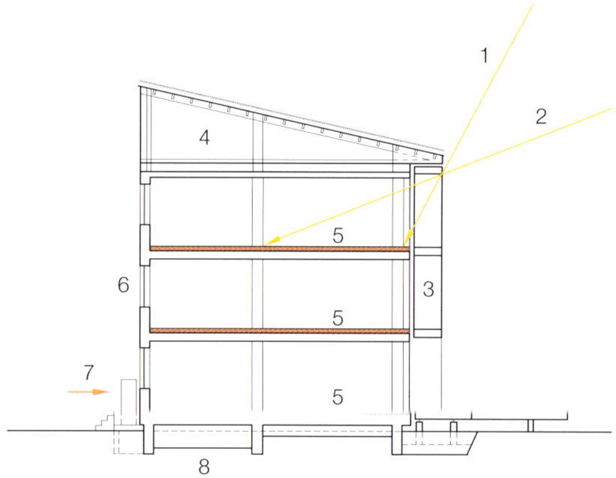

B 13.05 South-western elevation
B 13.06 Solar energy use concept
1 sun position in summer
2 sun position in winter
3 large glass surfaces facing south, shade provided by balconies
4 unheated attic
5 thermal storage mass
6 small windows facing north
7 supply air
8 ground loop
B 13.07 Technical schematic of energy provision
B 13.08 Comparison to reference building as per SIA standard
a building material share
b various lifecycle phases share
B 13.09 Energy evaluation

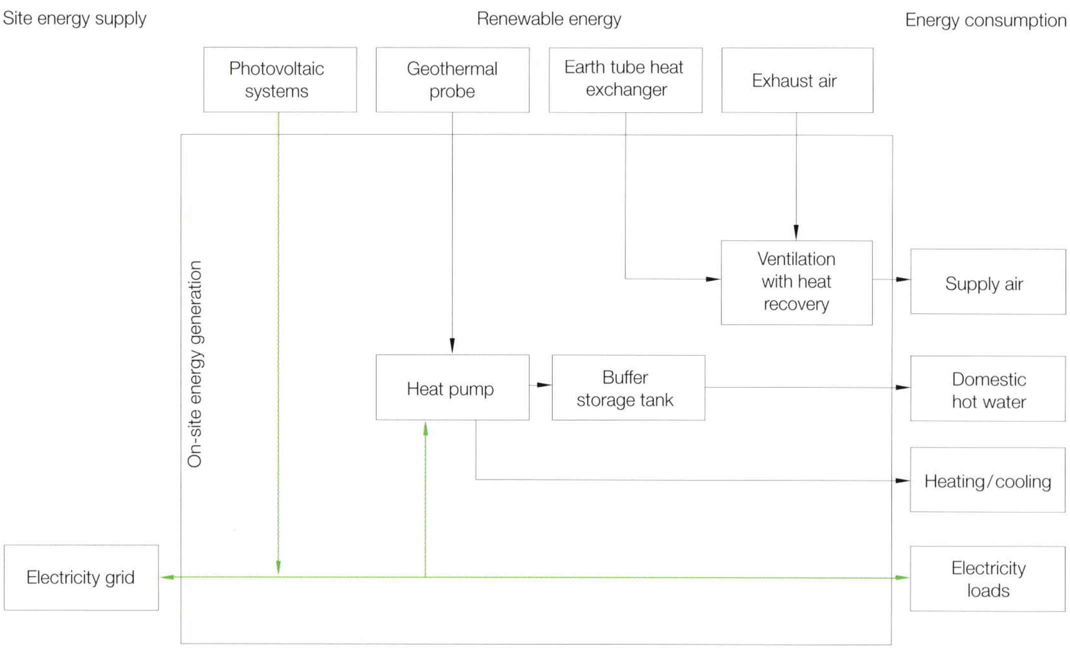

B 13.05
B 13.06
B 13.07

sunlight and protects interiors from overheating. In contrast to the graciously open design of the southern facade, relatively small wood frame, triple-glazed windows are placed at regular intervals along the other three facades insulated by 34 cm of mineral wool. Horizontal siding of strict design and made of Douglas fir with a ventilation gap situated behind it reinforces the impression of a discreet, enclosed building, typical for passive house designs. The floor slab rests on concrete foundation strips above a ventilation-free crawlspace and is insulated by 38 cm of glass wool infilled between wood joists.
A 12 m² green "living wall" on each floor serves as design element and regulates interior humidity. The open plan encourages communication, and its acoustic properties received special attention. A perforated sheet metal panel in combination with a layer of mineral fibre insulation is attached to the back of all cabinetry to improve sound absorption. In addition, suspended sound absorbing panels are placed above light fixtures (Fig. B 13.10, p. 124).

ENERGY EFFICIENCY The energy efficiency of the building is based on MINERGIE aspects such as compact design, excellent thermal insulation, triple-glazed thermal windows, and minimum thermal bridging. A ventilation system with a heat recovery factor of 91 % combined with an earth-air heat exchanger enables reduction of ventilation heat losses and calculated heating energy demands of 9.3 $kWh/m^2_{NFA}a$. A 25 m concrete channel is located approximately one meter below the floor plate within the crawlspace and adjusts the temperature of incoming air. This effectively prevents icing of the heat exchanger on the exhaust side, since supply air is always above freezing temperatures. The heat recovered from the exhaust air raises the supply air temperature to 20 °C before it enters office spaces at floor level via the air ducts integrated into the northern facade and the southern column row. It is vented through exhaust openings integrated into columns along the centre of the building. Windows usually remain closed due to noise from the nearby motorway. This is why a ventilation system is necessary to supply offices with fresh air. The fully glazed southern facade offers substantial solar gains in winter and maximises daylight intake. At the same time, the continuous balconies provide shade in summer. The thermal storage capacity of floors and the special glass elements compensate thermal peak loads (Fig. B 13.06).

ENERGY SUPPLY Due to intentional simplicity of utilities, the building only requires electricity for energy supply. Thus, this office building with heat pump and photovoltaic system is an "all-electric" building.

ELECTRICAL POWER SUPPLY The southern orientation of the building maximises its potential to generate electricity via the 44.6-kW_p photovoltaic system integrated into entire area of the shed roof with its 12° slope. The charcoal-coloured thin-film solar modules are similar to a scaly roof skin and eliminate the need for roofing tiles or metal roofing.

SPACE HEATING AND COOLING Under normal operating conditions, the 6.1-kW_{el} heat pump, together with two double U-shaped geothermal probes with a diameter of 40 mm each and installed 180 m below grade, serves as heat source with a capacity of 18.3 kW_{th}. This covers hot water and space heating demands via towel radiators and underfloor heating. The latter can also be used for cooling via the geothermal probes. This is possible via a bypass that acts as a heat exchanger between the ground loop and the heating distribution system without simultaneously operating the heat pump, a method also known as free cooling. There are no solar thermal collectors (Fig. B 13.07).

ENERGY BALANCE The photovoltaic system is rated to cover the annual electricity consumed by the heat pump, ventilation system, lighting, computers and auxiliary equipment. Due to cost, the photovoltaic system isn't financed by the owner. The energy provider of the district of Zurich (EKZ) is under contract to operate and maintain the photovoltaic system, as well as the geothermal heat pump. Solar electricity is handled directly by the EKZ power exchange and increases the renewable share of the electricity portfolio in the electricity grid. Marché is obligated to buy back one quarter of the solar electricity fed into the grid at market prices for solar electricity. Because of this commitment this percentage is also the maximum energy amount that could be balanced to offset the buildings energy consumption. In practice, these 25 % are sufficient to offset about 20 to 25 % of annual operating energy demands by primary energy. The office building's photovoltaic system is, thus, rather a part of the power grid than of the building's energy balance.
A lifecycle analysis for a projected fifty years indicates that total energy consumption will be reduced by 60 % (see p. 35). The analysis includes all energy consumed to manufacture the building materials used, to erect the building, for maintenance and repairs, for operating energy consumption, and for demolition and disposal, as well as the solar electricity generated on-site. By using regional building materials, untreated local evergeen wood, insulation made of 80 % recycled glass in the floor slab, and recycled concrete in the foundations and stairwell walls, embodied energy has been reduced to just below 30 % compared to reinforced concrete buildings that follow SIA standards. While the energy required to construct the building was reduced by almost 50 %, repairs to the wood cladding and upgrades of the photovoltaic system will result in higher maintenance costs than that of standard SIA buildings (Fig. B 13.08).

LESSONS LEARNED Although employees, unaccustomed to an open plan office, were sceptical at first, they soon embraced the concept. The fifty employees enjoy the excellent acoustic properties of the open plan office, its open atmosphere, and its creative design aspects. The solar facade's temperature regulation function also proved effective. In summer, interior surfaces of the solar facade are cooler than the wood-frame windows, and in winter they are warmer. The diffused light filtered through the facade's salt hydrate-filled glass panels doesn't impair use of computer monitors. However, an annual zero energy balance was neither achieved in 2008, nor in 2009. Metered consumption was higher than projected for the two years, by 7 and 10 $kWh_{end}/m^2_{NFA}a$ respectively. An uninterruptible

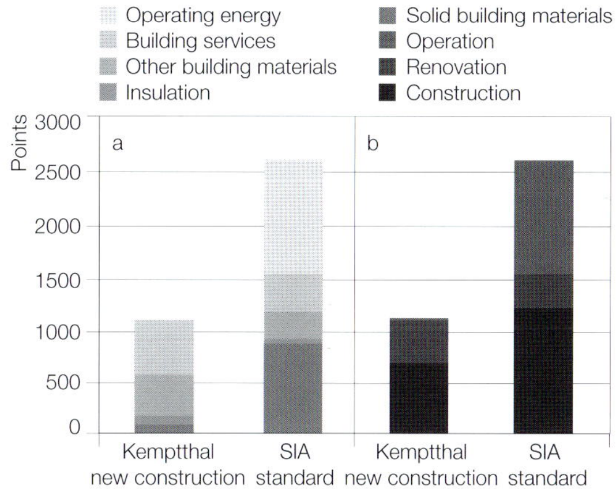

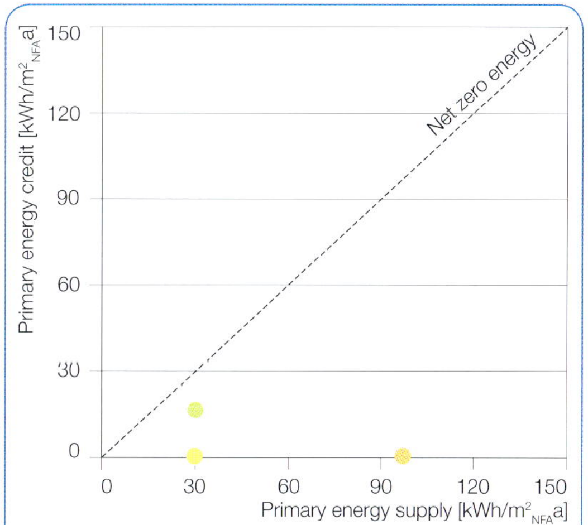

The corporate headquarters in Kemptthal didn't achieve net zero energy balance in 2008 and 2009.
- Metered annual total primary energy consumption (97 $kWh/m^2_{NFA}a$)
- Consumption, deducted monthly on-site demand coverage (31 $kWh/m^2_{NFA}a$)
- Remaining consumption after insufficient seasonal offset (15 $kWh/m^2_{NFA}a$)

Calculations indicate coverage shortfalls of annual demands on site of about 16 $kWh/m^2_{NFA}a$. Photovoltaic system operation by the energy provider of the district of Zurich not included.
Primary energy factors as per SIA 2031 (see Fig. 2.07, p. 31, Fig. 2.14, p. 35).

B 13.08
B 13.09

Ceiling construction:
composite cement particleboard 15 mm
cement screed 80 mm
underfloor heating 80 mm
polyethylene sheet
mineral wool impact sound absorption 30 mm
concrete pavers 400/400/50 mm
three-ply CLT panel 30 mm
joists, partial infilled insulation 260 mm
three-ply layer CLT panel 30 mm

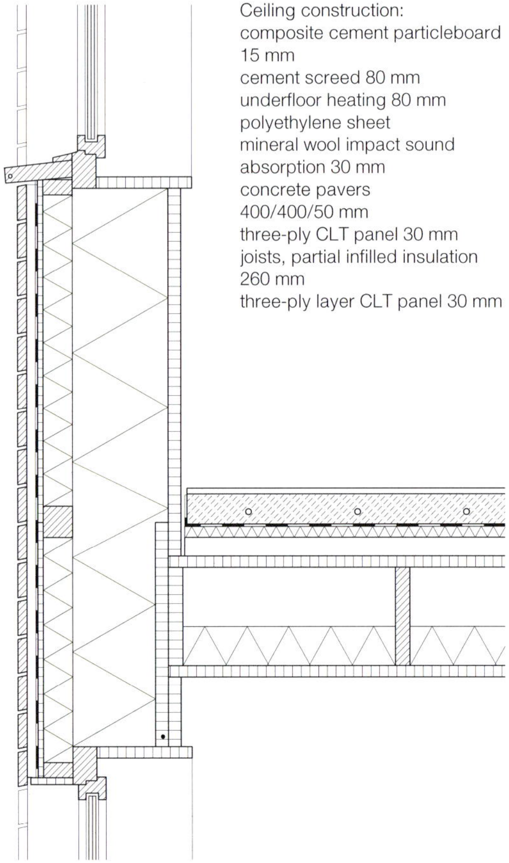

B 13.10
B 13.11
B 13.12

power supply (UPS) that wasn't originally planned increased building services consumption. In addition, lights stay on much longer than expected despite motion and light sensors. Conversion to manual operation is pending. The unforeseen need for a frequently used industrial dishwasher, several coffee machines, as well as copiers and printers that consume significantly more power than the manufacturers claim, indicate how limited the energy-related leeway actually is. The photovoltaic system yield was calculated at 40,000 kWh, which was precisely what it delivered in 2008. In 2009, almost 50,000 kWh were generated. If the best possible monocrystalline solar panels available at the time of construction had been used, the yield could be as high as 70,000 kWh. The decision to use thin-film cells was exclusively investment-related. The ambitious project shows that high architectural quality in combination with modern technology and low energy consumption is possible at the same cost of a conventional office building.

B 13.10 Open-plan office
B 13.11 Section through junction of floor slab and facade-, scale 1:25
B 13.12 Building and energy parameters (values refer to net floor area, NFA)

SITE	Kemptthal (CH)
Annual global radiation on site	1120 kWh/m²a
Annual mean temperature on site	8.9 °C
Context	rural
BUILDING ENVELOPE PARAMETERS	W/m²K
U-value, exterior walls	0.11
U-value, windows (including frames)	0.48–0.73
U-value, roof surface (cold roof)	0.08
U-value, floor slab	0.10
Mean U-value, building envelope	0.24
BUILDING EQUIPMENT PARAMETERS	
Photovoltaic system area	486 m²
System area per m²	0.38 m²/m²
Photovoltaic capacity	45 kW_p
Capacity per m²	35.20 Wp/m²
Thermal storage volume	500 L
Storage volume per m²	0.40 L/m²
GRID INFRASTRUCTURE AND ENERGY SOURCES	
Supply infrastructure	electricity grid
Energy source supply	electricity
Feed-in infrastructure	electricity grid
Feed-in energy sorce	electricity

DESIGN STRATEGIES, CONCEPTUAL FOCUS
Passive house components and concept, mechanical ventilation with heat recovery, geothermal heat pump, photovoltaics, ecologically sound building materials, active solar energy utilisation integrated into the building envelope

BUILDING PARAMETERS	
Net floor area, NFA	1267 m²
Gross floor area, GFA	1550 m²
Gross volume, V	5757 m³
dto., heated	4143 m³
Building envelope, A	2335 m²
Surface to volume ratio, A/V	0.56 m²/m³
Building costs (net, construction/ technical systems)	1964 €/m² (2007)
Number of units	1
Total number of users	50 persons
CONSUMPTION PARAMETERS (2008)	kWh/m²a
Space heating consumption	22
Site energy heat (incl. hot water)	7 (electricity)
Electricity consumption	31
Total primary energy consumption	97
Total primary energy generation	81

The Netherlands headquarters of the WWF (World Wide Fund for Nature) is located on the south-eastern periphery of the town of Zeist. Following the completion of the reconstruction measures in 2006 this abandoned agricultural laboratory from 1954 was regarded as the first carbon neutral administration building in the Netherlands. The building is efficiently conditioned by using exhaust heat which is stored in the ground by means of geothermal probes, as well as by a passive cooling system. Solar collectors cover a part of the heat required for the provision of hot water. The remaining hot water and heating demands are covered by a heat pump and a mini combined heat and power plant that runs on rapeseed oil. Over the course of a year the electricity produced by the grid-connected 300 m² photovoltaic system balances the entire amount of primary energy consumed – including all appliances.

DEVELOPMENT, DESIGN, AND STAKEHOLDERS

While searching for a suitable location for its new headquarters in 2002 the organisation was already fully aware of its pioneering role. Consequently, erecting a new building or demolishing an old one was out of the question. Instead of this, an existing building was to be found, renovated and, if necessary, expanded. Eventually, the abandoned research institute on a clearing in the woods of the Zeist nature conservation area was chosen. The

WWF HEADQUARTERS

Zeist, NL 1954/2006

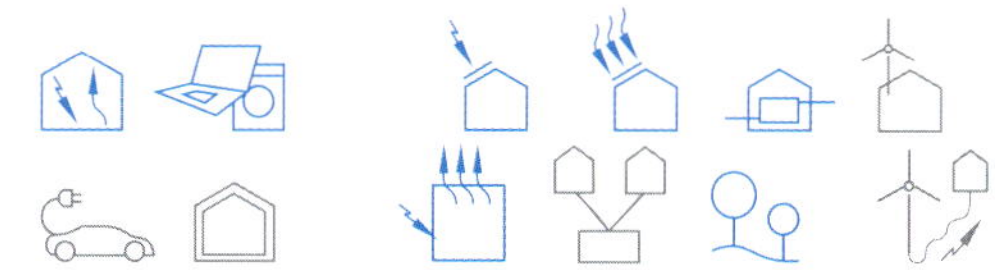

Client: WWF – World Wide Fund for Nature, Zeist
Architect: RAU architects, Amsterdam
Energy consultant: DGMR, Drachten
Building services: Arup, Amsterdam
Monitoring: Arup, Amsterdam
Main stakeholders: architect and client

B 14.01 View from south
B 14.02 Site plan, scale 1:1500

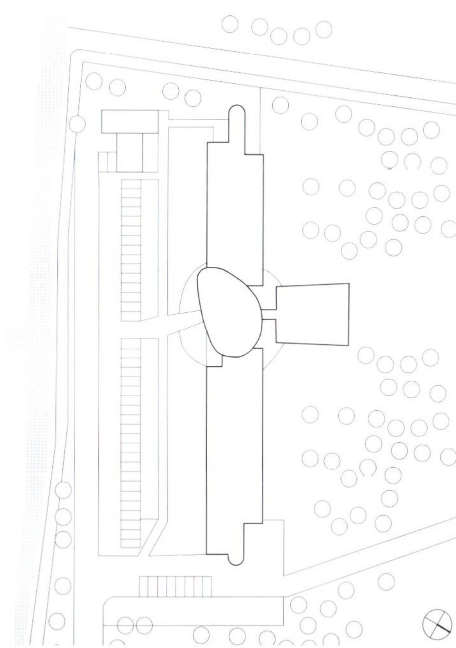

B 14.01
B 14.02

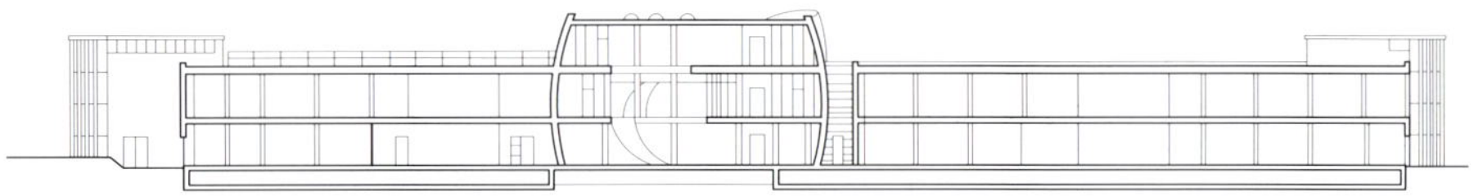

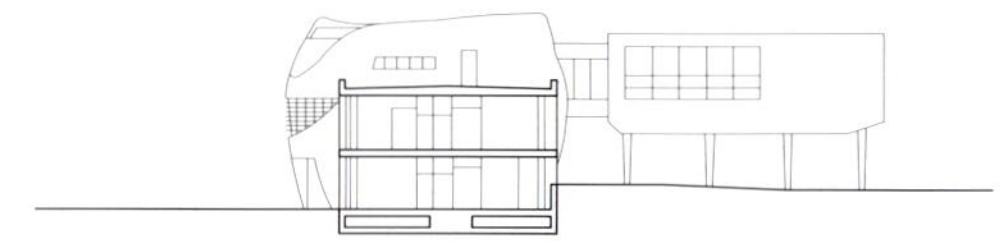

1 Entrance
2 Foyer
3 Information
4 Office
5 Restaurant
6 Meeting room
7 Storage/archive

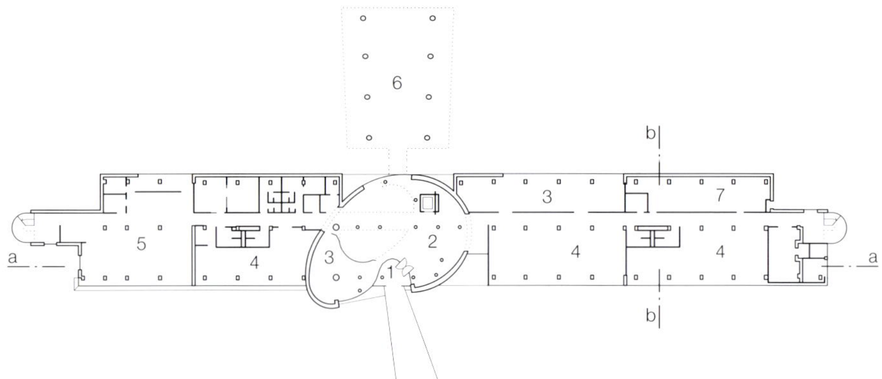

B 14.03
B 14.04
B 14.05

B 14.03 Section, scale 1:800
B 14.04 Section, scale 1:800
B 14.05 Ground floor plan, scale 1:800
B 14.06 Technical schematic of energy provision
B 14.07 Interior of the "blob"

existing building comprised of massive construction was in part very dilapidated and, in fact, demolition seemed to be the most logical option. But today, and through its new function, it offers important arguments for the preservation and careful handling of existing buildings as a resource. In a new building, the exceptional ceiling heights and the flexibility this offers, as well as the good lighting from the tall windows could not have been achieved. Also, the CO_2-neutral overall approach could not have been implemented, due to the limited budget. After a three-year planning and construction phase the headquarters was officially opened in September 2006.

ARCHITECTURE As part of the renovation, the meandering old building was reduced to an elongated block oriented almost exactly from north to south. An incision was made in the centre of what remained of the rigid, two-story original building and filled with a three-story "blob". (Figs. B 14.01 and 02, p. 125). This projecting volume with its organic shape contrasts with the rest of the building and together with the clearly legible main entrance makes an inviting gesture. The facade is clad in glazed ceramic tile supported by a wood frame and wood panel sheathing. These enclose a 24 cm deep air cavity and a 12 cm thick insulating layer of mineral wool. The inner face is comprised of a vapour barrier and a 5 cm thick adobe render layer. The lightweight timber construction achieves a U-value 0.32 W/m²K.
The organic appearance of the window openings underlines the effect of the "blob" and establishes a connection to the environment. The building contains a library, shop, the entrance hall with reception area, a coffee corner, and a central staircase that resembles a spatial sculpture and a communicative element (Figs. B 14.03–05). Glazed walls symbolise the openness of this nature conservation organisation and allow daylight to penetrate deep into all areas (Fig. B 14.07). The staircase emphasizes the three-dimensional qualities of the building, which had not been perceivable in the building prior to its renovation – neither on the exterior, nor the interior – due to the building's hard geometric delineation and absent spatial interrelations. A pedestrian bridge leads from here to a shingle-clad extension that stands on columns adjacent to the north side of the building. It houses the conference rooms and is also a preserved part of the old building. The non-publicly accessible office spaces in the two renovated wings of the building extending to the right and left of the central "blob" form a strong contrast. The open plan offices are clearly and flexibly organised by the 3 m grid of the existing reinforced concrete frame. The simplicity of the building's horizontal and vertical rganisation allow the office wings to be rented out in part or as a whole. Felt lining on the storage cabinets, perforated wooden facade panels covered with textile covers, as well as sound absorbing adobe ceilings improve the internal acoustics. The post and lintel construction of the curtain wall facade, clearly structured and based on a delicate grid, consists of triple glazed windows (U_g value 0.9 to 0.6 W/m²K), most of which are fixed, which alternate with opaque facade elements. The latter comprise lightweight wooden elements that are insulated with a total of 12 cm mineral wool and feature coloured glass

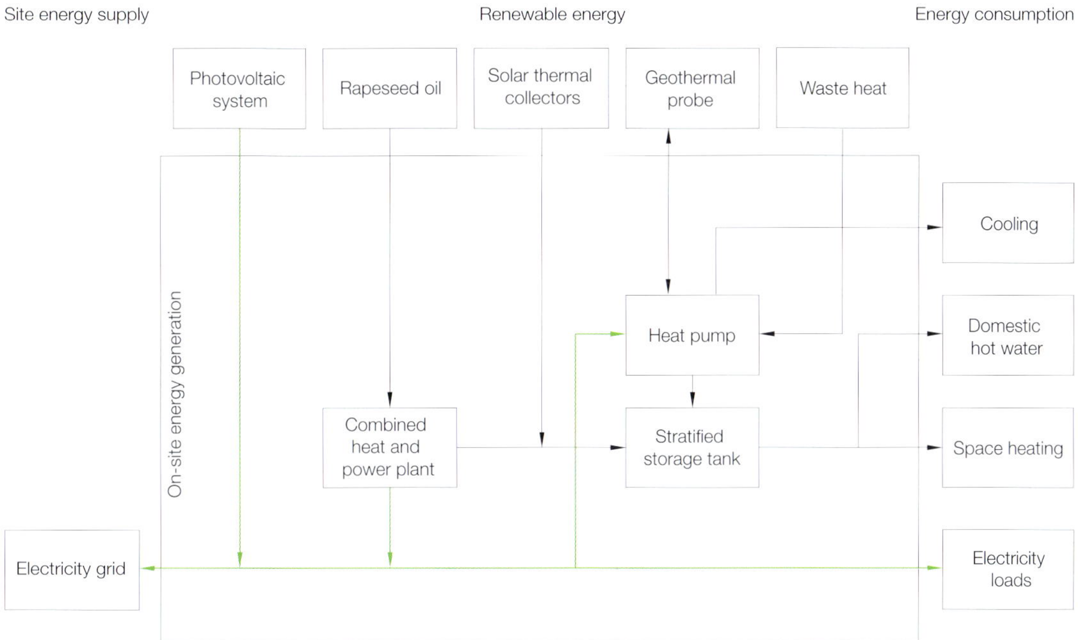

B 14.06
B 14.07

panels on their exterior. The U-value of the facade here is 0.32 W/m²K. The existing and new reinforced concrete floor slabs were thermally insulated and achieve a U-value of up to 0.13 W/m²K. The wood construction of the new ceiling in the entrance area has a total of 40 cm of polystyrene insulation and reaches a U-value of 0.12 W/m²K.

ENERGY EFFICIENCY As large areas of the facade are glazed, daylight can penetrate deeply into the offices, and they permit an almost unobstructed view of the surroundings. Horizontal wooden louvers shade parts of the south-facing facade in the office wings in the old part of the building. The high summer sun can therefore only enter the offices here indirectly, while the low winter sun can penetrate without obstruction. The fixed sun shade system emphasises the horizontal structure of the office wings. Four fixed blades protrude according to the opening portion of the facade up to 32 cm (upper façade area). On the south side, blinds are provided behind the triple glazing as additional protection against glare.
Openable windows combined with ventilation flaps in the lightweight timber construction elements ensure natural ventilation throughout the year. In the interior, ventilation casings with 40 cm of insulation are fixed in front of the ventilation openings. They house convectors that heat the cool air entering the building before it flows into the rooms through metal grilles. The circulation of the air is achieved by means of a central air exhaust plant. This takes heat from the exhaust air and stores it in geothermal probes in the ground. The heat that remains after seasonal storage is used, by means of a heat pump, to warm the building in winter. When it is necessary to cool the building, this process can be reversed in summer. The water for heating and cooling flows through a fine mesh of polypropylene capillary tubes embedded in the 5 cm thick layer of untreated adobe rendering. The building can be passively cooled at night by means of the exhaust air system and the cool air that enters in the evening. The adobe in conjunction with the concrete serves as short-term thermal storage. The large areas available for thermal exchange allow homogeneous temperature shifts and help avoid sudden temperature changes. In addition, the adobe retains moisture and improves the indoor climate.
The renovation of the WWF headquarters is based on Michael Braungart's "Cradle to Cradle" concept. In addition to reducing operating energy, this includes the use of sustainable building products and materials, and thus reduces the amount of energy expended in creating the building beyond its original use. Wood and bamboo certified by the FSC (Forest Stewardship Council) is used, alongside mostly natural and regional building materials. The clay of the ceramic tiles, like that of the internal plaster, is also from a nearby riverbed. The doormat is made of recycled car tyres; the floor is partly covered with recycled carpet.

ENERGY SUPPLY The building's heating demands are covered by a reversible heat pump (50 kW_{th} and 17 kW_{el}) in connection with geothermal ground probes. A total of 18 such ground probes are embedded up to 100 m deep and at a distance of 5 m apart from each other. A mini combined heat

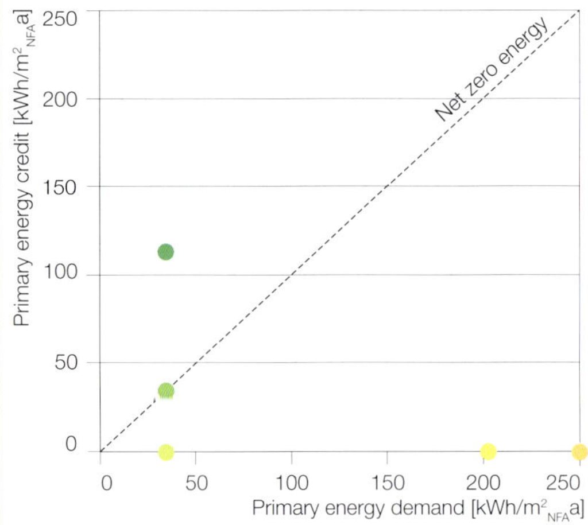

A consumption of 212 kWh_{prim}/m^2a is covered by monthly yields. The remaining 35 kWh_{prim}/m^2a of the total consumption is more than balanced by an additional surplus.

- Calculated annual total primary energy demand (247 kWh/m²a)
- Building-related demand (204 kWh/m²a)
- Monthly coverage of self-demand (212 kWh/m²a)
- Annual balance between remaining demand and feed-in
- Annual energy plus (114 kWh/m²)

The high monthly coverage results from the connection of the CHP and the large photovoltaic system.

Primary energy factors according to energy directive EN 15603 and GEMIS 4.5 (A 2.07, S. 31)

B 14.08
B 14.09

and power plant serves as back-up, runs on rapeseed oil, and supplies additional heat (45 kW_{th}) and electricity (25 kW_{el}). In addition, a 5 m² array of solar thermal collectors on the flat roof covers part of the hot water requirements.
A 40.5-kW_p photovoltaic array generates most of the electricity and the primary energy to cover heating needs. A total of 208 PV modules are positioned on the flat roof at an angle of 20 degrees.

ENERGY BALANCE The building is carbon neutral and has an equalized annual primary energy balance. In summer surplus electrical energy is fed into the public grid, in winter electricity is drawn from the grid. This building achieves an A++ certificate according to the Dutch energy certification system.
Both the total electricity consumption for operating the building and using electrical appliances, which amounts to 71 kWh/m²a, as well as using 121 kWh/m²a of rapeseed oil power can be balanced in terms of primary energy and emissions by the 104 kWh/m²a of electricity generated (Fig. B 14.08).

LESSONS LEARNED There were a number of initial problems in operating the building with regard to the interaction of the different building services systems. The software of the central building control system was adapted by reprogramming it in 2008. The incompatibility was due in part to the fact that, at the time this building was designed, several of the first-generation building services systems used had not been sufficiently tested and, therefore, could not be properly harmonised with each other. After the faults in this system of passive conditioning were eliminated, the projections and expectations in terms of comfort and the conditioning of the building could be met, and the users are now very satisfied. The building can be operated simply and economically.

B 14.08 Energy evaluation
B 14.09 Building and energy parameters (values refer to net floor area, NFA)

SITE	Zeist (NL)
Annual global radiation at site	990 kWh/m²a
Annual mean temperature at site	9.7 °C
Context	suburban

BUILDING ENVELOPE PARAMETERS	W/m²K
U-value, exterior walls	0.31
U-value, windows (including frames)	0.90
U-value, roof surface	0.12
U-value, floor slab	0.13
Mean U-value, building envelope	0.35

BUILDING EQUIPMENT PARAMETERS	
Area of solar thermal collectors	5 m²
Area per m²	0.002 m²/m²
Thermal storage volume	220 l
Storage volume per m²	0.07 l/m²
Photovoltaic system area	300 m²
System area per m²	0.09 m²/m²
Capacity	41 kW_p
Capacity per m²	12.20 W_p/m^2
CHP capacity	45 kW_{th} 25 kW_{el}
Capacity per m²	13.40 W_{th}/m^2 7.50 W_{el}/m^2

GRID INFRASTRUCTURE AND ENERGY SOURCES	
Supply infrastructure	electricity grid, delivery of rapeseed oil
Energy source supply	rapeseed oil, electricity
Feed-in infrastructure	electricity grid
Feed-in energy source	electricity

DESIGN STRATEGIES, CONCEPTUAL FOCUS
Mechanical ventilation with use of waste heat, combined heat and power with use of biomass, photovoltaics, solar thermal energy, ecological building materials

BUILDING PARAMETERS	
Net floor area, NFA	3360 m²
Gross floor area, GFA	3800 m²
Gross volume, V	14360 m³
Building envelope, A	2850 m²
Surface to volume ratio, A/V	0.20 m²/m³
Building costs (net, construction/ technical systems)	1050 €/m² (2006)
Number of units	1–3
Total number of users	200

DEMAND PARAMETERS (calculated)	kWh/m²a
Electricity demand	71
Total primary energy demand	247
Total primary energy generated	326

OFFICE BUILDING WITH APARTMENT

Villach, A 2002

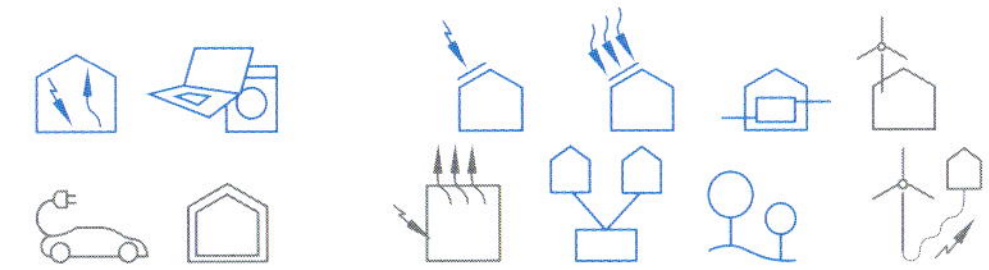

Clients: Miriam and Armin Themeßl, Villach
Architect: Anton Oitzinger,
Arbeitsgemeinschaft Erneuerbare Energie
– Institut für Nachhaltige Technologien (AEE INTEC), Villach
Energy planner: AEE INTEC, Villach
Building services: AEE INTEC, Villach
Carl Pfeiffer, Bad St. Leonhard; ÖkoFen, Niederkappel
PVT Austria, Wolfau
Monitoring: AEE INTEC, Villach
Main stakeholder: architect

The Institut für Nachhaltige Technologien of the Arbeitsgemeinschaft Erneuerbare Energie (Institute for Sustainable Technologies – AEE INTEC) Kärnten is responsible for the technical building services and ecological planning of new building projects. In 2000 when plans were made to erect the group's own office building in Villach, zero or plus energy concepts were still in their infancy. Despite this, the goal was to visibly implement the aims of the AEE by erecting an energy plus passive building as a demonstration object. The starting point of the concept was the integration of a large solar thermal system as a design element in the facade and to draw a major amount of the thermal energy for the building from this source. At the same time, the intention was also to use ecological and regionally available materials as far as possible. Thanks to the passive house building standard, periodic heat surplus from the solar thermal systems can be transferred to neighbouring buildings. The use of wood pellets, the transfer of heat, and the solar generation of electricity achieve a primary energy balance that is more than equalised, viewed over an entire year.

DEVELOPMENT, DESIGN, AND STAKEHOLDERS

In designing this office building, the architect not only emphasizes the professional focus of its occupants, but also created for AEE Intec (Institute for Sustainable Technologies) a part of its demonstration and research activity, visibly represented by the

B 15.01 View from south
B 15.02 Site plan, scale 1:1500

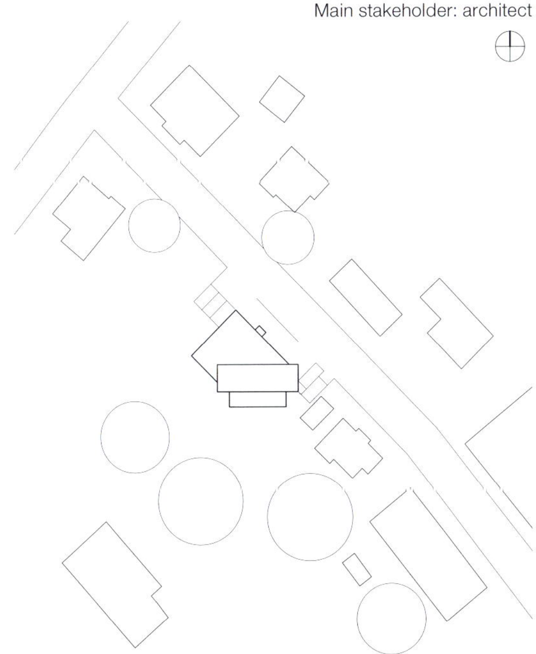

B 15.01
B 15.02

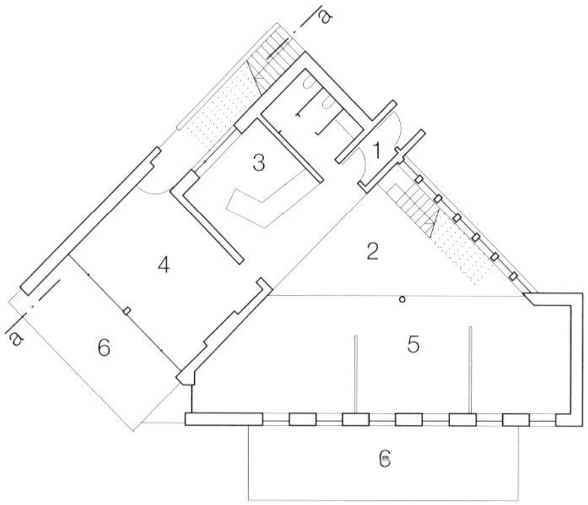

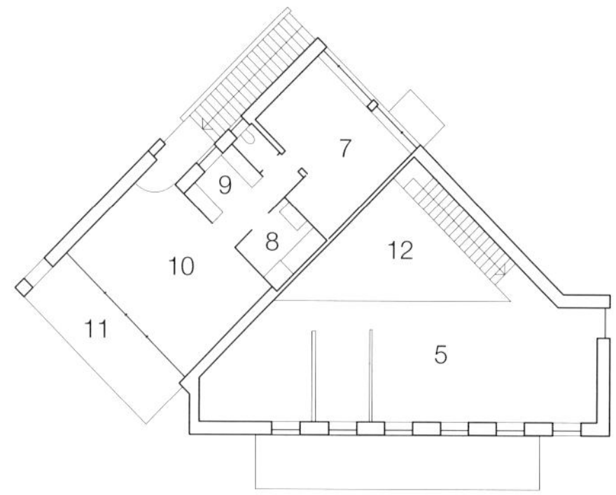

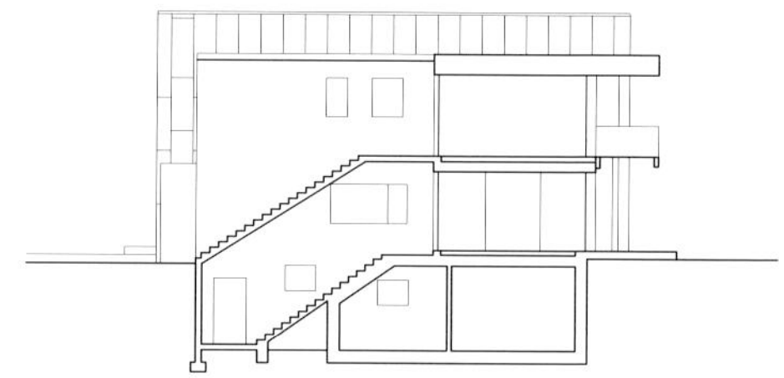

1 Lobby
2 Hall
3 Secretary
4 Meeting room
5 Office
6 Terrace

7 Bedroom
8 Bathroom
9 Kitchen
10 Living/dining
11 Balcony
12 Void

B 15.03 Ground floor plan, scale 1:400
B 15.04 First floor plan, scale 1:400
B 15.05 Section, scale 1:400
B 15.06 Technical schematic of energy provision

large area of solar collectors in the facade. The site is located in a mixed-use area on the periphery of Villach. Its longitudinal side runs parallel to the street and faces south-west (Fig. B 15.02, p. 129). The new administration building is divided into three areas (office wing, self-contained apartment, and events space) and into two building parts. The lower part follows the street and recalls the surrounding context, and a taller volume with a shed roof faces the south (Fig. B 15.01, p. 129). The southern facade provides the ideal orientation for active and passive use of solar energy by means of large areas with facade collectors and windows.The office is entered from the street side in the north. It opens towards the street through a large area of fixed glazing that was a subject of debate during the design phase. With dimensions of 2.3 × 6.0 m, this large glazed area is the sole exception to the otherwise consistently applied concept of keeping the building as closed as possible in order to reduce heat losses. However, the additional energy demand that results from 20 % higher transmission heat losses is balanced by the building's open and inviting gesture.

ARCHITECTURE The individual work spaces of the 20 staff members are interconnected, while still offering sufficient individual areas where people can work in privacy. Separate offices weren't required. The design uses flexible internal partitions and features an atrium at the centre of the building that connects the ground and first floors to create an interior with work zones and open communication areas (Fig. B 15.03). A meeting room (that can be subdivided) is located on the ground floor and enables access to the terrace and the outdoor area. During planning, the original intention was to exclude the basement from the thermal envelope. In the course of the construction work, however, the idea of using it as a lecture and events space arose. The basement was therefore insulated, integrated in the centralised ventilation concept, and provided with underfloor heating. A self-contained apartment on the upper floor is reached by an external steel staircase on the north-western side of the building (Fig. B 15.04). The living room on the upper level opens through a completely glazed facade onto a south-west-facing balcony.

CONSTRUCTION AND INSULATION Natural and regionally available materials were used to construct this passive building. Wood fibre panels cover the prefabricated wooden elements on both floors, as well as the facade collectors. Inside the building, unfinished OSB panels cover both the framing and a 45 cm deep cavity with blow-filled cellulose insulation. Structural framing members were selected with cross-sections as small as possible according to structural calculations and in an offset arrangement that creates a continuous thermal insulation layer and minimises thermal bridges. The construction of the walls, thus, achieves a U-value of 0.1 W/m^{2}K. All the internal walls and ceilings were plastered with regional adobe render reinforced with hemp chips and jute weave. The roof elements are constructed according to the same principle. The windows have triple glazing in a composite wood-aluminium frame with a U-value of 0.84 W/m^{2}K. Hemp rope and sheep wool were used to fill the joints between frame and building shell. The offices have an internal sunshade system.
The OSB panels form an airtight layer, since they are

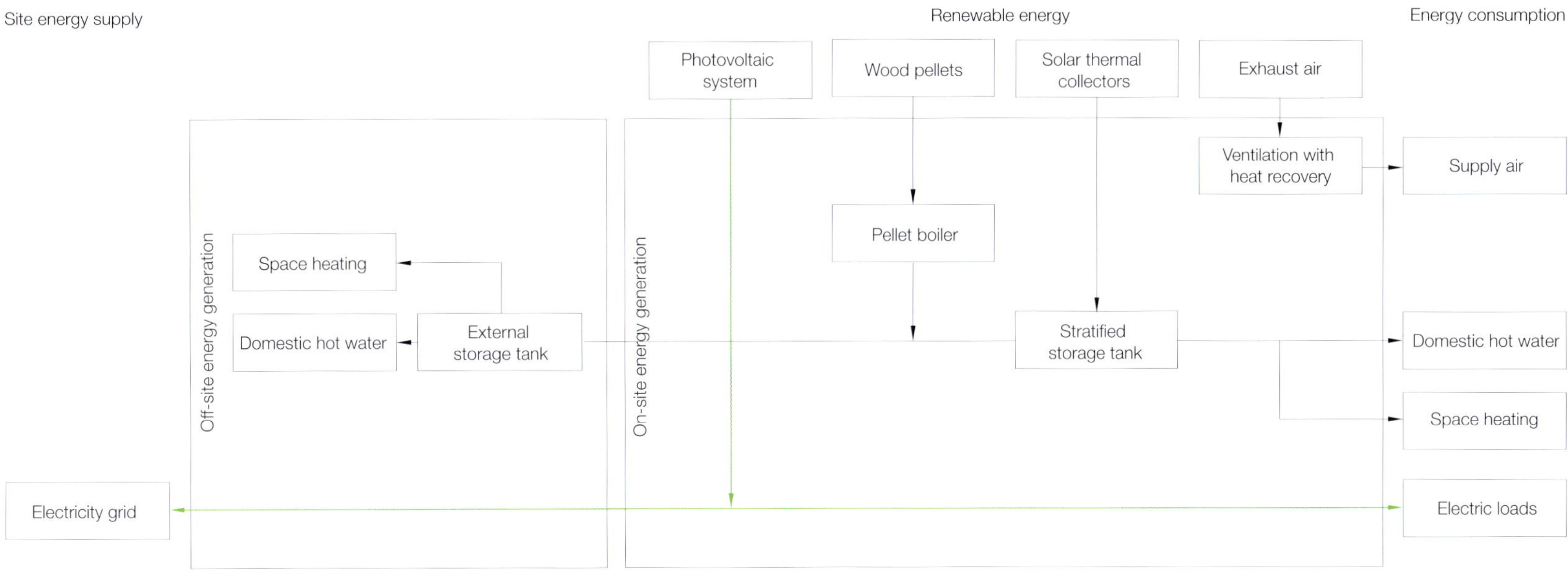

B 15.06

adhesively connected via vapour diffusing tape along butt joints and onto window frames. Butyl tape was used to seal the joints between the timber construction and the ceiling slab above the concrete basement. A blower door test produced an air change rate of 0,4 h^{-1} at 50 Pa negative pressure. Rainwater from the 200 m^2 roof surface is led via two drains into a cistern with a capacity of 8500 l. It provides grey water for the toilets, washing machine, and watering the garden.

ENERGY EFFICIENCY A central ventilation plant with heat recovery supplies the administration building with fresh air at a temperature of 19 °C. Depending on the season, an earth tube heat exchanger at a depth of 2 m consisting of three PE pipes (diameter 150 mm), each 40 m long, heats or cools the fresh air. The central ventilation plant can, thus, be operated without danger of icing at external temperatures down to -12 °C. Filter boxes in the intake rid the air of dust and pollen. In summer the pre-cooling performance is about 3 kW, in winter a temperature increase of up to 12 K can be achieved at a continuous volume flow of 600 m^3/h. Cellulose and paper pellets integrated in the construction insulate the spiral seam ducts in the floors and ceilings. The three areas of the building, office wing, apartment, and event space, can each be serviced separately. An additional heater for the supply air flow by means of a water-air heat exchanger with a heat capacity of up to 6 kW allows the ventilation plant to be also used as a monovalent heating system.
For extra comfort, in addition to the ventilation system, wall heating creates what are called warm areas with low-temperature heat with a flow temperature of 35 °C. This is a slower and less efficient method of heating than with conventional systems, but the users find the radiant heat extremely pleasant.

ENERGY SUPPLY The energy supply of the AEE office building is based entirely on renewable energy sources. The very large solar thermal collector in the vertical southern facade with an area of 78 m^2 produces heat in the "low-flow" system. This principle calls for heat available in the collectors to be layered in thermal storage at high temperatures without delay and mixing losses. The great advantage of the low-flow system is, as the name suggests, the lower amount that flows through it and, thus, minimises the expenditure on tubes and pipes and optimises temperature stratification. Low-flow enables reaching desired temperature levels quickly. A stratified tank with a self-regulating filling device is used for this purpose. The 3050 l steel tank stands in the utility room in the basement, within the thermal envelope. All the connections to this tank are beneath it, permitting seamless insulation without thermal bridges using 5 cm of rock wool. An OSB cladding on a surrounding timber frame construction allowed cellulose fibres to be blown into the entire space around the tank.
Until 2005 a bio-diesel combined heat and power unit produced 10.5 kW of heat and 5.3 kW of electricity. The remaining energy required for space heating and hot water was provided by a two tandem pellet boiler with 20 and 32 kW. The CHP was heat operated for two years but in 2005 had to be shut down for technical and economic reasons. The supply of bio-diesel was subject to technical

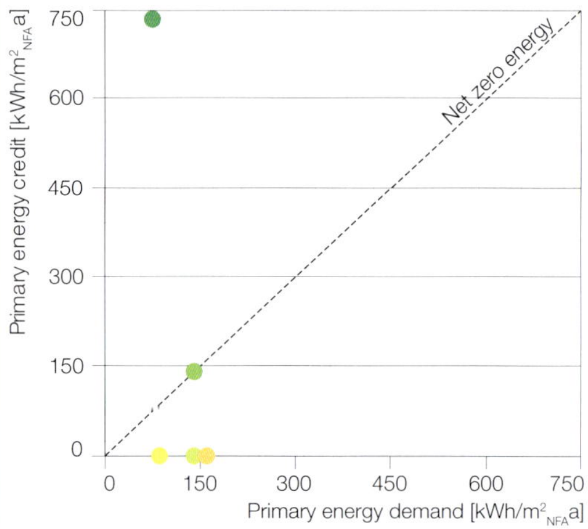

The following aspects and the enormous heat output into the local heating grid result in the Net ZEB-131-Standard:

- Measured annual total primary energy consumption including electricity for the apartment and the office wing as well as the production of heat for the heating network (149 kWh/m²a)
- Energy consumption of the technical plants for conditioning the building including local heating (89 kWh/m²a)
- Seasonal balance of consumption through chargeable feed-in of heat and electricity (17 kWh/m²a)
- Balance of remaining consumption through monthly surplus
- Annual energy plus feed-in of heat and electricity into the public grid and the local heating network (619 kWh/m²). The high figure is the result of replacing natural gas (primary energy factor 1.1) for the generation of heat in the connected buildings.

Primary energy factors according to GEMIS 4.5 (s. Fig. A 2.07, p. 31)

B 15.07 Energy evaluation
B 15.08 Primary energy balancing on the basis of measured consumption and generation data from 2008, primary energy factors according to GEMIS 4.5
B 15.09 Sankey diagram with primary energy values
B 15.10 Building and energy parameters (values refer to net floor area, NFA)

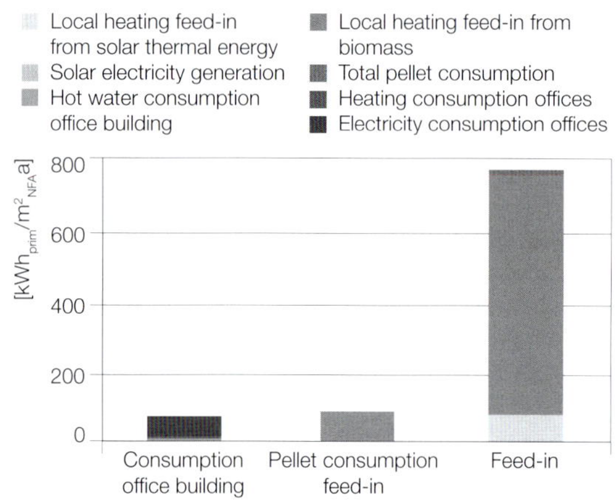

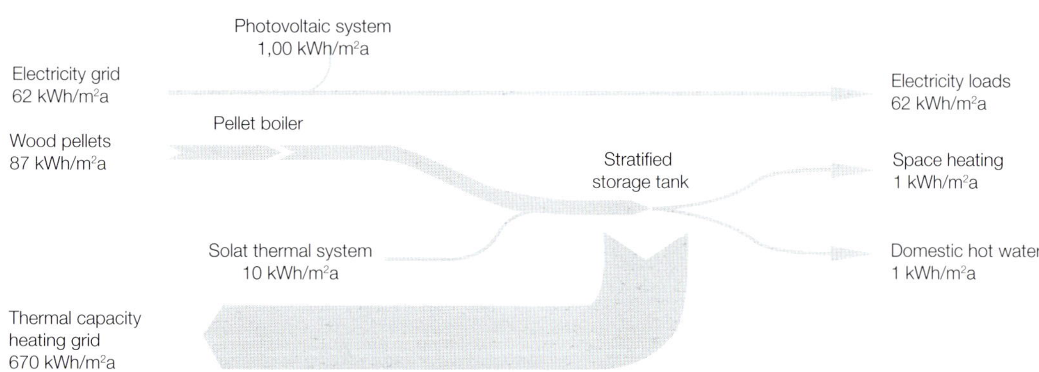

B 15.07
B 15.08
B 15.09

problems and was more expensive than had been calculated. In follow, a third pellet boiler with an output 56 kW was installed. In 2004 a 3.6-kW$_p$ photovoltaic system tilted at an angle of 30° was mount on the flat roof. It provides between 3.6 and 3.9 MWh of electricity per year, all of which is fed into the national grid (Fig. B 15.06, p. 131). As the building started operation before December 31st, 2004 a tariff subsidy applies according to the federal ecological electricity law (ÖSG). In Austria, the level of feed-in tariffs differs from state to state. In Carinthia, at the time the building was granted approval, the feed-in payment was 0.72 €/kWh for a period of 10 years.

MICRO NETWORK During the design phase the large solar collector suggested that a significant unused heat surplus could be expected during times of high solar intake. Consequently, in the immediate surroundings of the new building, suitable buyers for this heat supply were sought and they were later connected by means of a micro-heating network. Initially only three nearby pairs of semi-detached houses could be connected. In 2010 three apartment buildings with a total of 15 apartments and around 35 residents were added to the network. The calculated heating load of the connected buildings amounts to a total of 127 kW, including network losses.
The heat from the collector and the pellet boiler plant is brought together in a stratified tank (volume 3050 l) and prepared for further distribution. Three below-grade pipelines buried in the ground service the neighbouring buildings. The heat distribution system has a total length of 210 m. In three of the six buildings there are three further external tanks; each with a capacity of 1000 l. They are fed in winter between four and five times daily with a filling time of one to one-and-a-half hours, and at other times of the year between one and three times daily. Constant provision of heat would increase network losses. The heat energy for heating and hot water in the houses amounts to around 190,000 kWh/a. A joint heating circuit connects all but one of the users: one of the buildings has a very antiquated heat transfer system and is therefore supplied by a heat exchanger with separate hydraulics. Although underfloor or

surface heating is not installed in any of these buildings, various regulating measures ensure that the eturn temperatures of the district heating runs do not exceed 35 °C, as a low return temperature in the tanks improves the efficiency of the solar collectors.

ENERGY BALANCE Since the completion of this building in 2002 detailed records of the energy flows of generation and consumption have been kept that allow establishing a primary energy balance. This includes all the consumption sectors of the office building and the attached dwelling. Alongside space heating and hot water, office, operating, and household electricity are also included. Within the system boundaries of the building, the primary energy is calculated separately on the basis of the actual site energy consumption for pellets (hot water and space heating), and for grid power, using the primary energy calculation factors according to GEMIS 4.5 (Global Emissions-Model of Integrated Systems). As an efficiency measure, the energy provided by the facade collectors is not included in the balance. This method of producing heat that conserves resources is expressed in the building's reduced primary energy consumption. Taking electricity consumption into account raises this figure to 63 kWh_{prim}/m^2a.
The amount of pellets required to supply the neighbouring houses is also entered as primary energy expenditure, since pellets need to be purchased. The overall primary energy consumption is thus raised, which would not be necessary to operate the office building alone. The surplus heating energy is transferred as an energy plus to the local heating micro network and replaces the supply of heat derived from fossil fuels to the connected buildings, which would otherwise be necessary. The credit for the energy fed in is, analogous to the credit for feeding PV electricity, calculated using the factor for natural gas. The export and balancing of solar heating isn't decisive for the balance. However, the use of pellets with a primary energy factor of 0.2 as well as substituting natural gas (1.1) for the provision of heat in the connected buildings are of decisive importance. The average electricity consumption of the office building and apartment amounts to roughly 48 kWhend/m² per year. In the first years of operation the production of electricity by the heat controlled CHP covered this consumption without an additional solar electricity plant. After the CHP was disconnected part of the electricity consumption could be balanced by the primary energy credit of the surplus thermal energy that is fed into the local heating network, thus achieving a positive energy balance. By switching to increased production of thermal energy via wood pellets, the plus energy balance increasingly contains the exchange of energy sources. Wood pellets need to be supplied and heat or electricity is fed in (Figs. B 15.08 and 09).

LESSONS LEARNED By recording the operation of the building and analysing energy consumption, faulty systems or disproportionate consumption can be recognised and corrected. One example is the manually controlled gutter heating, which in 2007 required 15 kWh of electricity, but 69 kWh in 2008. After the back-up was accidentally switched on in 2009, consumption soared to a figure of 568 kWh until the meter was read for the first time. This problem was identified and corrected. In 2007 the failure of a power inverter remained unnoticed for a short time. However, the annual yield of the solar electricity plant sank only to 2200 kWh compared to the typical figure of 3700 kWh.
Growing staff numbers led to a constant increase in the consumption of electricity for IT and lighting. Increasing user awareness and the appropriate management of the IT equipment counteracted this, so that from 2008 onwards consumption has remained at a comparatively constant level.

SITE	Villach (A)
Annual global radiation at site	1180 kWh/m²a
Annual mean temperature at site	9.2 °C
Context	suburban
BUILDING ENVELOPE PARAMETERS	W/m²K
U-value, exterior walls	0.10
U-value, windows (incl. frames)	0.88
U-value, roof	0.10
U-value, basement ceiling / ground floor slab	0.19
Mean U-value, building envelope	0.10
BUILDING EQUIPMENT PARAMETERS	
Area of solar thermal collectors	76 m²
Area per m²	0.26 m²/m²
Thermal storage volume	3050 l
Volume per m²	10.50 l/m²
Photovoltaic system area	29 m²
System area per m²	0.10 m²/m²
Photovoltaic system capacity	4 kWp
Capacity per m²	12.30 W_p/m^2
GRID INFRASTRUCTURE AND ENERGY SOURCES	
Supply infrastructure	electricity grid, delivery of wood pellets
Energy source supply	wood pellets, electricity
Feed-in infrastructure	electricity grid, local heating network
Feed-in energy source	electricity, heat

SOLUTION STRATEGIES, FOCAL POINTS OF CONCEPT
Passive house concept, facade collectors, central ventilation plant with heat recovery and earth tube heat exchanger, local heating micro network, photovoltaics, regionally available building materials

BUILDING PARAMETERS	
Net floor area, NFA	292 m²
Gross floor area, GFA	394 m²
Gross volume, V	1341 m³
Building envelope, A	614 m²
Surface to volume ratio, A/V	0.45 m²/m³
Building costs (net, construction/ technical systems)	1050 €/m² (2002)
Number of units	2
Total number of users	20
CONSUMPTION PARAMETERS (2008)	kWh/m²a
Space heating consumption	6
Water heating consumption	3
Site energy consumption for heat (including hot water)	10
Electricity consumption	48
Total primary energy consumption (incl. heat generation, local heating network)	149
Total primary energy generation	768

B 15.10

PIXEL BUILDING

Melbourne, AUS 2010

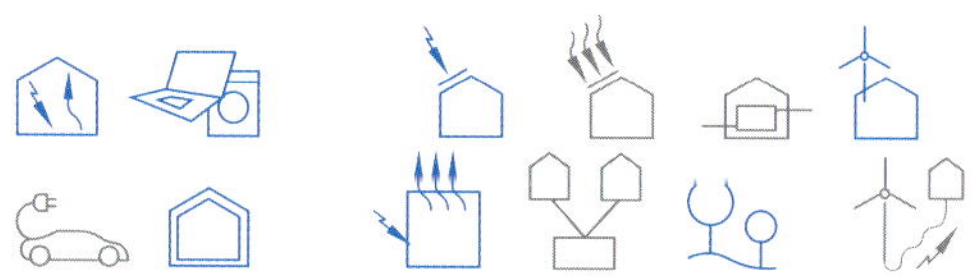

Client: Grocon, Melbourne
Architect: Studio 505, Melbourne
Energy consultant: Umow Lai, Victoria
Building services: Umow Lai, Victoria
Monitoring: Grocon, Melbourne
Main stakeholder: client

B 16.01 Site plan, scale 1:4000
B 16.02 Ground floor plan, scale 1:600
B 16.03 Section, scale 1:600
B 16.04 View from north-west

1 Office
2 Entrance
3 IT/services
4 Corridor/photocopier
5 Storage
6 Staircase
7 Sunshade
8 Wind turbine
9 Solar trackers

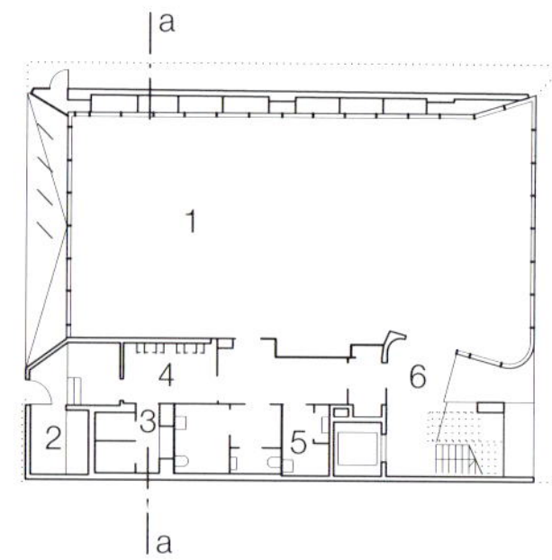

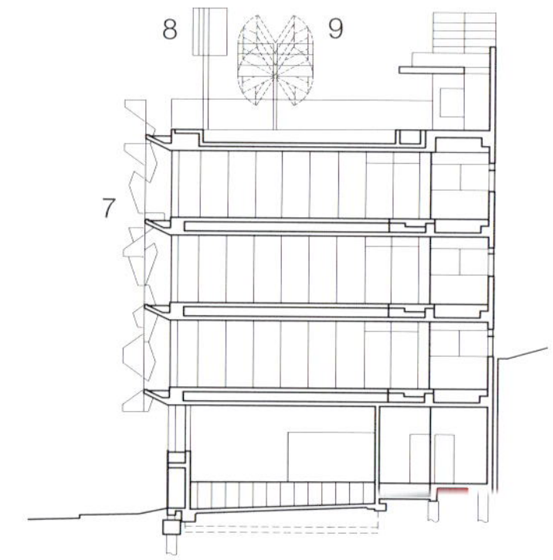

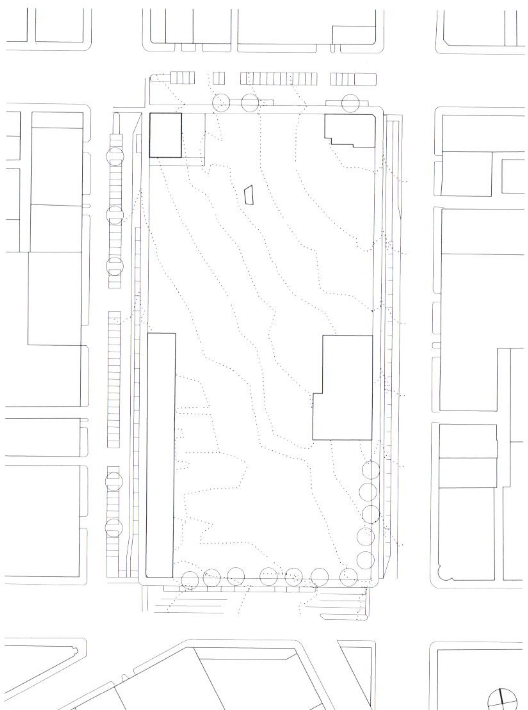

B 16.01
B 16.02
B 16.03
B 16.04

On the site of a former brewery in Melbourne, Australia, an office building with the illustrative name Pixel was erected in 2010. It is used by the developers of the new urban district being created there as a planning and information location and through its special sustainability and positive CO_2 balance serves as a demonstration object for a number of new buildings. Passive strategies such as sunshade and daylight optimisation reduce the need for cooling and heating, while recyclable materials minimise the amount of CO_2 produced in making the building. The photovoltaic arrays and the micro wind turbine power generators on the roof are dimensioned to offset all climate gas emissions from the gas absorption heat pump and the other building service plants. It is planned to create a surplus to balance the emissions resulting from the building's construction.

DEVELOPMENT, DESIGN, AND STAKEHOLDERS
The central location of this site offered immense development potential, turning the urban renewal project into one of the most ambitious construction projects in Australia and attracting great attention to the Pixel office building as a demonstration object (Fig. B 16.01). By using the building as an example to initiate sustainable concepts and energy-efficient new buildings locally, the planning team aims at implementing a variety of strategies to reduce CO_2, and increase energy efficiency and sustainability. Due to the lack of an assessment basis for climate neutral buildings in Australia, during the design phase theoretical concepts on balancing from the United States were used as a basis for the emissions balance. Upon completion in 2010 the building achieved the maximum number of points for "Green Star Office Design" from the "Green Building Council of Australia" (GBCA) and is regarded as Australia's "greenest" building. The project team also aims to achieve the highest ratings under the American LEED and the British BREEAM certification systems.

ARCHITECTURE The Pixel building features a total floor area of 840 m² distributed on four levels and stands on the north-western boundary of the site that covers nearly 20,000 m². The column-free ground floor offers space for a sales area, while the three upper floors contain office areas that can be used in a variety of ways (Figs. B 16.02 and 03). They are accessed from a staircase core bordering the east-facing firewall. The other three facades comprise large areas of glazing.

CONSTRUCTION AND INSULATION The reinforced concrete floor slabs rest on the walls of the solid staircase core and on three precast concrete piers outside the insulated building envelope of the western facade. The windows, which have thermal glazing (U_w-value 1.80 W/m²K), recede here by about a metre to provide architectural sun shade for the facade that receives a lot of sun during working hours. The massively built core contains all the utility spaces, an elevator, and the single escape staircase. Those interested in what is happening on the building site can visit a viewing area on the roof of the building. The staircase landings are generally separated from the office areas only by a glass wall, thus providing these spaces with a vivid impression and offering insights into the development of this new district. The solid eastern wall, interrupted in only a few places by single glazed windows, (U-value 5.8 W/m²K), is insulated internally with 5 cm of glass wool and has a U-value of 0.55 W/m²K. Beneath 30 cm of soil for the extensive planting, the top of the reinforced concrete roof slab is insulated with 10 cm of extruded polystyrene (U-value 0.31 W/m²K).
Most of the materials used for the building's structure and envelope can be reused after its demolition. An innovative type of concrete was developed for the concrete elements, and the "Centre for Design" of RMIT in Melbourne has certified that this mix offers an almost 50 % reduction in energy expenditure and CO_2 emissions compared to standard concrete, while still providing comparable compressive strength. It further contains a high proportion of recycled aggregates.

ENERGY EFFICIENCY The primary structure of concrete and the large window fronts form a simple building volume, the face of which is defined by the structure of the sun shade panels that gives this building its name (Fig. B 16.04). The numerous coloured four or five-sided composite aluminium panels are arranged in front of the facade at different angles so that the view outside is not obstructed. Sufficient daylight can enter through the full-height windows while preventing direct solar radiation from heating up the rooms excessively. The ways in which the individual plates are overlaid and tilted, and the angles at which they are fixed reflect their particular orientation (Fig. B 16.06, p. 136). The largest number of plates is in front of the very sunny western facade. Here, they are only slightly turned away from the building and form an almost closed surface. In front of the southern facade, which receives less sun, there are relatively few such panels to provide shade. On the northern side, the panels are fixed in an almost vertical position. In the southern hemisphere, this is where solar intake occurs due to the high sun position. The colourful metal plates also protect the staff from glare. The shade system was optimised with the help of daylight simulations. The few windows in the eastern facade feature an internal, centrally controlled glare protection system that closes automatically between 8 and 11 a.m. to exclude the intensive solar thermal load of up to 150 W/m². The windows in the loading area of the ground floor are shaded either by the tall surrounding buildings or internal blinds.
In addition to the intelligent exploitation of daylight, an energy-efficient lighting system (fluorescent tubes) in the office ensures low heat loads. They are dimmed in accordance with the amount of daylight and connected to presence sensors. In all other rooms apart from the offices LED lighting is used.
To passively cool the building during the night, the windows of the upper floors open automatically on cool nights, allowing cold air to flow across the solid ceiling slabs (there are no suspended ceilings) and withdraw the heat stored during the day.

ENERGY SUPPLY The wind turbines on the roof are also a unique feature that make the theme of renewable energy clearly visible. The vertical axis wind turbines developed specially for the inner city location and the local wind situation in Melbourne are not affected by the frequent change in wind direction and allow an even generation of

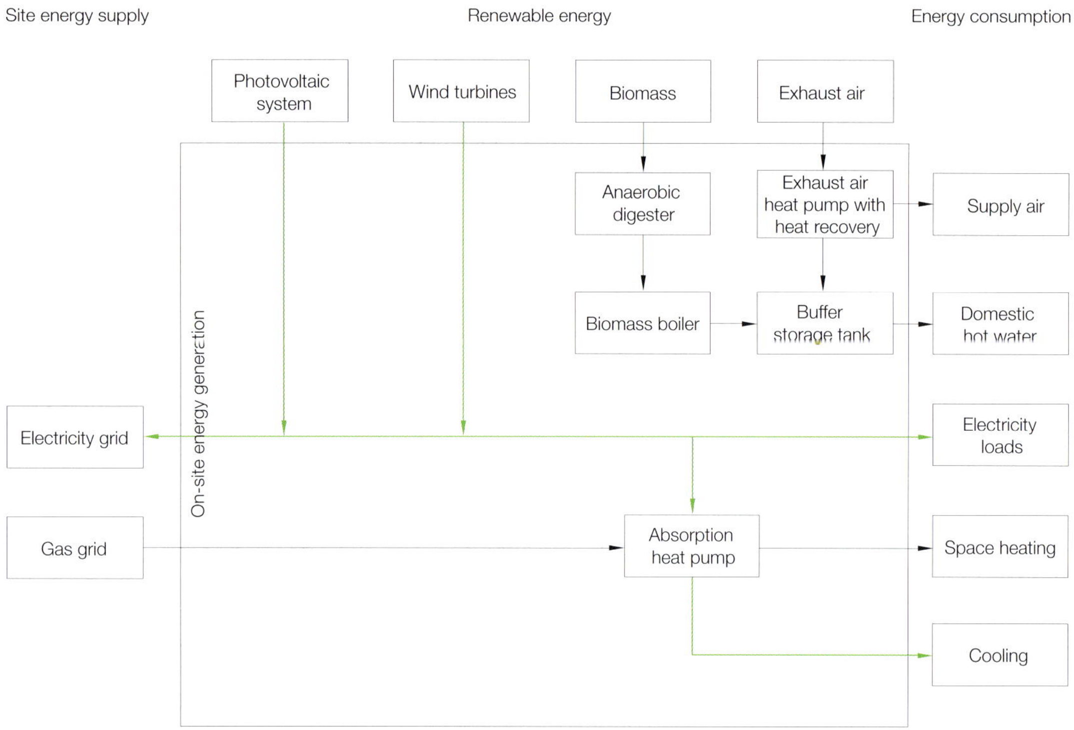

B 16.05
B 16.06
B 16.07

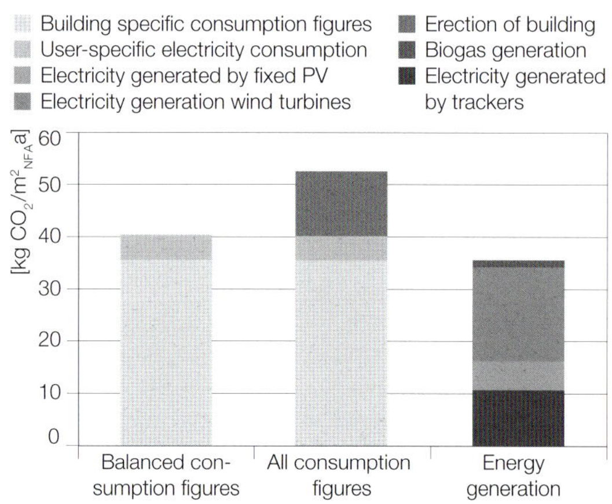

1.7 kW of electricity at a wind speed of 8 m/s. To exploit the high solar thermal load of almost 1600 kWh/m²a, a photovoltaic array measuring 38 m² produces electrical energy. 18 of a total of 30 photovoltaic modules, each with a performance of 211 W_p, are mount on three solar trackers (dual axis system that follows the sun). The other modules are positioned on the roof of the staircase tower that is inclined slightly towards the north.

The centralised ventilation system with heat recovery (75 %) by means of a plate heat exchanger is coupled with an air-air heat pump serving as air conditioning system. A network of ducts in the raised floors of the office areas and floor outlets equipped with diffusers introduce pre-conditioned air into the building. The four floors can be serviced separately with different air and temperature levels by means of a volume flow regulator. The supply air is introduced at a temperature of at least 18 °C (at full cooling capacity of 363 kW) and maximum 34 °C (at full heating capacity of 28 kW). The relative air humidity is supposed to not exceed 60 % .

In addition, a reversible absorption heat pump with a maximal thermal capacity of 68 kW cooling and 141 kW heat feeds the water coils located on the underside of the solid concrete ceilings with cold or warm water. This thermally activates the mass of the concrete slabs. The heat pump runs on natural gas (101 kW).

A burner uses biogas produced from faeces by an anaerobic digester to provide hot water in connection with a 300 l buffer tank. This enables water to be heated when there is sufficient biogas, and detaches production, which can't be precisely predicted on account of the irregular amount of biogas, from consumption (Fig. B 16.05).

B 16.05 Technical schematic of energy provision
B 16.06 Shade provided by rigid sun shade elements on the north facade
B 16.07 Calculated annual energy consumption parameters including embodied energy and energy generation
B 16.08 Energy evaluation
B 16.09 Building and energy parameters (values refer to net floor area, NFA)

ENERGY BALANCE The dominant factor in the energy consumed at the Melbourne location is clearly the cooling system. The gas used to heat the building amounts to almost 6 $kWh/m^2_{NFA}a$ in comparison to 42 $kWh/m^2_{NFA}a$ required for cooling. The total power consumption for operating the building amounts to 22 $kWh/m^2_{NFA}a$. Due to the low emissions factor for natural gas (0.21 kg CO_2/kWh) in comparison to grid electricity (1.34 kg CO_2/kWh), the introduction of gas in the emission balance is advantageous, despite the low performance value of an absorption heat pump. The electricity comes, for the most part, from brown coal power stations. The entire CO_2 equivalent of the energy consumption amounts to roughly 26 kg $CO_2/m^2_{NFA}a$, whereby the function-related consumption for the offices, calculated at 12.5 kg $CO_2/m^2_{NFA}a$, is not included. This contrasts with a power generation figure of 12 $kWh/m^2_{NFA}a$ from the photovoltaic arrays and 14 $kWh/m^2_{NFA}a$ from the three wind turbines. The CO_2 emissions credit from the grid connected plant amounts to a total of approximately 36 kg $CO_2/m^2_{NFA}a$. The production of biogas contributes 2 $kWh/m^2_{NFA}a$.

The negative emissions balance does not allow the emissions generated in producing the building, which amount to 239 kg CO_2/m^2_{NFA} (Figs. B 16.07 and 08) to be offset. However, simulations reveal that the generation by means of wind and solar energy is well suited to the needs of Melbourne's electricity grid. The high cooling loads of the building mean that there are peak loads around midday on warm summer days.

LESSONS LEARNED The intensive search for suitable technologies to reduce the amount of energy needed to cool and heat the building, as well as the time required to develop the wind turbines meant that the planning and construction phases extended over a period of 29 months.

In the initial phase of operating the building the combination of the ventilation and cooling measures in the floor slabs and in the adjoining raised floors required a considerable amount of fine tuning. But this ultimately resulted in a pleasant internal climate. The individually controlled systems do not interfere with each other and it is possible to respond to the different spatial situations on the individual floors.

Uplights fitted later in the offices have provided an even distribution of light that was not achieved initially. However, directing light at the ceiling caused problems with the light sensors of the daylight control system. The uplights were therefore integrated in an adapted lighting control system. Initially the tracking system of the photovoltaic panels reacted solely to brightness, and at night rotated the panels to face the nearby lights of the city, thus using additional energy. Timers are to be added to the system so that the trackers move only during the daytime.

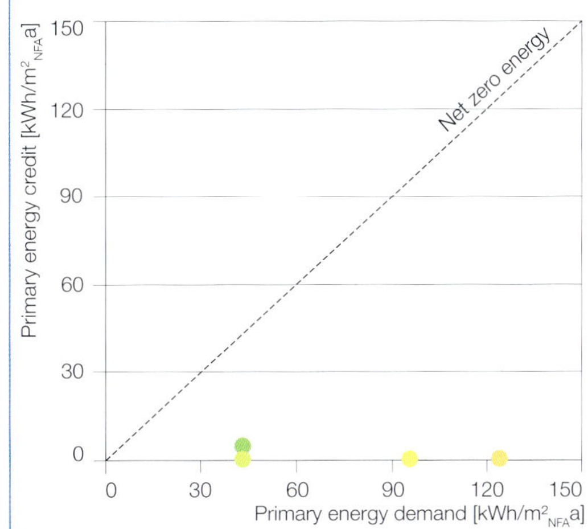

By including all calculated demand values an annual zero energy balance is not achieved:

- Measured annual total primary energy consumption including user electricity (123 kWh/m^2a)
- Building-specific primary energy demand (94 kWh/m^2a)
- Self-demand coverage by monthly calculable biogas, wind, and solar electricity yields (81 kWh/m^2a)
- Annual balance by electricity surpluses not entered in the monthly primary energy balance (3 kWh/m^2a)

Use of specific primary energy factors for the location Melbourne

SITE	Melbourne (AUS)
Annual global radiation at site	1600 kWh/m^2a
Annual mean temperature at site	13.5 °C
Context	urban
BUILDING ENVELOPE PARAMETERS	W/m^2K
U-value, exterior walls	0.55
U-value, windows (incl. frames)	1.80
U-value, roof	0.31
U-value, ground floor slab	0.13
Mean U-value, building envelope	1.06
BUILDING EQUIPMENT PARAMETERS	
Photovoltaic system area	38 m^2
System area per m^2	0.05 m^2/m^2
Photovoltaic capacity	6 kW_p
Capacity per m^2	7.50 W_p/m^2
Wind capacity	5 kW_{el}
Capacity per m^2	6.00 W_{el}/m^2
GRID INFRASTRUCTURE AND ENERGY SOURCES	
Supply infrastructure	electricity grid, gas grid
Energy source supply	natural gas, electricity
Feed-in infrastructure	electricity grid
Feed-in energy source	electricity

DESIGN STRATEGIES, CONCEPTUAL FOCUS
mechanical ventilation with heat recovery, external sunshade system as signature element, thermal activation of concrete core, gas-fired absorption heat pump, photovoltaic units, micro wind turbines, biogas production and use, reusable and recycled building materials

BUILDING CHARACTERISTICS	
Net floor area, NFA	837 m^2
Gross floor area, GFA	1136 m^2
Gross volume, V	3567 m^3
Building envelope, A	1310 m^2
Surface to volume ratio, A/V	0.37 m^2/m^3
Building costs (net, construction/ technical systems) (2010)	5100 €/m^2
Number of units	4
CONSUMPTION PARAMETERS (simulation)	kWh/m^2a
Space heating demand	5
Water heating	1
Site energy consumption for heat (incl. hot water)	6
Final energy cooling	42
Electricity demand	22
Total primary energy demand	123
Total primary energy generation	84

B 16.08
B 16.09

COMPANY HEADQUARTERS

Berlin, D 2008

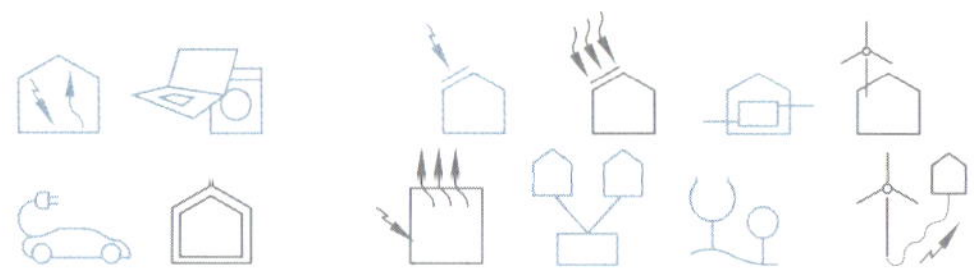

Client: Solon, Berlin
Architect: Schulte-Frohlinde Architekten, Berlin
Energy consultant: EGS-plan Ingenieurgesellschaft für Energie-, Gebäude- und Solartechnik, Stuttgart
Building services: EGS-plan, Imtech, Hamburg
Monitoring: IGS – Institut für Gebäude- und Solartechnik, Braunschweig
Main stakeholder: client

The headquarters of the Solon Company in the Adlershof Science and Technology Park in Berlin, which was completed in 2008, comprises both the company's administration and production facilities. As one of Europe's leading solar businesses, Solon is committed to the principles of sustainable business and the consistent use of renewable energies. These principles are also made apparent by this building. The planning team used traditional building materials and high-tech communication technologies and covered energy requirements by a mix of biogas-power heat/cold-coupling and photovoltaic arrays.

DEVELOPMENT, DESIGN, AND STAKEHOLDERS

At an early stage Solon put together a planning team that formulated the goals for the project in conjunction with the client. The aim was a sustainable, energy-efficient building with a good internal climate, highly efficient use of space and spatial flexibility. The office workspaces were to receive plenty of light during the day, to have windows to allow individual ventilation, and wireless IT.

Ambitious energy goals for the entire property were determined during planning:

- The office building was to have an annual primary energy requirement for heating, cooling, ventilation, and lighting of less than 100 $kWh/m^2_{NFA}a$ – and thus, surpass the requirements of the German Energy Saving Directive (EnEV) by more than 50 %.
- A further aim was to achieve largely CO_2 neutral operation for the entire complex from 2010 onward in combination with a biogas-fuelled combined heat and power unit. The energy requirements of the production plant aren't included in these calculations.

Architects, energy consultants, and specialist planners worked closely together in developing the concept, and examined in detail the characteristics of the location, the requirements of production, and of the various users of the building. Already at this stage the manufacturers of individual systems and components were consulted, with the aim of integrating the potential of timber construction as an

B 17.01 View from south-east
B 17.02 Ground floor plan, scale 1:1500
B 17.03 Sections, scale 1:1500
B 17.04 Southern elevation

B 17.01

unusual choice for office and commercial buildings and the innovative information and communications technologies in the overall concept in the best way possible.

ARCHITECTURE The office and production complex is located on a site covering 36 000 m². The office building with its floor area of 11,200 m² provides workplaces for a staff of 350 (32 m² gross floor area/workplace).
The production and administration buildings of Solon form two parts of an ensemble interconnected by bridges. Whereas the production building resembles a large shed that meets the requirements of production technology, the office building is designed as a spatial landscape that circumscribes five internal courtyards beneath a roof that slopes to the south (Fig. B 17.03). The ensemble borders the street in a straight line, while forming a curved perimeter towards the park (Fig. B 17.02).
An open circulation axis that leads to the central glazed atrium of the office building runs between the two parts of the building. The combination of glass, steel, and wood in the external envelope creates an open and inviting atmosphere. This is continued on the accessible green roof where people can meet to work together.
The ground floor of the production building features various production plants for the manufacturing of photovoltaic modules as well as spaces for storage and research and development. On the upper floors, there are offices and meeting rooms along the external south-western and south-eastern facades that have direct visual contact with the production areas on the lower levels. On the ground floor there is also a restaurant and an area for special events. The five courtyards embedded within the ensemble are each designed according to an individual theme and are accessible to the staff members.

ELECTROMOBILITY In the park behind the building there is a solar filling station for eight electric scooters belonging to the company. The battery charge indicator and charging sockets are integrated into a wall and connected to the batteries of "solar trackers", PV elements that follow the course of sun and are placed along the park-like site. This demonstrates the (future) potential of solar mobility, and the energy generated by the "solar trackers" is incorporated in the building's balance. They are, therefore, also connected to the building's power network.

ENERGY EFFICIENCY The office building was optimized in an integrated planning process that aimed at achieving an extremely low heating and primary energy demands for operating the building. The large areas of glazing in the building envelope with an average U-value of 0.75 W/m²K and an effective external sun shade system protect the building from the effects of the weather. The reinforced concrete floor slab features 12 cm of continuous perimeter insulation. In addition, perimeter insulation is placed 80 cm below grade with a U-value of 0.30 W/m²K. Polystyrene insulation with a thickness of 16 cm is placed on the 20 cm strong prestressed concrete slab of the roof. Together with the green roof, this achieves a U-value of 0.23 W/m²K.

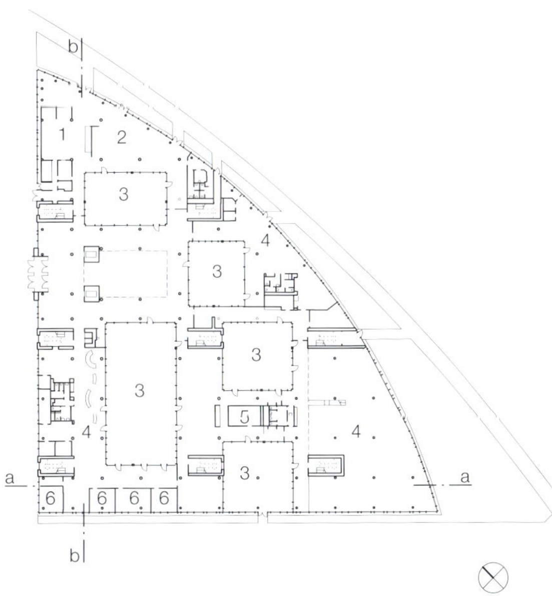

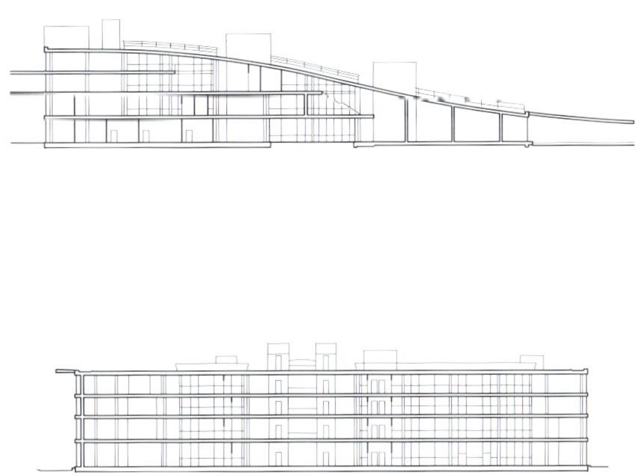

1 Kitchen
2 Staff restaurant
3 Atrium
4 Offices
5 Services
6 Meetings

B 17.02
B 17.03
B 17.04

FACADE An important planning goal in developing the building envelope was to reduce embodied energy as well as externally induced cooling and heating loads. By means of close collaboration between architect, energy consultant, and construction companies, a completely prefabricated timber element facade was developed. This integrates all necessary building elements: triple glazing, vacuum insulation panels, motorised sunshade system, ventilation panels with opening contact, and small radiators. A facade element consists of two grid modules spanning 1.35 m. The width and position of the opaque and transparent areas in the full-height elements alternate, giving the elevations a particular rhythmic structure. Simulation calculations suggested a figure of 60 % as the optimal proportion of transparent areas in the external envelope. In the courtyards, the openable elements feature full-height glazing and an overall proportion of transparent areas of 90 %. While the heavy triple-glazed elements are fixed, the opaque areas contain openable ventilation panels in front of which wooden louvers are fixed to provide protection against rain and intrusion. The lower, non-openable part includes a convector with window contact for quick regulation of room temperatures. The facade profiles are designed to minimise thermal bridges and are made of glue-laminated European fir. In addition to the external sun shade, there is also colour-neutral sun protection glazing with a g-value of 0.34 and a light transmission value of roughly 0.57. The triple-glazed windows are filled with argon gas, feature thermo-elastic spacers, and achieve a U-value of 1.2 to 1.6 W/m²K. On the north-western facades, low-E glazing is used on the ground and first floors, since they are subject to a lesser amount of solar intake. Their Uw-value is 1,2 W/m²K and their g-value reaches 0.60.
The vacuum insulation panels (VIP) that find use here have a total thickness of only 47 m and achieve a U-value of roughly 0.25 W/m²K. Especially for this project, the VIP manufacturers considerably improved long-term performance under increased temperatures. In laboratory tests of the Forschungsinstitut für Wärmeschutz in Munich (FIW), a thermal conductivity of below 6 W/mK over a period of 30 years was calculated.
In front of the area of the facade with solar protection glazing, there are also sun protection blinds that run in tracks and are stable in windy conditions. They can be rotated, feature a ventilation gap behind them, and are automated, while achieving a reduction factor (Fc) of about 0.2. To improve exploitation of daylight, the blinds in the upper third of the windows can be opened, and thus allow light to enter, while the lower part remains in the cut-off (closed) position to provide shade.

BUILDING SERVICES The high thermal quality of the building envelope reduces the heating and cooling demands of the office building, and thus

B 17.05 Mobile energy supply by E-Shuttle
B 17.06 E-Shuttle charging station
B 17.07 Interior with meeting "island"
B 17.08 Energy concept

B 17.05

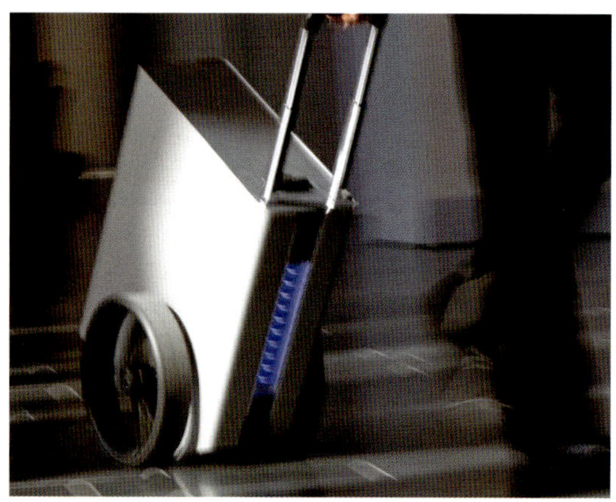

B 17.06

allows the use of simple heating and cooling technology. By thermal activation of the concrete floor slabs using a pipe system (thermally activated building parts – TABs,) the building is heated and moderately cooled. To cover peak heating loads during exceptionally cold weather and to allow individual temperature regulation, convector heaters are integrated in the facades. In front of the fully glazed courtyard facades, underfloor convectors were used.

The simulations produced a heating energy requirement of around 25 kWh/m^2_{NFA}a, of which 80 % is provided by the TABs and only 20 % by the convectors. In summer the 30 kWh/m^2_{NFA}a cooling energy requirements are met to about 85 % by activation of the concrete core and about 15 % by conditioning supply air.

A mechanical ventilation system with heat recovery – designed for hygienic air change of roughly 1.2 to 1.4/h – ensures a pleasant indoor climate. Supply and waste air is handled in the central zone by means of ventilation ducts that are suspended from the ceiling and are otherwise unadorned. The mechanical ventilation is operated only when there are heating or cooling requirements. In other cases, ventilation is exclusively by means of the ventilation panels in the facade. A window contact conveys information about the open window to the building automation system, which then closes the heating valve in the same facade element (Abb. B 17.08).

LIGHTING The building is provided with artificial lighting by a combination of background and individual lighting. Ceilings offer merely background lighting. The average electrical load of the ceiling lighting is 8.4 W/m^2 and produces an average of approximately 350 lux. Additional workplace lamps are planned for the workplaces to ensure the projected luminance or intensity of light.

RAUMTALK The software "Raumtalk" was used for room automation. This system is easy to use, can react flexibly to changes in the constellation of spaces, and also aids the monitoring of energy consumption and of the building operation. All appliances that are connected to the Raumtalk installation communicate on the basis of open IP (Internet Protocol) standards. In addition to the usual applications, IP-enabled functions in the areas of security, facility management, multimedia, infotainment etc. can be integrated. The systems are coupled via the open OPC standard (Object Linking and Embedding for Process Control – OPC). The Solon staff can adjust light, sun protection, and heating by means of freely programmable control elements on PC monitors and touch panels. Otherwise, buttons and sensors that employ battery-free radio technology are used.

To detach electricity generation and consumption, at certain selectable workplaces flexibly specially designed E-shuttles are available (Figs. B 17.05 and 06). Their battery units are charged at central docking stations. As mobile units, they allow staff to work independently of the power grid inside and even on top of the building.

ENERGY SUPPLY To cover the basic heating load and to supply the entire complex with electricity, a 530-kW_{th} gas combined heat and power unit is used. Once the nearby biogas system is connected to the gas network following an investment by Solon, it will deliver heat from power-heat coupling with a primary energy factor of 0.7.

HEAT SUPPLY A joint services control centre supplies the office and the production buildings with heating and cooling. To facilitate maintenance, the CHP was moved to the control centre of the nearby power station of the local energy provider (Blockheizkraftwerks-Träger- und Betriebsgesellschaft – BTB). The waste heat of the CHP is used in winter for space heating and in summer to run an absorption chiller. Through a planned biogas plant in the rural surroundings (to which Solon will make a financial contribution) it is intended to produce as much gas as is used in the biogas CHP.

The high cooling requirements for production are met by the absorption chiller and additional electrically-operated compression cooling engines. In summer hybrid cooling towers on the roof provide a large amount of the process cooling and the cold

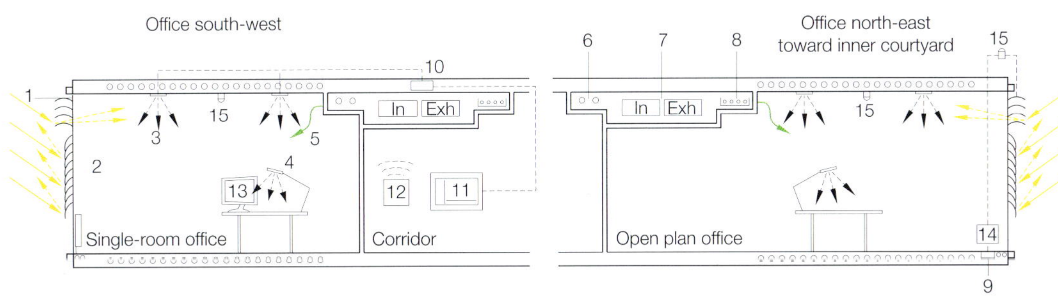

1 Sun protection glazing, external sunshade with daylight function
2 Window ventilation
3 Background lighting, grouped
4 Desktop lighting
5 Mechanical ventilation
6 Distribution of concrete core cooling
7 Distribution of supply and waste air Mechanical ventilation
8 Electrical distribution
9 Underfloor convector
10 Data cabling
11 Touch panel
12 Radio switch for light and blinds
13 SoftControl-PC for light, blinds, temperature
14 Controller
15 Sensor (light, temperature)

B 17.07
B 17.08

water for cooling the production processes and the office spaces.

POWER SUPPLY To ensure a redundant supply for production, a second electricity network was made available by the BTB headquarters. A photovoltaic array on the roof of the building with a total capacity of 230 kW_p generates around 258,000 kWh/a of solar electricity and feeds its surplus into the national grid. The plant points out the integrated concept behind the planning. To make the green roof useable, the photovoltaic arrays (produced by Solon itself) do not, as in comparable projects, cover the entire roof, but only the roof perimeter (Fig. B 17.09).

ENERGY BALANCE The Institute for Building and Solar Technology (IGS) of the TU Braunschweig developed a comprehensive planning and monitoring concept that was integrated in the building automation system. This work is part of the joint research project Energie-Navigator with the faculty of software engineering at the RWTH Aachen in the framework of the research focus energy oriented operation optimisation (EnBop) of the Federal Ministry of Economics and Technology (BMWi).

The Internet-based work platform Energie-Navigator ensures overall functional quality management from planning to the operation stage. In the first step, the functional requirements are defined by planning by means of the software-supported, so-called active function description. This serves as a precise model for programming controls. After this the installer supplements data point addresses in the functional description, and thus connects planning requirements with operating data produced later, while the software automatically analyses and reports any discrepancies or differences. These data can also be used for intensive adjustments and for optimising operations. For innovative and complex buildings, this concept promises new standards in quality assurance and management.

Nearly two years after start of operations, the first results produced in 2009 showed that the office building generally achieves the energy target values aimed for during planning. At present the annual primary energy consumption (balance boundaries according to DIN 18599) is around 98 $kWh/m^2_{NFA}a$.
An estimated energy balance for the entire property in accordance with the energy directive (EnEV) balancing limits (excluding plants used predominantly for production) was produced for the period from October 2009 to September 2010. The final energy consumption for heating and power is balanced against energy production from regenerative sources from the biogas CHP as well as the solar power plant. When combined, a clear energy plus is achieved.

LESSONS LEARNED The thermal spatial comfort is evaluated by means of what is known as "spot-monitoring", which was also developed at the IGS.

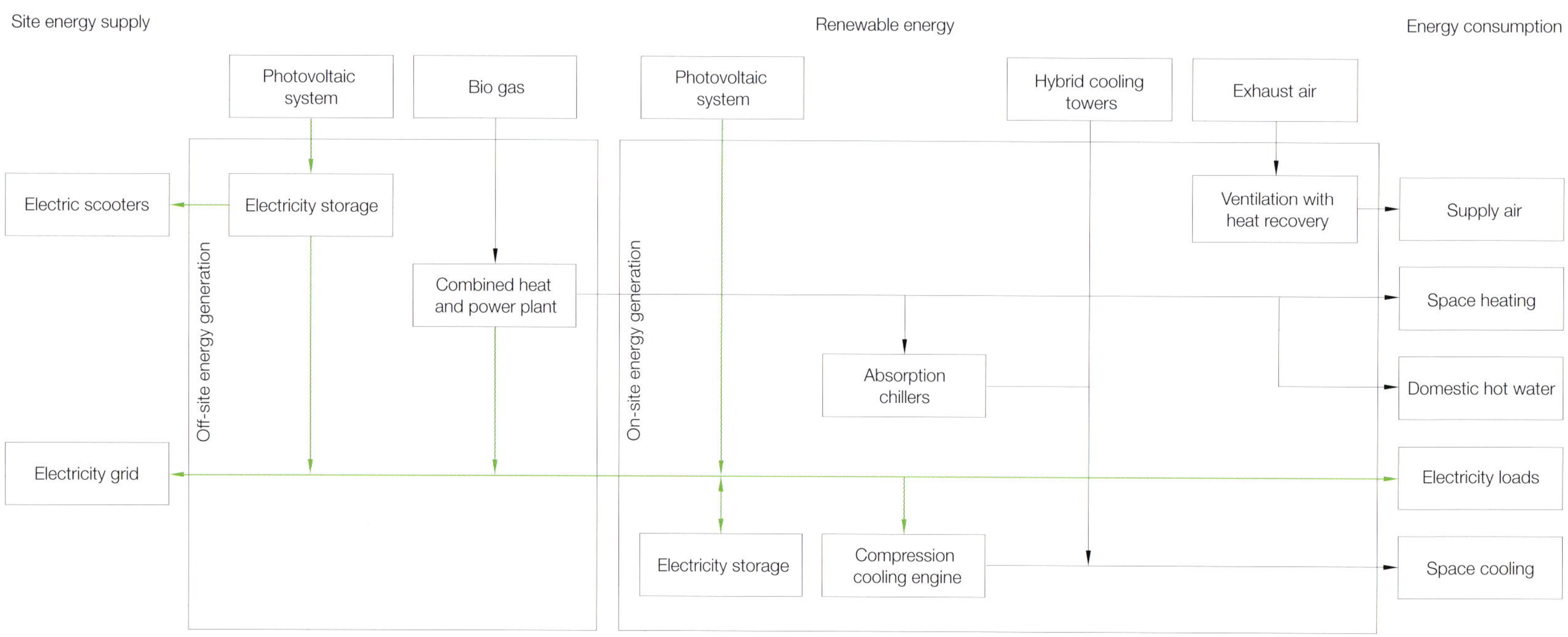

B 17.09

This connects measurements taken at workplaces with short-term surveys of the individual users. To assess the internal climate over an entire working day, in rooms chosen as examples measurements are taken three times daily during different weather conditions and over a period of one year. The first measurements from December 2009 produced positive results. Almost all measurements were in the top category A according DIN EN ISO 7730:2006. There were reservations only in the area of air quality. As initially only window ventilation was tested in winter and the use of ventilation systems was avoided (unlike what was intended in the original concept), the CO_2 values were around 1200 ppm – not untypical for buildings ventilated by openable windows – and therefore were in category C. In the course of optimising the building operation, depending on the weather conditions the ventilation systems were employed in a focused way in the building so as to achieve continuous air exchange.

Through the involvement of an interdisciplinary planning team at an early stage and careful analysis of the design and its functional, organisational, and technical possibilities, an optimal concept was developed. After almost a year of intensive adjustment the building arrived at its routine operating mode. Not only can it prove its qualities in terms of comfort and energy-efficiency through measurements taken, but has also been enthusiastically accepted by the staff.

B 17.09 Technical schematic of energy provision
B 17.10 Monthly site energy supply and solar electricity yield without electricity yields from the CHP unit (end energy)
B 17.11 Building and energy values (the values relate to the net floor area)

SITE	Berlin (D)
Annual global radiation at site	1050 kWh/m²a
Annual mean temperature at site	9.3 °C
Context	urban
BUILDING ENVELOPE PARAMETERS	W/m²K
U-value, exterior walls	1.20
U-value, windows (incl. window frames)	1.20
U-value, roof	0.23
U-value, roof lights (incl. frames)	1.60
U-value, ground floor slab	0.30
Mean U-value building envelope	0.75
BUILDING EQUIPMENT PARAMETERS	
Photovoltaic system area	1792 m²
System area per m²	0.18 m²/m²
Photovoltaic capacity	230 kW_p
Capacity per m²	23.60 W_p/m^2
CHP capacity	530 kW_{th}m 360 kW_{el}
Capacity per m²	54.30 W_{th}/m^2 36.90 W_{el}/m^2
GRID INFRASTRUCTURE AND ENERGY SOURCES	
Infrastructure supply	electricity grid, local heating grid
Energy source supply	local heating, electricity
Feed-in infrastructure	electricity grid
Feed-in energy source	electricity

DESIGN STRATEGIES, CONCEPTUAL FOCUS
mechanical ventilation, vacuum insulation, combined heat and power utilising biomass, local heating network, on-site electrical network, investment in biogas plant, photovoltaic system, quality control, and operation optimisation.

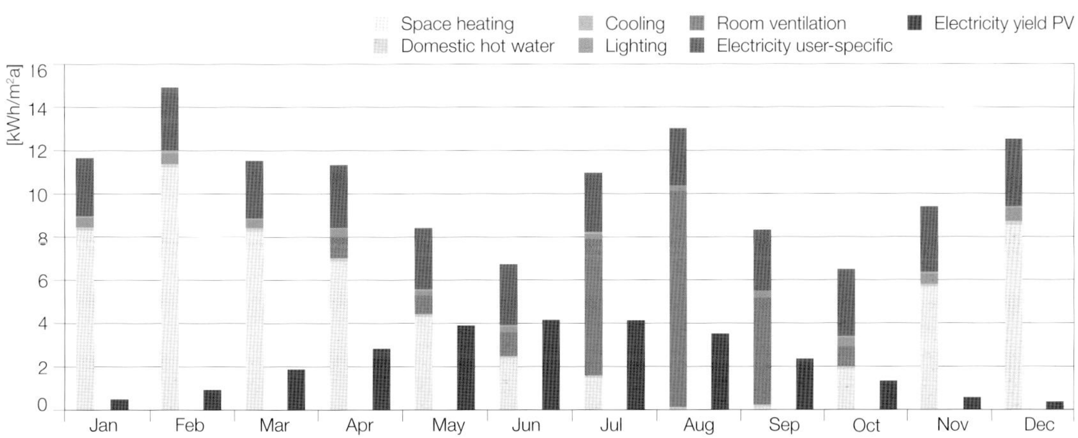

BUILDING CHARACTERISTICS	
Net floor area, NFA	9760 m²
Gross floor area, GFA	11,218 m²
Gross volume, V	48,905 m³
Building envelope, A	15,649 m²
Surface to volume ratio, A/V	0.32 m²/m³
Building costs (net, construction/ technical systems)	2200 €/m² (2008)
Number of units	1
Total number of users	approx. 350
CONSUMPTION PARAMETERS (2010)	kWh/m²a
Space heating consumption	54
Water heating consumption	2
Site energy heating (incl. hot water)	56
Electricity consumption	40
Cooling consumption	25
Electricity generation	27

B 17.10
B 17.11

ZERO EMISSIONS FACTORY

Braunschweig, D 2002
(2nd and 3rd construction phases: 2008–2010)

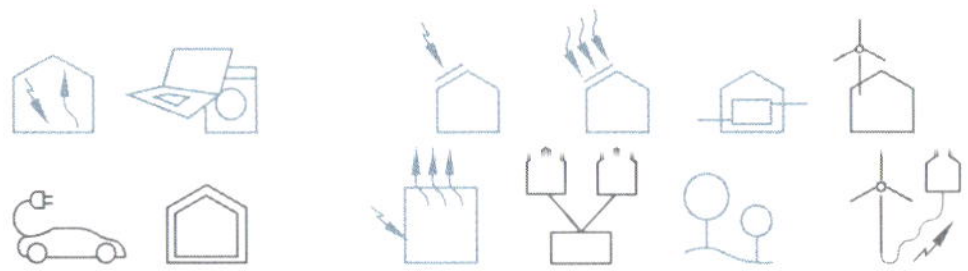

Client: Solvis Energiesysteme, Braunschweig
Architect: Banz + Riecks Architekten, Bochum
Energy consultant: Fraunhofer Institut für Solare Energiesysteme ISE, Freiburg
Building services: solares bauen, Freiburg
Monitoring: Fachhochschule Braunschweig/Wolfenbüttel
Main stakeholders: architect and client

Solvis, a manufacturer of energy systems, erected its new headquarters between 2001 and 2002 in the industrial district at Braunschweig's port near the Mittellandkanal. The intention was that, while producing no emissions, the building should combine comfortable working conditions with flexible production facilities. Exposed solar technology and the use of environmentally friendly building materials reflect the company's sustainable philosophy, also in the two later construction phases dating from 2008 and 2009. The building's striking external structure carries the solar collectors and photovoltaic plant and contrasts with the plain volume of the building itself. In the annual emissions balance, energy consumption levels were reduced by the use of passive house components as well as efficient cooling, lighting, and office equipment. These receive renewable energy from photovoltaic arrays, solar collectors, and biomass combined-heat-and-power systems. A similar concept was carried out in 1999 for the company headquarters of the Solarfabrik in Freiburg (see Solarfabrik, project list p. 178f. and Fig. B 2.04, p. 115). Over the period of one year the photovoltaic systems and combined heat and power unit offset the consumption of rapeseed oil.

DEVELOPMENT, DESIGN, AND STAKEHOLDERS

Planning and execution were based on an invited competition that was won in September 2000 by the architecture office Banz + Riecks. The planning team included solares bauen, an engineering group for energy planning, the Fraunhofer-Institut für Solare Energiesysteme, the structural engineers, and fire control experts as well as staff members of Solvis. They integrated potential and ideas and aimed at a high standard of efficiency regarding use of floor space, costs and energy.
The ambitious planning was aided by the subsidy program Energy Optimised Building (EnOB) of the Federal Ministry of Economics and Technology (BMWi). The stated goals of this program are a maximum figure of 40 kWh/m²a for heating energy

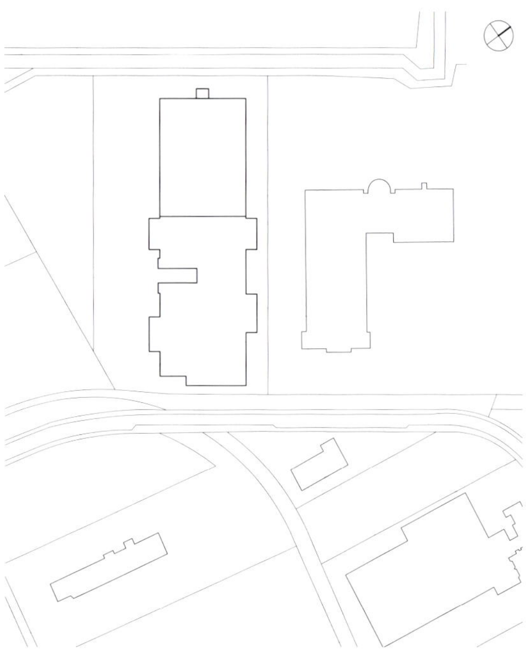

B 18.01
B 18.02

and 20 kWh/m²a for the energy consumed by the building services. Additionally, the building's primary energy requirements should be covered to 100 % by solar energy.

ARCHITECTURE After completion of the first construction phase the volume of orders and staff numbers rose so substantially that it soon became necessary to extend the building. Consequently, the zero emissions factory is structured into three construction phases.

FIRST CONSTRUCTION PHASE Clear lines and horizontal natural wood facades determine the character of the factory building. It is interrupted by an office wing on the eastern facade, which is accentuated by red window elements, and by the front buildings for deliveries, which are also red (Fig. B 18.02). These are directly adjoined with the production and storage facilities, can accommodate an entire trailer truck, and allow loading and unloading to be carried out in a warm environment. The temporary opening of the insulated rapid opening doors for driving in and out as well as a control that locks one set of doors when others are open avoids high ventilation heat losses in winter.
The U-shaped plan of the office and administration building intersects the rectangle of the production hall to create a courtyard that is adjoined by a cafeteria as a central meeting point. Offices and lounges on two stories surround the courtyard.
A third lightweight story set on top of the building (3rd construction phase) accommodates the training rooms as well as the research and development department. The incision of these areas in the production section, and a corridor (Solvis route Fig. B 18.07, p. 146) as internal longitudinal axis that traverses the production area connecting the two office wings, ensures short routes that support communication and productivity. Everything is close by, which underscores how this staff-managed company sees itself. On the ground floor, all the ancillary functional rooms and sanitary facilities are integrated in the Solvis route. On the first floor, all ductwork is exposed.

CONSTRUCTION The function of the longitudinal axis that traverses the building and the adjacent administration areas in reinforced concrete are: thermal storage mass, night-time passive dissipation of heat from the building mass, and fire protection. As high vertical loads had to be dealt with and the aim was to limit the volume that required heating, the roof construction of the production hall is connected to external steel structures (A-frames) placed on top of the reinforced concrete walls (Figs. B 18.04 and 05). The loads of the laminated timber beams that directly carry the roof skin are transferred on both sides into the A-frames via external tension and compression rods. About two-thirds of the roof loads are transferred from here to the reinforced concrete core of the building's central long axis. The light-weight timber construction spans 27.50 m,

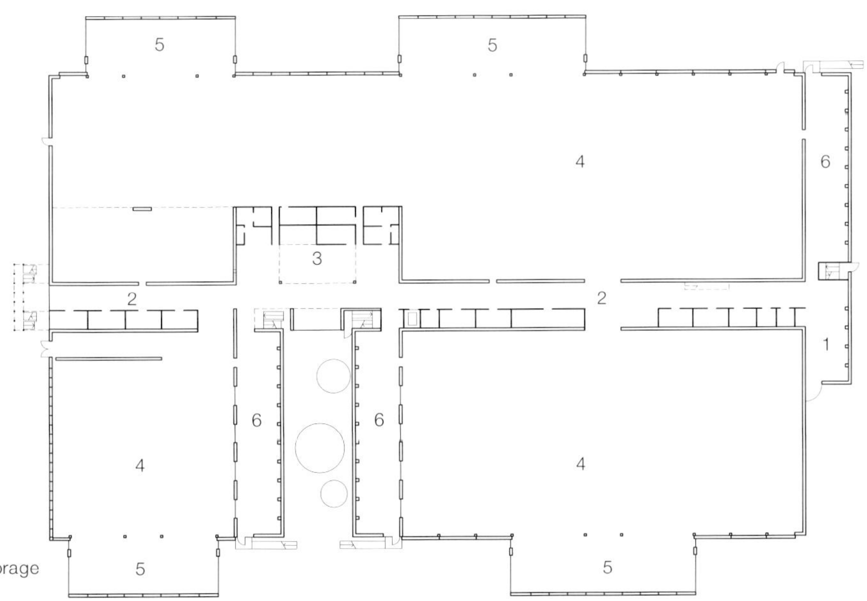

1 Reception
2 Solvis route
3 Lounge area
4 Production and storage
5 Delivery buildings
6 Office area

B 18.01 Site plan, scale 1:7500
B 18.02 View from south-west
B 18.03 Ground floor plan, scale1:1000 (1st construction phase)
B 18.04 Section through administration and courtyard, scale 1:1000
B 18.05 Section, scale 1:1000

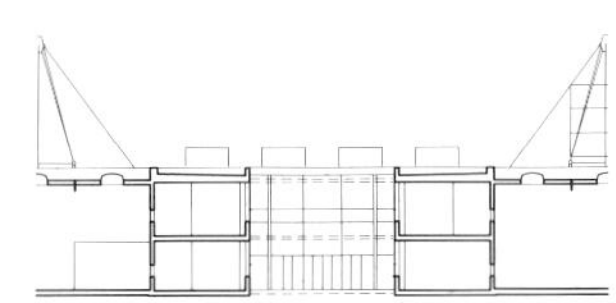

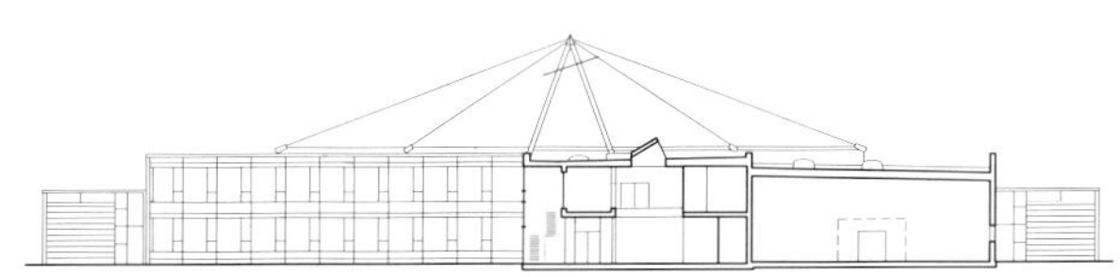

B 18.03
B 18.04
B 18.05

B 18.06 Manufacturing facilities in the column free production area, exploitation of daylight by means of skylights
B 18.07 Solvis route
B 18.08 Technical schematic of energy provision

B 18.06

B 18.07

B 18.08

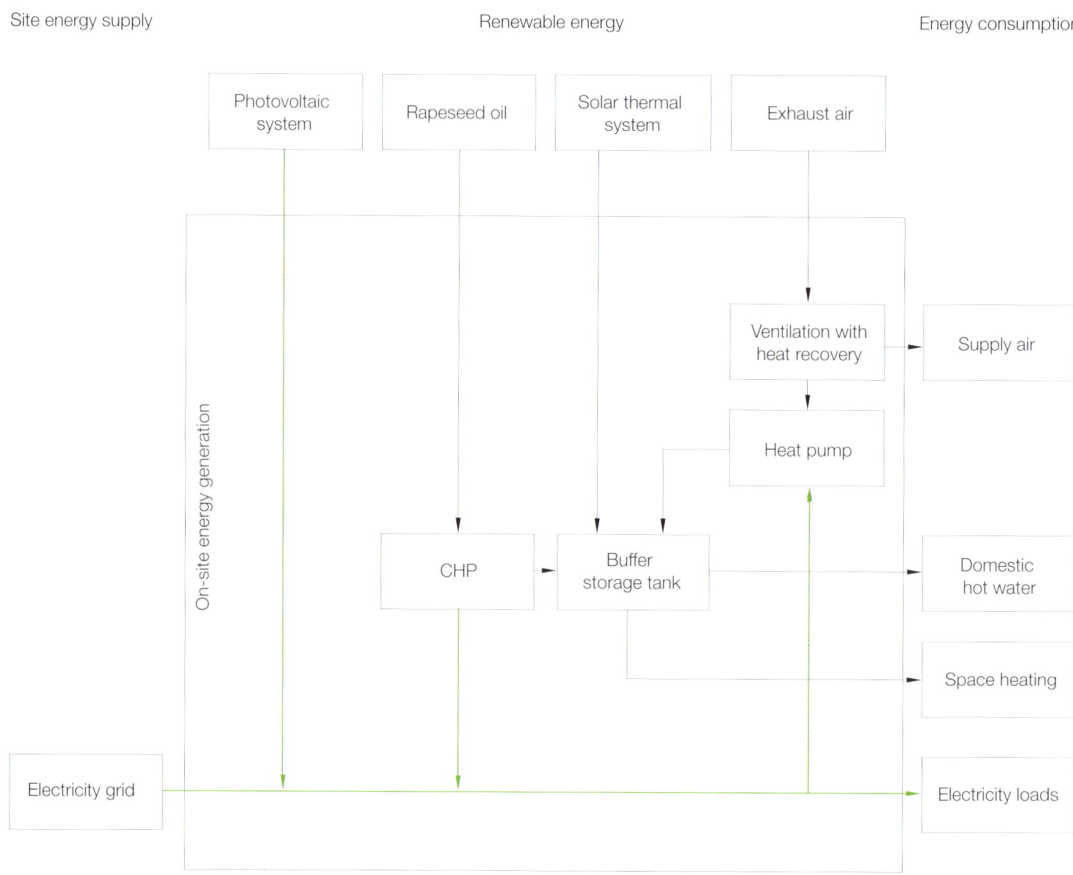

thus reducing the structural height of the beam cross-sections and the corresponding facade area by 1.20 m, while maintaining the usable clear height of the hall. The result is a flexible production area without internal columns and a reduction of the building volume by 15 %, and consequently, of the amount of energy used for heating and ventilation. The striking pylon structures also carry the large thermal and photovoltaic solar arrays and visibly display them as an effective kind of advertising (Fig. B 18.09, p. 148).

ENERGY EFFICIENCY The entire building is extremely compact in shape and the first phase achieves an A / V ratio of 0.37. Along the two external longitudinal column lines, column structures are placed with a curtain timber element facade or as timber frame construction with 24 cm of mineral fibre insulation. The roof and walls of the production hall consist of highly insulated light-weight timber building elements with 30 cm of thermal insulation. 12 cm of polystyrene foam below-grade insulation, triple low-e glazing in the office area, as well as double low-e glazing in the production area complement the thermal insulation concept. The building has a high air-tightness value of 0.22 h^{-1}.
While achieving a CO_2 neutral energy supply was a main goal, structural considerations limited the size of the area available for solar active systems, and thus, restricted the amount of electricity to be generated. Therefore, the energy concept focuses on avoiding air conditioning and using an energy efficient ventilation system instead. The ventilation concept for the offices differs from that for the production area, where the ventilation functions by means of a controlled supply and waste system with heat recovery via cross-flow heat exchangers. In addition to heating and ventilation, the ventilation system in the office area must also provide night-time cooling (air change 3/h by means of a mechanical exhaust system). Here, fresh air flows across intake elements in the parapet area of the facades. When no cooling is required, they can be closed by means of a central control to ensure the air-tightness of facade. The fresh air intake flows through the corridor areas into the air venting zones in the restrooms and coffee kitchens, where it is vented by a system of two downstream heat pumps. These extract heat from the waste air before it is directed outside, and feed two buffer tanks that are connected to the hot water network and the low temperature surface heating system for the offices. In summer mode the heat pumps switch off at a buffer tank temperature of 45 °C. Then the buffer tanks are fed primarily by the solar thermal plant. The heat pumps are always available as back-up.
An average daylight quotient of 4.5 % at workplaces and a daylight autonomy of 77 % in combination with skylights, and an automatic, daylight-linked dimming system for lighting in the halls and a demand-related control of artificial light in the offices save energy and reduce thermal loads. In addition, the amount of energy used for outdoor lighting is minimised by the use of LED lights and optimised technology is employed for the pumps, ventilators, servers, and the rest of the infrastructure in the areas of IT and communication. An external two part sunshade system with daylight louvers in the upper area helps prevent summertime heating in the offices. Opening wooden panels with vacuum insulation serve as ventilation openings in the office facades, the glazed areas are fixed. The various uses of wood in large areas of the wall and roof construction signal the efficient use of resources to the outside world.

ENERGY SUPPLY Solar collectors, photovoltaic arrays, and a rapeseed oil-fired combined heat and power unit cover the remaining energy requirements by means of regenerative energies. A total solar coverage of over 30 % is achieved. 550 m² of photovoltaic modules (52 kW_p of polycrystalline cells and 1 kW_p of semitransparent amorphous cells) generate 49 MWh of solar power alongside the 115 MWh_{el} of power from the rapeseed oil CHP (reference year 2005). The 200 m² of solar collectors feed the uninsulated sprinkler tanks that are situated in the building. As part of the fire protection concept they contain 500 m³ water for the sprinkler system and also serve as short-term storage or low temperature radiant heating in the heating period.
The waste heat from the IT centre and from the heating boiler in the development area of Solvis is led to the buffer tanks of the CHP. These are insulated steel pressure accumulators that stand in the production hall and supply the heating register of the ventilation system, the surface heating in the offices as well as ensuring the supply of hot water. The CHP is operated in a heat controlled mode. To allow the minimum operating period of two hours per start that the manufacturer calls for, first of all the two buffers are loaded and then the heat is led to the sprinkler tanks. If peak loads are exceeded the CHP can be operated in a power controlled mode in order to ensure a favourable power tariff. In this case emergency cooling is by means of the sprinkler tanks. (Fig. B 18.08).

ENERGY BALANCE As production was increased and working hours were extended during the period measured, the planning goal of an equalized complete annual energy balance and a zero emissions balance was not achieved. (Fig. B 18.10, p. 148). The increased consumption of power is covered only by increased use of grid power, as the consumption for computers and lighting rose by up to 60 %. Following the extension of the production hall it became possible to meet the energy demand locally, thanks to additional photovoltaic arrays. A cooling machine fitted later to cool the server room caused additional consumption of power, while the waste heat in summer remained partly unused. In 2004 the additional amount of energy used amounted to roughly 5 kWh/m²a, which rose to more than 11 kWh/m²a in 2005. In this period the CHP did not produce the amount of power that had been hoped for, as the heat already produced by the boilers in the research and development department reduced the CHP's operating time and therefore the amount of electricity it generated. For a CO_2-neutral supply of energy, it would be necessary to reduce the specific electrical energy demand from 20 to 12.5 kWh/m²a, or to develop renewable electricity production further. The photovoltaic array was increased to the maximum size structurally possible, but the hall's lightweight construction was not designed for greater loads and, therefore, a further increase is not possible.
According to primary energy consumption based on DIN V 18599 without the production facilities and work aids such as the computer and coffee

B 18.09
B 18.10

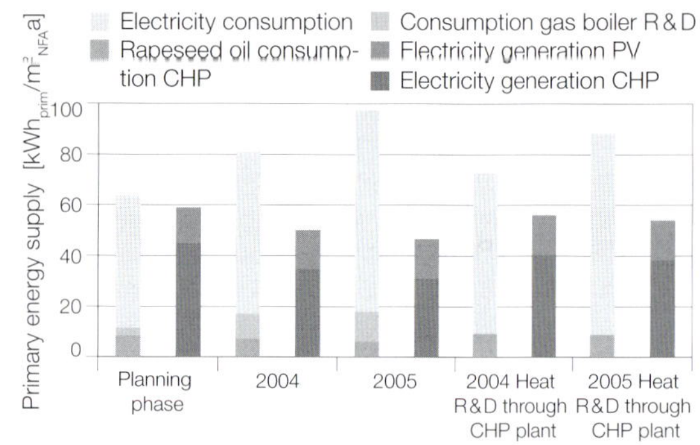

kitchens a CO_2-neutral energy supply is achieved, whereby 75 % less energy is consumed than in comparable industrial facilities. This means that the building avoids 800 tonnes of CO_2 annually. Here, the high quality thermal insulation and the heat recovering achieve roughly 70 % of savings (Fig. B 18.10).

EXTENSION BUILDINGS Within six years after completion the staff numbers had increased from 120 to around 300 and training rooms were required.

SECOND BUILDING PHASE In response to these needs, the building was extended from 2008 by around 5000 m^2 of multi-functional production and warehouse space on the north-western area of the site. The building extension used the same concept as the first phase to achieve the same flexibility and variability. In the area of the extension to the hall, the central long wall of the Solvis route was broken up into individual piers to transfer the vertical loads from the external steel structure and to further increase flexibility. The new building contains the warehouse areas, while the original building now accommodates the manufacturing of the entire range of products with a new automated production line (Fig. B 18.06, p. 146).

THIRD BUILDING PHASE A third construction phase consisting of the addition of a story from 2009 onwards included the expansion of the training and user centre as well as further offices with a floor area of around 1000 m^2. On account of the loads assumed during planning of the first building phase, the additional floor had to be a lightweight structure, and therefore timber construction was chosen. Self-supporting, horizontal, 18 cm thick gluelam roof slabs span the entire area of the additional floor between the external walls, a span of 22.50 m, without any internal columns. The training areas receive daylight through horizontal solar protection glazing situated between the primary beams placed at 2.50 m off centre. The thermal envelope to the internal courtyard is a prefabricated timber element facade based on a modular dimension of 1.25 m. The facade elements are individual components that accommodate all technical functions such

as glare protection, sun protection, and supply lines. The vertical loads in the area of the facade facing the inner courtyard are transferred by narrow wood composite bulkheads measuring 10 cm × 36.50 cm and placed at 2.50 off centre.
The energy concept is also retained. Expanded long-term buffers with a total capacity of 100,000 l buffer the periodic additional need for heat, so that even after completion of the extension it was not necessary to supplement the CHP. The photovoltaic array now measures 3200 m². Most of the modules are installed on the company building itself, roughly 24 kW_p on the roof of the bike shelter, and 20 kW_p on eight solar trackers. Additional solar energy capacity with just a few new energy users now allows a complete zero emissions balance.

LESSONS LEARNED All the Solvis employees discussed the design approach with the architects, which helped achieve wide acceptance of the completed building. The employees are satisfied with the building, although measurements have revealed comfort problems relating to poor heat recovery figures from the ventilation plant and in the area of heat protection in summer. Working hours considerably longer than originally planned led to an increase in the internal heat input and reduced the periods of passive dissipation of heat. Consequently, individual office areas heat up more than planned (during 9 % of working hours the internal ambient temperature is above 25 °C).
The technical measurements taken up to 2005 included the energy generated, fed, and used for electricity and heat as well as building and plant-specific parameters. This revealed a higher electricity consumption by the ventilation system than assumed during planning, as well as lower working figures for the heat pumps. Nevertheless, the results were, in part, clearly below the target maximum values of 40 kWh/m²a for heating energy requirements and 20 kWh/m²a electricity for building services. The primary energy consumption of the technical plant for heating, hot water, cooling, and lighting can, for the most part, be covered by regenerative energy sources or by credits for the energy fed into the public grid.
On account of its innovative impulses and the proof it provides of the feasibility of theintegrated concept, the new building is a model for solar building in ecological, economic, and architectural terms. It has already received several prizes in the area of energy efficient building.

B 18.09 Photovoltaic system on the primary steel structure
B 18.10 Analyses of energy consumption
B 18.11 Building and energy parameters (values refer to net floor area, NFA)

SITE	Braunschweig (D)
Annual global radiation at site	980 kWh/m²a
Annual mean temperature at site	8.7 °C
Context	suburban
BUILDING ENVELOPE PARAMETERS	W/m²K
U-value, exterior walls	0.20
U-value, windows (incl. frames)	1.10
U-value, roof surface	0.16
U-value, roof lights (incl. frames)	1.80
U-value, ground floor slab	0.27
Mean U-value, building envelope	0.27
BUILDING EQUIPMENT PARAMETERS	
Area of solar collectors	240 m²
Area per m²	0.03 m²/m²
Thermal storage volume	100,000 l
Storage volume per m²	12.20 l/m².

Photovoltaic system area	560 m²
System area per m²	0.07 m²/m²
Photovoltaic capacity	45 kW_p
Capacity per m²	5.50 W_p/m^2
CHP capacity	166 kW_{th} 105 kW_{el}
Capacity per m²	20.20 W_{th}/m^2 12.80 W_{el}/m^2
GRID INFRASTRUCTURE AND ENERGY SOURCES	
Supply infrastructure	electricity grid, delivery
Energy source supply	rapeseed oil, electricity
Feed-in infrastructure	electricity grid
Feed-in energy source	electricity

DESIGN STRATEGIES, CONCEPTUAL FOCUS
Passive house components, mechanical ventilation with heat recovery, vacuum insulation, CHP with use of biomass, photovoltaic array, solar thermal energy, passive cooling

BUILDING PARAMETERS (2005, 1st building phase)	
Net floor area, NFA	8215 m² (1500 m² admin., 6715 m² production)
Gross floor area, GFA	9600 m²
Gross volume, V	54,700 m³
Building envelope, A	19,706 m²
Surface to volume ratio, A/V	0.37 m²/m³
Building costs (net, construction/ technical systems)	757 €/m² (2002)
Number of units	office wing/production
Total number of users	approx. 150
CONSUMPTION PARAMETERS (2005)	kWh/m²a
Space heating consumption	28
Site energy consumption for heat (including hot water)	42
Electricity consumption	31
Total primary energy consumption	98
Total primary energy generation	47

B 18.11

SCHOOL RENOVATION

Wolfurt Mähdle, A 2010

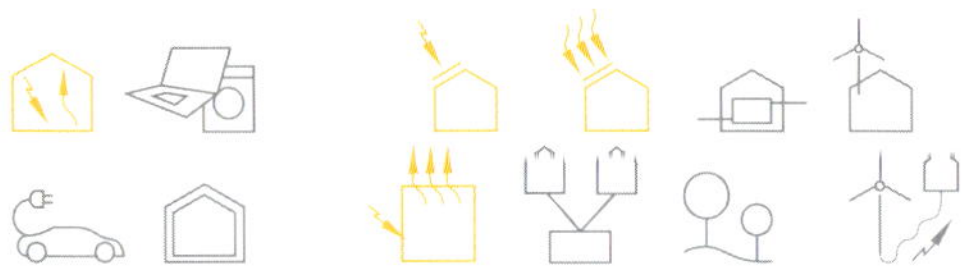

Client: Gemeinde Wolfurt
Architect: Architekt Gerhard Zweier, Wolfurt
Energy consultant: Bauphysik Spektrum, Dornbirn
Building services: Planungsteam E-Plus, Egg, Vorarlberg
Monitoring: Planungsteam E-Plus, Egg, Vorarlberg
Main stakeholders: architect and client

After 35 years of use the elementary school in Wolfurt, Austria was redesigned and renovated in 2009 to meet current and future demands of everyday school life. The renovation included wide-ranging thermal energy improvements to the building envelope aimed at bringing it up to passive house standards, and an increase in the building's compactness, something normally difficult to achieve with an existing building. Adhering to passive house standards as closely as possible is a requirement for the building-related zero energy balance. The hot water and heating energy demands of the school were reduced by about one quarter thanks to a large area of solar thermal collectors, while a ground source heat pump covers the remaining energy requirements. In the annual balance, a photovoltaic array on the roof of the school building offsets the electricity consumption of the heat pumps and other technical plants in the school and of a neighbouring fire station for the town of Wolfurt, erected at the same time as the renovation of the school.

DEVELOPMENT, DESIGN, AND STAKEHOLDERS

The renovation was carried out by the Marktgemeinde Wolfurt Immobilienverwaltungsgesellschaft commissioned by the authority responsible for the school, the town of Wolfurt. The basis for the passive house concept was a resolution passed in 2008 by Wolfurt's town council stating that all new community buildings as well as comprehensive renovation projects in Wolfurt are to be erected to or brought up to passive house standards. In intensive project meetings the town, school administration, teachers, and architects defined the new philosophy of the school as a "feel-good" school. The goals here included a wide range of spaces combined with high level of comfort thanks to good air quality, appropriate amounts of daylight, and high thermal comfort, and as well as low energy consumption. In addition, the "Ökoleitfaden Bau" ecological guidelines for building construction commissioned and issued by the Umweltverband Vorarlberg (Vorarlberg environmental association) for ecological and resource saving material and building concepts, promotes the use of ecological materials. These guidelines, formulated by the inter-university Research Centre Graz (IFZ) and the Austrian Institute for Building Biology (IBO) describe ecological parameters of recommended products with information on where they

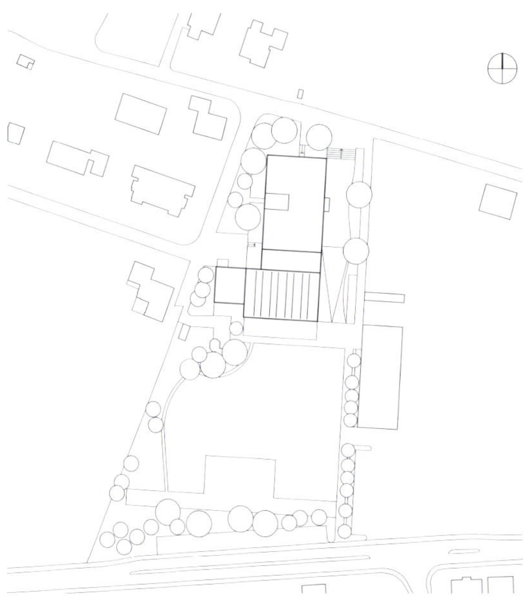

B 19.01
B 19.02

can be sourced, cite reference projects, and provide specification texts. They also supply information on the application, use, and disposal of the building materials. This allows ecological quality to be considered in the specifications and material selection phases and reduces the amount of additional time required for planning.
Apart from a 20% subsidy from the State of Vorarlberg, the project was financed largely by the town itself. The exemplary implementation of play and leisure areas for various generations was assisted by a subsidy of 50% from the State of Vorarlberg. The entire remodelling of the school building was carried out during the summer holidays in 2009 – which were extended to eleven weeks especially for this purpose – without any additional interruptions to the school year. Only the gym hall and the building services centre were adapted prior to that.

ARCHITECTURE By closing the former open space between the school and the gym hall and widening the school building by 4 m, additional 580 m^2 of usable space were gained. At the same time the building was made considerably more compact and the A/V ratio was increased from 0.43 m^{-1} to 0.36 m^{-1}. In terms of external appearance, this led to a building that displays a restrained, contemporary impression (Fig. B 19.02). Thanks to these measures and moving the restrooms to a former teaching aids room, each of the nine classrooms now has a directly allotted group room, and the school as a whole has a new multi-functional school hall (Fig. B 19.04). In addition, the former wardrobes in the corridor in front of the respective classrooms were combined in a central wardrobe in the new entrance area of the school, which adjoins the sheltered and generously sized school yard. This space serves as an antespace and was lowered to the level of the gym hall and connected to the sports and play areas by a wheelchair accessible ramp (Fig. B 19.03). The corridor area in front of the classrooms made available by the redesign can be used by teachers and pupils from the four classrooms on each story as a "learning studio" with open access to teaching materials. The organisational concept was developed in collaboration with the teachers. All the teaching materials are kept in plastic boxes in four different sizes, which are stored in a simple, flexible system of drawers. Therefore, materials can be accessed in the learning studio, in the classroom, or in the group room. There are also two long work tables in the learning studio where people can also work while standing. At both ends, there are quiet areas for resting. All-in-all, the new spatial concept provides a communicative and open building that can be used flexibly and facilitates modern educational concepts such as "open learning" or independent learning and meets the future demands of an all-day school.

CONSTRUCTION AND INSULATION The materials and colours used in the interior support the educational concept through their haptic qualities and the warm, pleasant nature of the surfaces and tones. The materials, the selection of which was influenced by the "Ökoleitfaden Bau", are predominantly natural, and therefore ecologically sound: wooden doors, linoleum flooring, and acoustic ceilings made of wood-wool panels. The double wood shingle cladding of the facades is supported by a wood frame that is infilled with 30 cm of mineral wool insulation (14 + 16 cm) and fixed to the existing reinforced concrete walls that are either 18 or 25 cm thick This wall construction

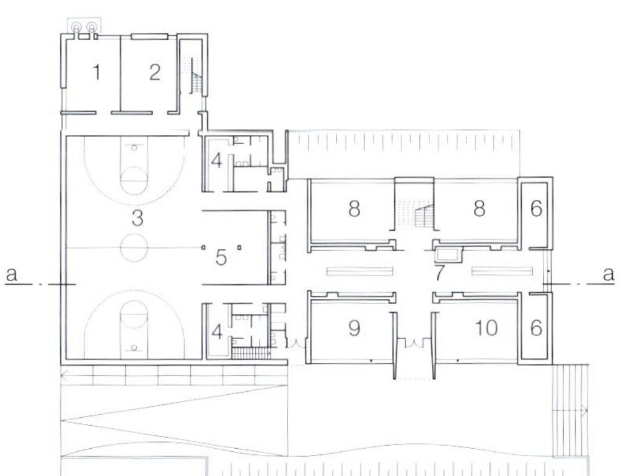

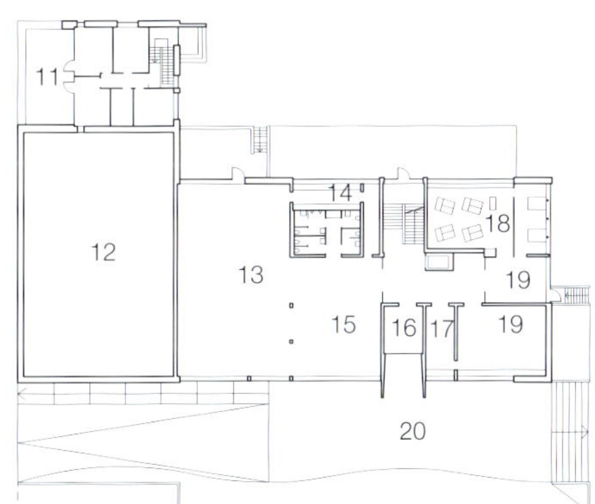

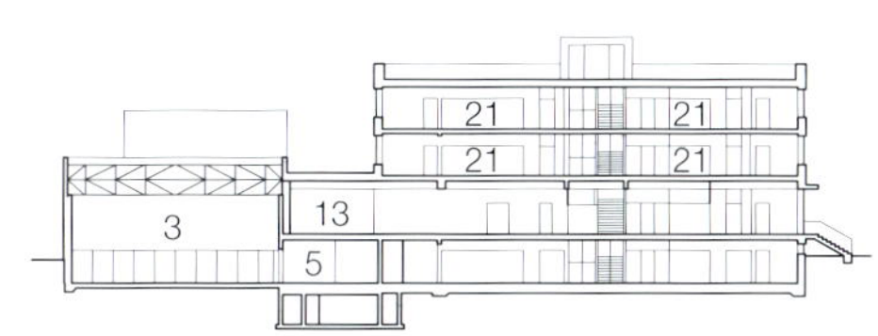

B 19.01 Site plan, scale 1:2500
B 19.02 View from north-east
B 19.03 Ground floor plan, scale 1:1000
B 19.04 First floor plan, scale 1:1000
B 19.05 Section, scale 1:1000

1 Boiler room
2 Storage
3 Gym hall
4 Locker room
5 Appliance store
6 Storage
7 Wardrobe
8 Arts & crafts
9 Music classroom
10 Art classroom
11 Janitor's apartment
12 Void to gym hall
13 Assembly hall
14 Coffee kichen
15 Library
16 Meeting room
17 Director
18 Pre-school classroom
19 Teachers
20 School yard
21 Learning atelier

B 19.03
B 19.04
B 19.05

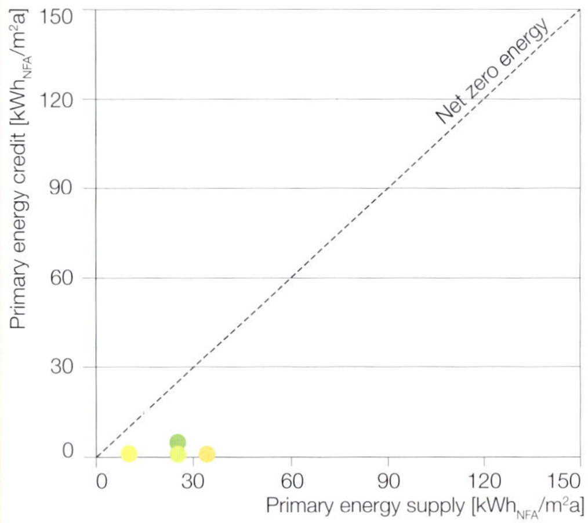

Taking into account user-specific consumption not conceptually included in the energy balance, a zero energy balance is not achieved. The solar electricity plant is designed just to cover consumption by building services. In the first year of measurements this goal was narrowly missed due to increased use of heat to dry the building, fluctuating user behaviour, initially uninsulated heating pipes, and low solar thermal yields due to a sensor problem and shading by a big tree.

- Total annual primary energy consumption 34 kWh/m²a
- Building specific primary energy consumption (10 kWh/m²a)
- Primary energy consumption after self-demand coverage by monthly yields (25 kWh/m²a)
- Remaining feed-in (4 kWh/m²)

Primary energy factors according to GEMIS 4.5 (see Fig. A 2.08, p. 32)

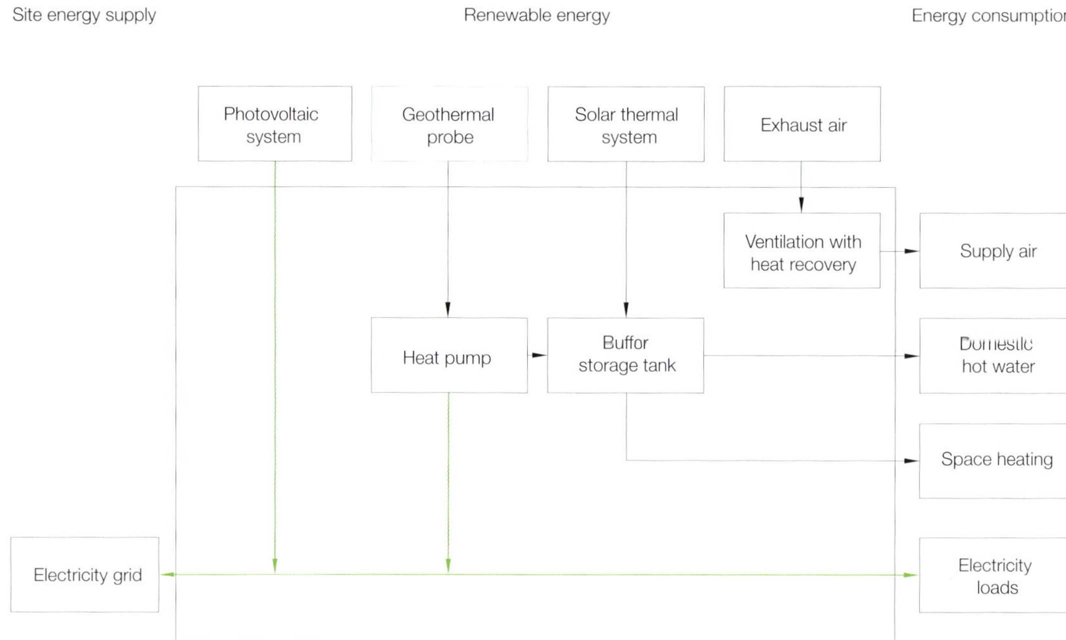

B 19.06 Energy evaluation
B 19.07 Technical schematic of energy provision
B 19.08 The 80 m² solar energy system on the southern facade

B 19.09 Building and energy parameters (values refer to net floor area, NFA)

B 19.06
B 19.07
B 19.08

achieves a U-value of 0.12 W/m²K. The new facade was partly produced in prefabricated elements to reduce construction time on site as far as possible. A post-and-lintel construction replaces the old double-glazed windows and the strip windows incorporate triple-glazed openable casement windows in wood frames. On the existing roof construction consisting of 18 cm of reinforced concrete and 14 cm of insulation, an additional 16 cm of EPS (expanded polystyrene hard foam) insulation was laid out (U-value 0.13 W/m²K), whereas the existing, almost entirely uninsulated floor slab could not be additionally insulated. Here, XPS (extruded polystyrene hard foam) insulation covers the foundation perimeter below grade and the base of the building above grade.

ENERGY EFFICIENCY The entire school was given a controlled supply and exhaust ventilation system with a heat recovery level of 85 %. A centrally located ventilation plant supplies the gym hall and the school hall. The classrooms feature decentralised ventilation that supplies fresh air directly via the facades. This made expensive ventilation ducts within the building largely unnecessary. The air change rate is 0,45 h^{-1} on average, while the specific power consumption (of the ventilation) is a low figure of 0.38 Wh/m³. The ventilation appliances in the classrooms are controlled by presence sensors. They can also be operated manually by the users. The solid concrete construction of the building means that there is sufficient thermal storage mass to buffer heat loads and to avoid over-heating of the classrooms. To prevent large thermal loads that could result from the larger areas of glazing, all windows received external blinds as sun protection. Their louvers can be tilted according to user needs and sun position, and thus also serve to prevent glare. A BUS system controls the sun protection in accordance with the external temperature, the room temperature, the position of the sun, the time of day, and the wind load. Several times a day (during recess) the controls switch to the default position. However, similar to the ventilation system, pupils or teachers can manually override it in every classroom. All switching levels are centrally visualised, monitored, and operated. This allows maximum passive solar yields to be achieved in winter at times when there are no classes. The skylights in the gym hall feature sun protection glazing.

In addition to energy-efficient lighting, all building services such as pumps and ventilators are also energy-efficient. As the computer equipment was taken over from the existing building, no influence could be exerted on its energy consumption. However, there are only relatively few computers. A hot water circulation pipework runs only between the showers in the gym hall area and the storage tank on the same level to avoid pipe losses. Water-saving fixtures and waterless urinals were fitted throughout the building.

ENERGY SUPPLY The energy supply system, which was deliberately kept simple, is based on an 80 m² solar thermal collector plant (Fig. B 19.08). This is integrated in the southern facade of the school and feeds a 6000 l combination buffer storage tank. As well as heating domestic hot water, this serves as a heating buffer storage. With high quality insulation, heating losses are kept low, as only one storage tank is used for all applications. A 56-kW_{th} ground water pump covers the major part of the heating energy demand. The heat source is an 18 m deep drawing well (one borehole) that is also used by the heat pump of the neighbouring Wolfurt fire station. Both buildings have their own heat pump, and in both the ground water returns and seeps away through a 12 m deep return well.

The electricity required to run the building services for the two all-electric buildings is generated by a 188 m² 26-kW_p photovoltaic array on the school roof (Fig. B 19.07). However, the two buildings are independent of each other in terms of energy, with the exception of the shared ground water well.

ENERGY BALANCE About 23 % of the calculated heat demand for space heating and hot water is covered by the solar thermal plant. Thanks to the photovoltaic array placed on the school roof, which is also used by the new fire station building, the entire electricity consumption of the building services amounting to around 26,000 kWh/a (school) can be offset in the annual balance. The solar thermal plant feeds generated energy into the national grid and is dimensioned to cover the energy demands of the two buildings for ventilation, heating, and hot water in the annual balance. The other energy consumption parameters are not included in the conceptual energy balance.

LESSONS LEARNED Supported by weekly discussions on the design with the teachers and school administration, planning proceeded in a positive way and architectural, educational, and energy requirements were consistently incorporated and integrated. The excellent communication also helped in adhering to the very short construction period of only eleven weeks (school building) and eight months (infrastructure). Consistent monitoring of the technical equipment during operation of the building as well as analysis of the results and the wishes of the users made it possible to explain effects on the consumption of energy and to adjust the system. The technical fine-tuning and harmonisation of the space heating system, ventilation plant, and sun protection measures extended over a period of nearly one year.
In comparison to the old building, the improved air quality and especially the optimised acoustics were very positively received by pupils and particularly teachers. Furthermore, the changes to the energy concept became part of the curriculum. In conjunction with the high visibility profile of the solar thermal plants, the intention here was to motivate pupils to use energy responsibly.

SITE	Wolfurt Mähdle (A)
Annual global radiation at site	1090 kWh/m²a
Annual mean temperature at site	8.9 °C
Context	rural
BUILDING ENVELOPE PARAMETERS	W/m²K
U-value, exterior walls	0.13
U-value, windows	0.85
U-value, roof surface	0.12
U-value, roof lights (incl. frames)	1.00
U-value, base plate	0.95
Mean U-value, building envelope	0.36
BUILDING EQUIPMENT PARAMETERS	
Area of solar thermal collectors	80 m²
Area per m²	0.02 m²/m²
Thermal storage volume	6000 l
Storage volume per m²	1.80 l/m²
Photovoltaic system area	188 m²
System area per m²	0.06 m²/m²
Photovoltaic capacity	26 kW_p
Capacity per m²	7.70 W_p/m^2
GRID INFRASTRUCTURE AND ENERGY SOURCES	
Supply infrastructure	electricity grid
Energy source supply	electricity
Feed-in infrastructure	electricity grid
Feed-in energy source	electricity

DESIGN STRATEGIES, CONCEPTUAL FOCUS
Renovation according to passive house concept, increased compactness, decentralised ventilation plant with heat recovery, facade solar collectors, ecological construction materials

BUILDING CHARACTERISTICS	
Net floor area, NFA	3367 m²
Gross floor area, GFA	4096m²
Gross volume, V	15,006 m³
Building envelope, A	5426 m²
Surface to volume ratio, A/V	0.36 m²/m³
Building costs (net, construction/ technical systems)	950 €/m² (2007)
Number of units	school and sports hall
Total number of users	180
CONSUMPTION PARAMETERS (2010)	kWh/m²a
Space heating consumption	18
Water heating consumption	4
Site energy consumption for heat (including hot water)	7
Electricity consumption	26
Total primary energy consumption	34
Total primary energy feed-in	11

B 19.09

UNIVERSITY BUILDING

Saint-Pierre, La Réunion island, F 2009

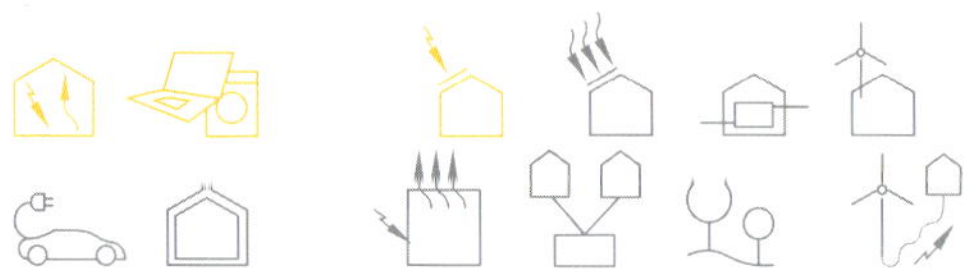

Client: Université de la Réunion, Saint-Pierre, La Réunion
Architect: Thierry Faessel-Bohe, Saint-Pierre, La Réunion
Energy consultant: Imageen, Saint Denis Cedex, La Réunion
Building services: Imageen, Saint Denis Cedex, La Réunion
Monitoring: Université de la Réunion, Saint-Pierre, La Réunion
Main stakeholder: client

The University of La Réunion's EnerPos building (French acronym for energy plus) is situated on the Saint Pierre satellite campus in the island's south. It was opened in January 2009 and is one of the few zero energy buildings in a subtropical climate. Located 800 km east of Madagascar in the Indian Ocean, La Réunion Island, in geopolitical terms, is a French overseas department. Situated in the southern hemisphere, its subtropical climate and global solar intake of about 2,000 kWh/m^2 make it an ideal location for buildings supplied with solar energy. Due to consistently high temperatures, there is hardly any need for heating. However, air-conditioning is a must, especially during the hot and humid summer between December and March, with temperatures averaging at 25 °C and humidity over 70 %. During the dry season between June and September, average temperatures reach approximately 22 °C (Fig. B 20.09, p. 157). Typically, air conditioners are used to cool and dehumidify interiors, yet consume large amounts of electricity. An extensive sun protection concept and providing natural ventilation supported avoiding air conditioners to a large degree in the EnerPos building. A photovoltaic system connected to the electricity grid more than offsets any annual residual electricity demand for building equipment and user-related consumption.

DEVELOPMENT, DESIGN, AND STAKEHOLDER

Already during its development, the EnerPos building served as a research project. It allowed researching the implementation of new planning instruments and developing genuine methods for the design of zero energy buildings in hot climates. Since planning began in 2005 within a research project of the three French universities La Réunion (LPBS), Chambery (INES/LOCIE), and Toulouse (PHASE), the goal of minimising energy demand was pursued with determination. Eliminating air-conditioners in the seminar rooms and providing natural ventilation and plenty of daylight are intended to cut total electricity consumption to less than 70 kWh/m$^2_{NFA}$a. This is less than half the demand of comparable buildings in La Réunion (160 kWh/m$^2_{NFA}$a). Early on, the research team coordinated by La Réunion University's Laboratory for Physics of Buildings and Systems (Laboratoire de Physique du Bâtiment et des Systèmes) used simulation instruments to research strategies for the reduction of energy consumption and for passive cooling. Both thermal and daylight simulations provided data that were evaluated according to energy demand, available daylight, and interior comfort. The results contributed to optimising architectural design and building construction.

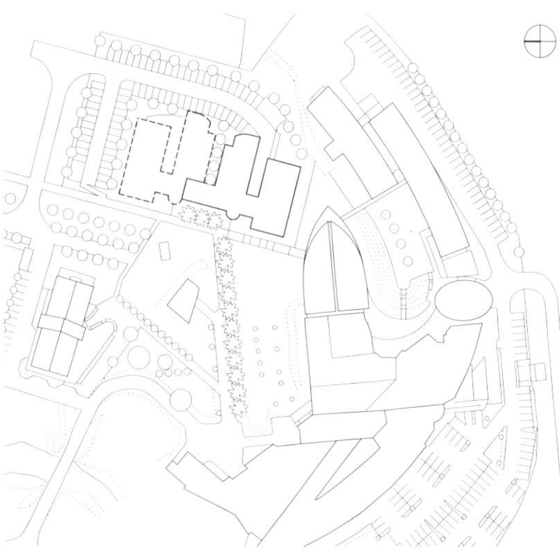

B 20.01
B 20.02

ARCHITECTURE The two-story building is divided into two parallel strips by a landscaped inner courtyard. The gross floor area of 1425 m² encompasses an administration area with, among others, seven offices at ground level, five seminar rooms on the ground and upper floors, and an underground garage. The inner courtyard plays an important role for access and recreation (Figs. B 20.03–05). Windows comprise 30 % of facades and are oriented towards the north and south, in order to take advantage of winds from the sea. A three-metre-wide strip of indigenous vegetation following the building perimeter prevents ambient air from heating up too much as it enters the building. Parking spaces are located beneath the building to prevent asphalt surfaces from radiating too much heat too close to the building. In addition, unpaved surfaces improve drainage during tropical downpours.
Covered walkways oriented towards the inner courtyard provide access to the building and benefit from ambient daylight, while providing shade for each other as well as adjacent windows.

ENERGY EFFICIENCY The building structure consists of reinforced concrete. Northern and southern walls feature no insulation. Due to minor temperature differences between interior and exterior, heat transmission gains or losses are negligible. However, for uninsulated walls, shading and low heat absorption are of decisive importance. The exterior surface temperature should not exceed the exterior ambient temperature, in order to prevent transmission heat flow. In the southern hemisphere, facades oriented to the north require protection from direct sunlight. This is accomplished by a cantilevered roof extending 2.40 m and fixed wood louvers as sun protection. The spacing between louvers was based on simulations that optimised both shading effect and natural lighting of interior spaces. The analysis showed that during class hours from 8 a.m. to 6 p.m. two ot the seminar rooms required no artificial lighting.
The shade-less west and east facades feature exterior wood cladding and are insulated with 4 cm of mineral wool matching interior ceiling heights. All wood used is indigenous and is untreated, due to its particular durability and limited number of rainy days in the region.
The photovoltaic systems are mounted on top of corrugated sheet metal roof decking elevated above and providing shade for the two building volumes below. Ceilings feature a 10 cm polystyrene light insulation layer. Elevating the seemingly weightless roof structures improves ventilation beneath the photovoltaic systems, which contributes to solar yield in relation to strong intake and high exterior temperatures. The roof inclination supports drawing air from the inner courtyard between the two building volumes via the typical sea-borne wind currents, and thus, facilitates passive ventilation of interiors. Based on the layout of the building volumes with their inner courtyard and covered access walkways, all class rooms have two exterior walls and are naturally ventilated via single-pane louvered windows that can be manually adjusted to control air flow. The louvered glass offers improved security compared to traditional openable windows. They are also used in the administration area and are placed within walls between offices and hallways, where they provide natural cross-ventilation through the building (Figs. B 20.06 and 07).
The offices and seminar rooms are equipped with a total of fifty-five ceiling fans to provide air movement similar to cross-ventilation when there is no wind. They can be individually adjusted in the offices and in groups of two or four units in the seminar rooms. Each three-speed ceiling fan serves 10 m² and consumes a maximum of 70 W. Natural ventilation, partly enhanced by fans, increases interior air flow. This causes people to dissipate more body heat through convection and evaporative cooling,

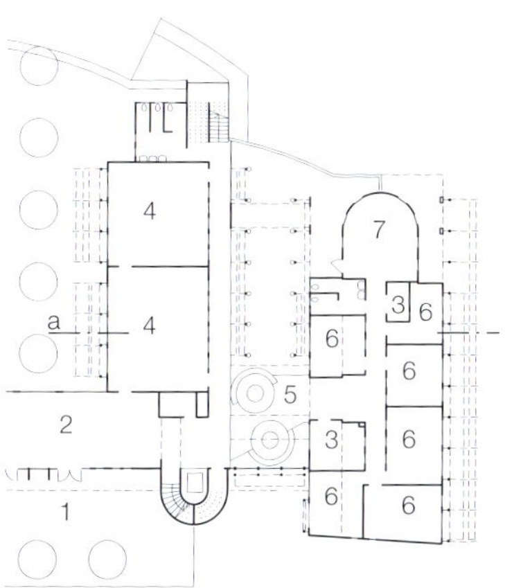

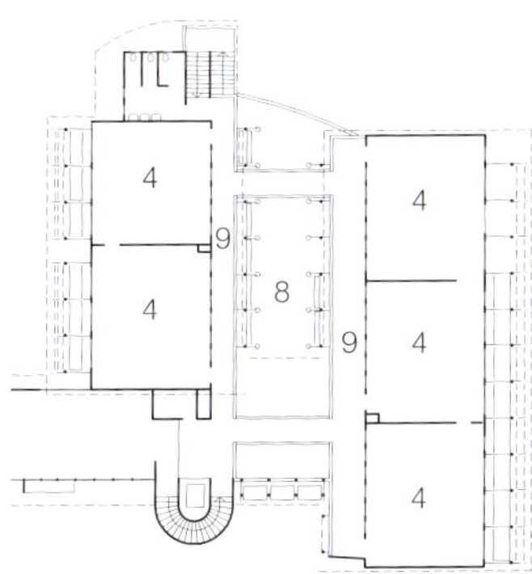

1 Antespace
2 Foyer
3 Utility/storage room
4 Seminar room
5 Forum
6 Office
7 Meeting room
8 Void, forum
9 Covered walkway

10 Photovoltaics and fixed sun protection
11 Summer sun
12 Winter sun
13 Fixed sun protection: wood louvers
14 Louvered glass windows: 30 % transmission factor
15 Cross ventilation
16 Ventilator

B 20.01 Site plan, scale 1:3000
B 20.02 View from south-west
B 20.03 Ground floor plan, scale 1:750
B 20.04 Upper floor plan, scale 1:750
B 20.05 Section, scale 1:500

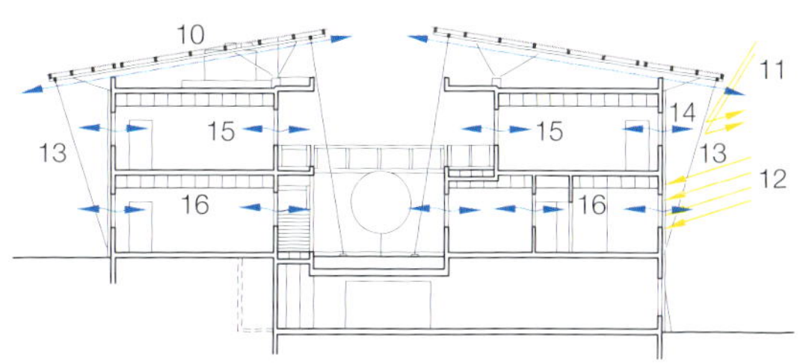

B 20.03
B 20.04
B 20.05

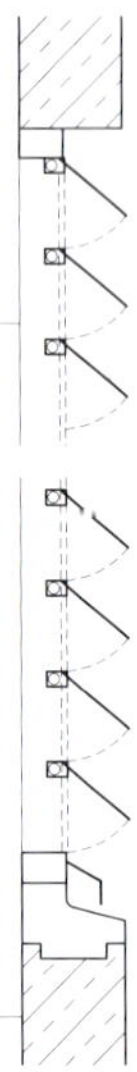

1 Window, eleven movable single pane louvers 700 × 1400 mm
2 Wall construction: 140 mm reinforced concrete, no insulation

B 20.06
B 20.07

making them feel more comfortable even when temperatures are higher than usual. The sparingly used LED-type lighting in office spaces consumes only 4.4 W/m^2 and features an illuminance of 300 Lux. In the seminar rooms, lights are automatically turned off two hours after lectures are scheduled to end.

ENERGY SUPPLY A central air-conditioning system serves to cool and dehumidify individual office spaces and computer classrooms during particularly hot days without wind. Due to natural ventilation and ceiling fans, use of air conditioners in computer rooms has been limited to a runtime of six weeks per year and currently isn't even required in remaining spaces. The air conditioning systems of comparable buildings on the university campus typically are in operation nine months per year.

A photovoltaic system covering 350 m^2 and with a rating of 50 kW$_p$ is integrated into the roof structure of the two buildings, sloped at 9°. One half of the roof surfaces is oriented to the north, the other half to the south. Because of the minor roof slope and high sun angle, yield difference is relatively small at only 8 % (Fig. B 20.08).

ENERGY BALANCE All electrical loads are monitored. Initial measurements indicate that the total electric power consumed is 32 kWh m$^2_{NFA}$a, which is significantly less than the target of 70 kWh/m$^2_{NFA}$a. The photovoltaic systems generate about 70,000 kWh/a (1,428 kWh/kW$_p$). The measured solar yield is just under 80 kWh/m$^2_{NFA}$a, considerably higher than consumption. Due to consistently high solar intake, low seasonal differences in electricity consumption, and the fact that the buildings are mainly used during daytime, electricity consumption and generation are highly synchronous (Fig. B 20.11). There is no space heating or hot water heating demand. However, a shower for cyclists wasn't included. Since electricity is the only energy source, utilities and metering concept are kept simple.

OVERALL EXPERIENCE The extensive use of simulation tools served primarily to optimise the passive elements and integrate fundamental architectural principles for a net zero energy building in a subtropical climate, such as natural ventilation and sun protection. Load matching was of secondary importance. The result is that the building consumes one-fifth of the energy of comparable standard buildings. Air conditioning systems are usually not required, even during the hottest days of the year.

Recorded data on air temperature, relative humidity, and interior air velocity, as well as a user survey based on nearly 2,000 questionnaires report high comfort levels of interiors and excellent feedback.

Measurements taken during the first two years of service show that the groups of ceiling fans in the seminar rooms are slower than their individually controlled counterparts in the office spaces. Switching is currently under revision. Arranging the interior lighting according to zones in the seminar rooms is considered a potential further improvement. Daylight measurements indicate three zones. The daylight near windows not facing the inner courtyard is significantly stronger than on the opposite side of the room. The amount of light in the centre of the room lies between the above two situations. Therefore, three rows of lights that can be switched separately will be installed.

B 20.06 Section, interior wall with louvered windows between hallway and office, scale 1:20
B 20.07 Louvered window, administration area hallway
B 20.08 Technical schematic of energy provision
B 20.09 Climate data for La Réunion: average humidity and temperature
B 20.10 Comfort parameters, building interior, observed temperature and interior humidity values during typical hours of service
B 20.11 Energy evaluation
B 20.12 Building and energy parameters (values refer to net floor area, NFA)

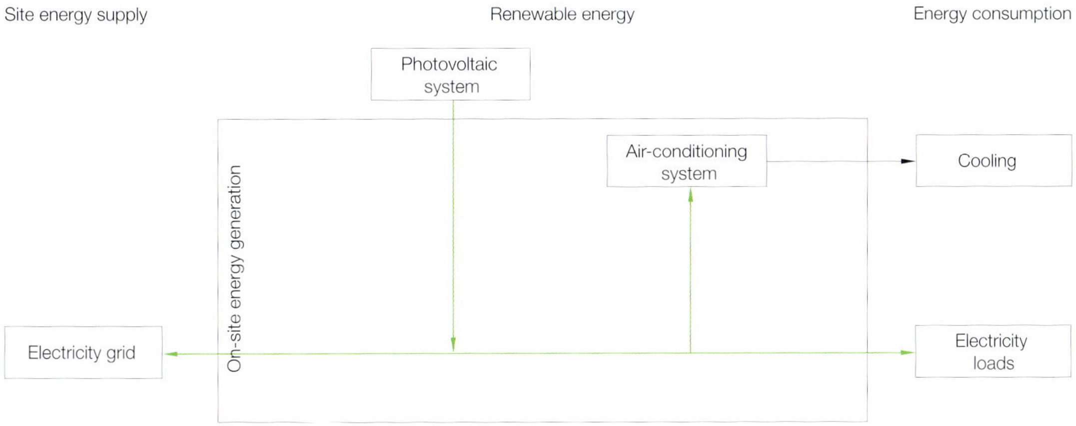

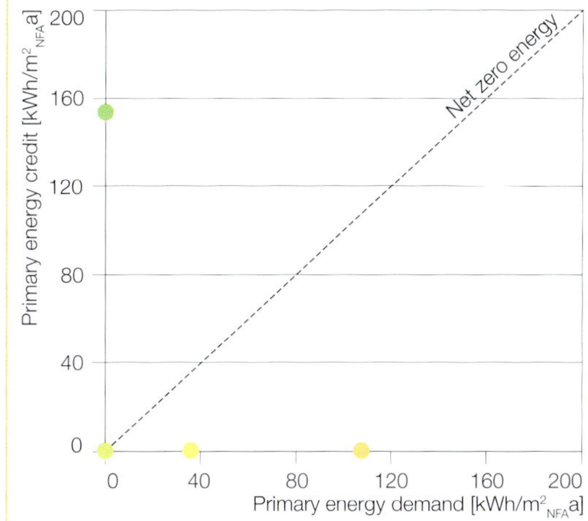

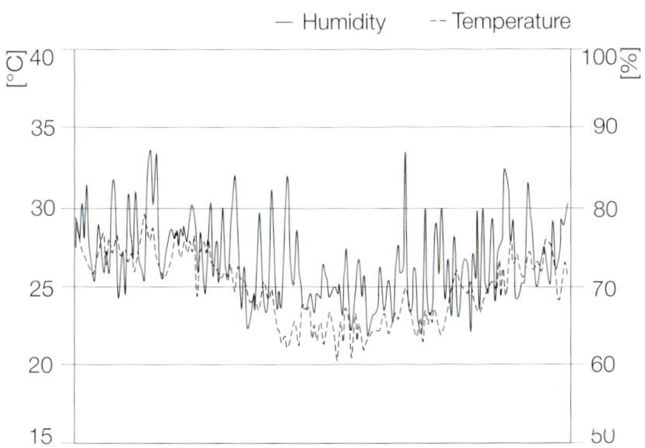

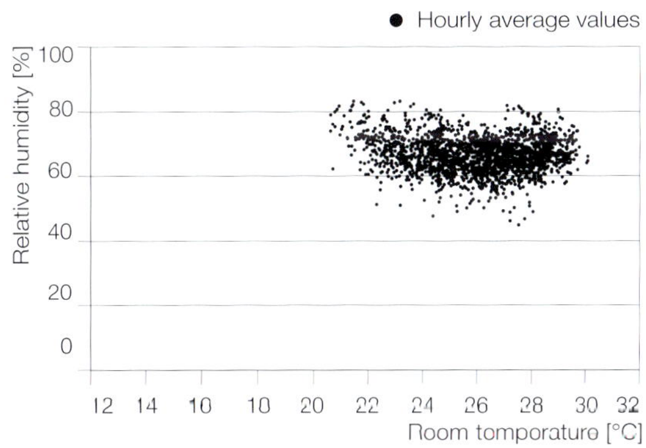

The very large solar yield covers total consumption each month; also, a very large surplus is generated. Clearly, user consumption is the dominant factor, and its monthly compensation is due to consistent year-long solar intake and absence of space heating or hot water demand.

- Metered annual total primary energy consumption (106 kWh/m²a)
- Building-specific primary energy consumption (36 kWh/m²a)
- Seasonal compensation of total consumption via creditable monthly energy yield (106 kWh/m²a). Coverage of residual energy consumption is not required.
- Annual energy surplus, grid feed-in (154 kWh/m²)

Primary energy factor for electricity (3.3), specific to La Réunion

SITE	Saint-Pierre, La Réunion
Annual global radiation at site	1,929 kWh/m²a
Annual mean temperature at site	25 °C
Urban context	suburban
BUILDING ENVELOPE PARAMETERS	W/m²K
U-value, east and west exterior walls	0.67
U-value, north and south exterior walls	3.30
U-value, windows (including frames)	6.00
U-value, roof	0.43
U-value, floor	0.26
Mean U-value, building envelope	2.90

BUILDING EQUIPMENT PARAMETERS	
Photovoltaic system area	350 m²
System area per m²	0.45 m²/m²
Photovoltaic capacity	50 kW_p
Capacity per m²	64.00 W_p/m^2
GRID INFRASTRUCTURE AND ENERGY SOURCES	
Supply infrastructure	electricity grid
Energy source supply	electricity
Feed-in infrastructure	electricity grid
Feed-in energy source	electricity

DESIGN STRATEGIES, CONCEPTUAL FOCUS
Photovoltaic system, sun protection concept, natural ventilation, ceiling fans, use of natural daylight, efficient artificial lighting

BUILDING PARAMETERS	
Net floor area, NFA	781 m²
Gross floor area, GFA	1425 m²
Gross volume, V	3847 m³
Building envelope, A	1234 m²
Surface to volume ratio, A/V	0.32 m²/m³
Building costs (net, construction/technical systems)	1664 €/m² (2009)
Number of units	2
Total number of users	170
CONSUMPTION PARAMETERS (2010)	kWh/m²a
Electricity consumption	32
Total primary energy consumption	106
Total primary energy generation	260

B 20.08
B 20.09
B 20.10
B 20.11
B 20.12

DAY CARE CENTRE

Monheim, D 2009

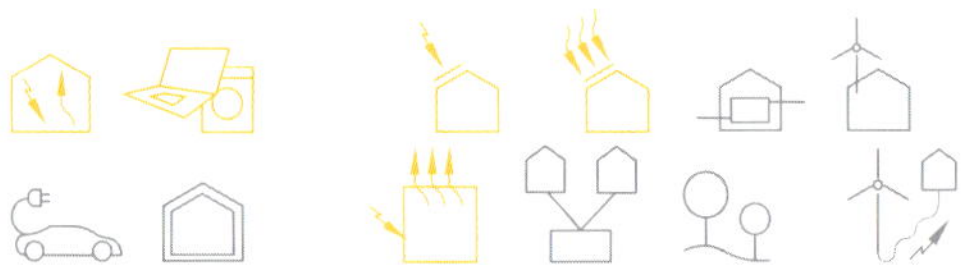

Client: Bayer Real Estate, Leverkusen
Architect: tr.architekten, Cologne
Energy consultant: IPJ Ingenieurbüro P. Jung, Cologne
Building services: E+W Ingenieurgesellschaft, Leichlingen
Monitoring: Bayer MaterialScience, Leverkusen
Main stakeholder: client and architect

B 21.01 Site plan, scale 1:2000
B 21.02 Northeast elevation
B 21.03 Ground floor plan, scale 1:750
B 21.04 Sections, scale 1:750

To help its employees manage family life and career, Bayer built a day care centre called Sprouts (die Sprösslinge) at its Monheim site near Leverkusen. The facility serves five groups of altogether sixty children between six months and six years of age. Based on an expert appraisal of proposals, the design submitted by tr.architekten was selected and completed in 2009.
As a planning goal from the very beginning, the passive house standard was further developed and defined throughout the planning process towards a carbon-neutral building in the context of the EcoCommercial Building Program. This program promotes planning and realisation of sustainable, environmentally friendly, economically feasible projects, based on integrated material selection and strategies that meet future challenges of continuous changes in urban developments and finite energy resources. Built according to these principles, the day care centre's total annual energy consumption is balanced, and thus, climate neutral; a debut for day care centres in Germany.

DEVELOPMENT, DESIGN, AND STAKEHOLDERS
The project's development is based on an integrated working model: During the competition phase, the team already intended to create a building that meets contemporary standards of function, efficiency, convenience, and design, while implementing ideas to cover residual energy consumption via renewable resources. This way, being environmentally sound can be defined as an integral rather than an additive aspect of the planning process. Design concepts were discussed in close cooperation among the various disciplines involved and evaluated using computer modelling and dynamic simulation. This enabled comparing design alternatives early on as an immanent part of the planning process. Consumption was reduced to a minimum by optimising thermal insulation and form of the building envelope, maximising utilisation of daylight, and optimally coordinating technical requirements with user-specific demands. Heating demands are covered by a solar thermal system and a geothermal heat pump. A photovoltaic system provides required electricity for building services and equipment.

ARCHITECTURE Despite the restrained design of openings on the building's entrance side, the day care centre offers an inviting appearance (Fig. B 21.02). A rectangular, longitudinal building volume extends deeply into the site and houses ancillary rooms, separates the driveway from

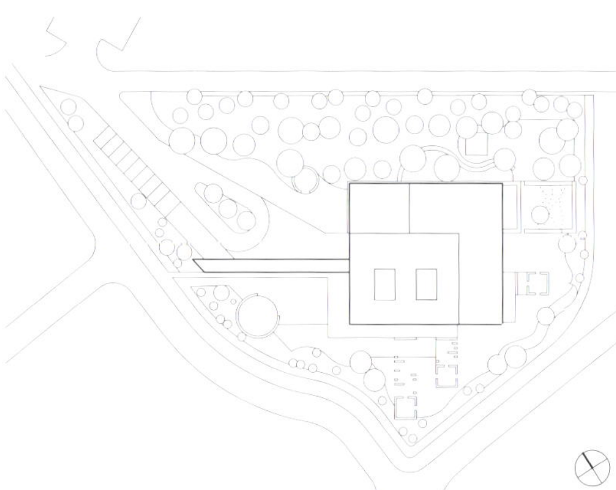

B 21.01
B 21.02

the outdoor area, and leads visitors from the antespace into the generous and bright centre of the building, the so-called piazza. This central space forms a common area and links access between the adjacent functional areas and group rooms (Figs. B 21.03, 07 and 08, p. 160).

In its northern corner, the rectangular volume of the child care facility comprises two stories. Personnel-related areas are located on the second floor. All areas for children are on the ground floor, which features spatial variety due to differing ceiling heights (Fig. B 21.04). Areas with public character such as entrance and wardrobe feature greater ceiling heights than the play or group areas, occasionally revealed on the exterior by the building's geometry. Transparent partitions instead of solid walls and large, triple-glazed windows with insulated wood frames and thermally separated spacers in the south-western facades offer views towards the exterior. Maximising passive solar energy gains via the envelope, form, and orientation of the building was already a conceptual precondition. The floor layout was optimised during construction to improve the compact character of the building with simultaneous increase of daylight use.

CONSTRUCTION AND INSULATION Construction is based on prefabricated, highly insulated wood frame components meeting passive house standards. The 40 cm strong facade is insulated by 24 cm of polyurethane infilled in stud walls and 10 cm of mineral fibre in the external composite thermal insulation, resulting in a heat transfer coefficient of 0.135 W/m²K. The wood beam roof onstruction is also prefabricated. It is infilled with 40 cm of polyurethane insulation and covered with tapered rigid insulation with a combined depth of 50 cm and a heat transfer coefficient U-value of 0.09 W/m²K. With 18 cm of polyurethane rigid insulation above a reinforced concrete floor slab and 10 cm insulation along the perimeter as well as below slab, the ground floor heat transfer coefficient is 0.1 W/m²K.

Prefabrication of wood construction components results in excellent workmanship and reduced construction time of only five months. The airtight envelope was subjected to a blower door test, indicating an n_{50} value of 0.18/h.

ENERGY EFFICIENCY The particular roof geometry enables natural daylight intake in the central and inner areas via skylights facing north without excessive heat intake. Combined with skylight-integrated, highly efficient lighting controlled according to daylight and sunlight deflection, this leads to an optimised lighting concept according to energy and user demands. Movement sensors and light meters reduce artificial lighting demands. An exterior sunscreen with an shading value of 0.2 and fixed, angled Retro blinds installed in the gap between skylight panes help prevent excess heat intake (Fig. B 21.05, S. 160). Installing the blinds in the gap between panes protects them from any manipulation and prevents heat from entering the building.

All rooms receive fresh air via a central passive house ventilation system through vents that are discreetly integrated into walls, wardrobe cabinetry, inbuilt furniture, and floors. Ventilation is designed for energy savings by reduced pressure loss within ducts and high efficiency ventilators with performance rates of 0.71 W/(m³/h). Exhaust air heat is recovered and used to pre-heat supply air. To reduce the required air flow to a minimum, air is supplied into and extracted from the building based on the cross flow principle. Fresh air is mainly routed into group spaces and flows into the centre of the building via cross flow vents. Air

1 Vestibule
2 Reception
3 Piazza
4 Ancillary room
5 Group room
6 Exercise room
7 Mother-and-child room
8 Wardrobe
9 Storage room
10 Kitchen
11 Utility room
12 Greenhouse
13 Rainwater storage

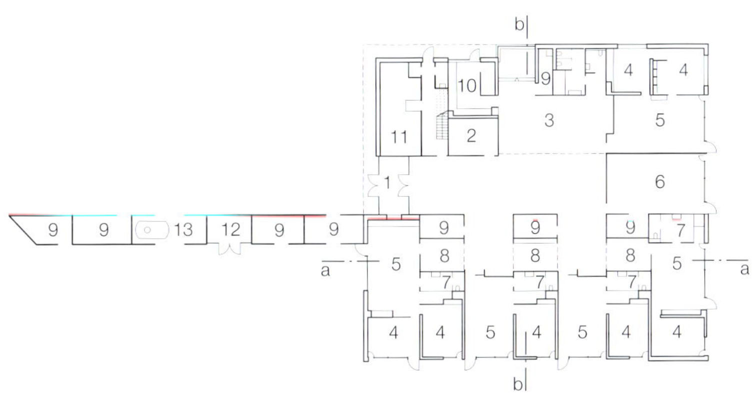

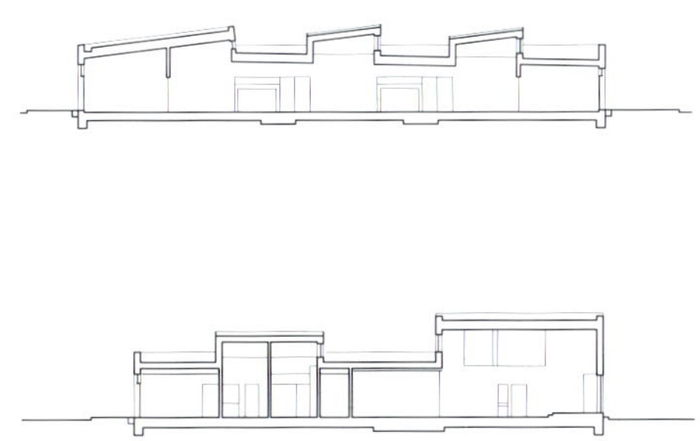

B 21.03
B 21.04

B 21.05
B 21.06

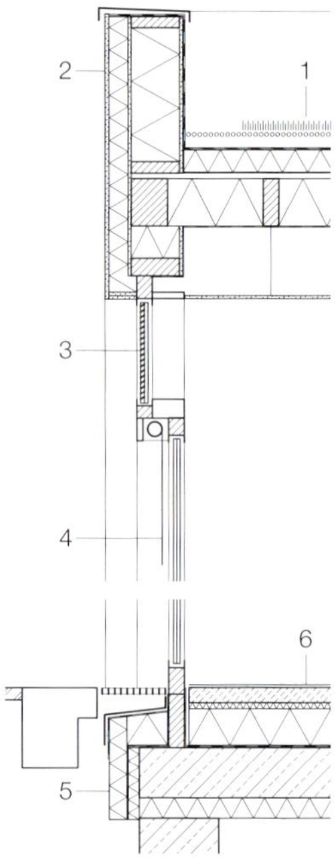

1 extensive planting 80 mm
roofing membrane
insulation 120 mm
vapour barrier
roofing 20 mm
roof trusses 240 mm
infilled insulation 240 mm
vapour barrier
acoustic tile ceiling 175 mm
2 composite thermal insulation 120 mm
OSB panel 15 mm
studs 240 mm
insulation 240 mm
OSB panel 15 mm,
polyethylene layer
gypsum board 12 mm
3 fixed glazing
reflective louvres
4 sun protection blinds
5 perimeter insulation 150 mm
6 flooring 10 mm
cement screed 75 mm
polyethylene vapour barrier
impact sound insulation 30 mm
thermal insulation 180 mm
sealant layer, floor slab 250 mm
perimeter insulation 100 mm

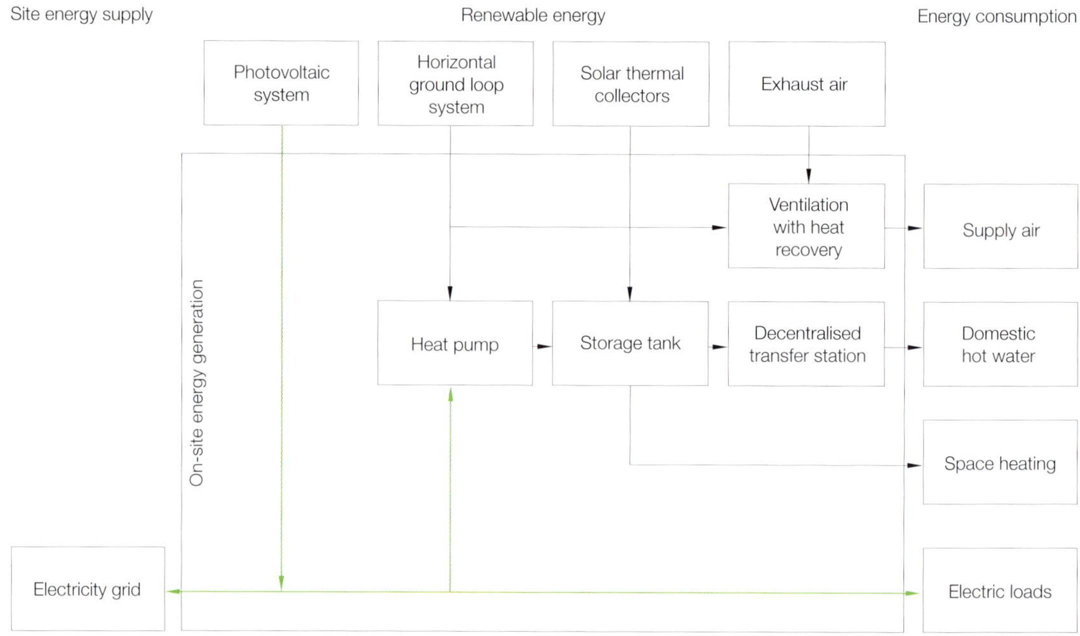

B 21.07
B 21.08

B 21.09

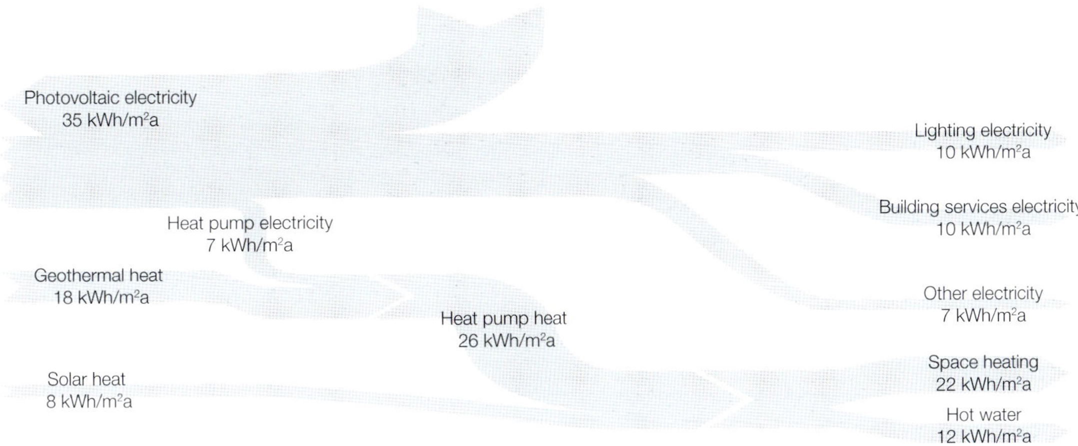

B 21.05 Facade section, scale 1:50
B 21.06 Technical schematic of energy provision
B 21.07 Entrance area to central room

is extracted via vents or restrooms and storage rooms and discharged to the exterior. This ensures that all areas of the building receive adequate fresh air even at low supply air volumes, which reduces the energy consumed by the building ventilation system. Considerably higher air flows would have been required if a standard air supply and return system had been installed for all rooms. In winter, a heat exchanger in the area of the geothermal probe prevents freezing of the ventilation system. Brine originating in the geothermal area is used indirectly to de-ice the heat exchanger.

To avoid fire protection problems, the building's single fire zone is subdivided into smoke zones. Automatic smoke control dampers prevent smoke distribution in case of a fire. The pressure drop across the cross flow opening elements is between 5 and 8 Pa to ensure that the air flows only through the cross flow opening elements and not through gaps in the doorways or other leakage areas.

B 21.08 Central room used as common area
B 21.09 Sankey diagram with planning data for electricity (top) and heat (bottom)

ENERGY SUPPLY Heat is supplied by a 28-kW_{th} brine/water heat pump that uses the geothermal potential on-site via four vertical heat loops reaching 98 m deep. Solar heat is fed in at high temperatures by 22 m^2 of solar collectors. The roof-mount individual vacuum tube solar collectors are positioned and arranged to maximise coverage of space heating according to day care centre demands except for vacation periods. The solar thermal system feeds one of two 1000 l buffer storage tanks located in the utility room within the insulated building envelope. Its priority is to heat tap water, but it also serves to support space heating in winter. When solar yield is insufficient, the heat pump performs this task at temperatures of 50 °C. Decentralised domestic water heaters heat water. These use heat exchangers and operate similarly to tankless water heaters; they are fed from the storage tank and are located in each restroom. At low storage tank water temperatures of 50 °C and reduced recirculation times, this provides required legionella protection and improves efficiency of the solar thermal system. An underfloor heating system is used to heat rooms. Low pre-heating temperatures of approximately 25 °C in the heating system along with vertical ground loops specified for high temperature output improve the heat pump's annual performance rate. In summer, the underfloor heating system can be used to provide cooling with a heat exchanger via the vertical ground loops if required.

The photovoltaic system consists of modules of multi-crystalline cells covering an area of 344 m^2 rated at 49 kW_p. It is embedded in the complex roof geometry and hidden from views at street level. Electricity is fed into the grid and covers total electric power consumption of heat pump, lighting, auxiliary systems, office equipment, and kitchen (Fig. B 21.06).

ENERGY BALANCE The calculated annual energy demand of the building for space heating, ventilation, lighting, hot water, and operation of all electric equipment is 61 kWh/m^2a (34 kWh/m^2a of heat energy for space heating and hot water and 27 kWh/m^2a for electricity). Following the first twelve months of service and research conducted by Bayer MaterialScience, the observation was made in November 2009 that metered consumption was more than 10 % lower than predicted by simulations. However, photovoltaic system yield exceeded forecasts by more than 20 %. Thus, the goals of covering 100 % of building energy needs by regenerative sources and creating a carbon neutral building were exceeded substantially.
The annual emission balance indicates savings of nearly fifty tonnes of CO_2, nine tonnes more than expected (Figs. B 21.09, p. 161, 10 and 11).

OVERALL EXPERIENCE The children and personnel state that they experience the building in a very positive way. Interior quality and indirect sunlight create a pleasant atmosphere. Building use is straightforward and intuitive. Especially temperature levels in the building are described as very comfortable. This is in part due to gentle, draft-free air movement, as well as temperatures of surfaces that convey residual heat into rooms almost unnoticeably via minor temperature differ-

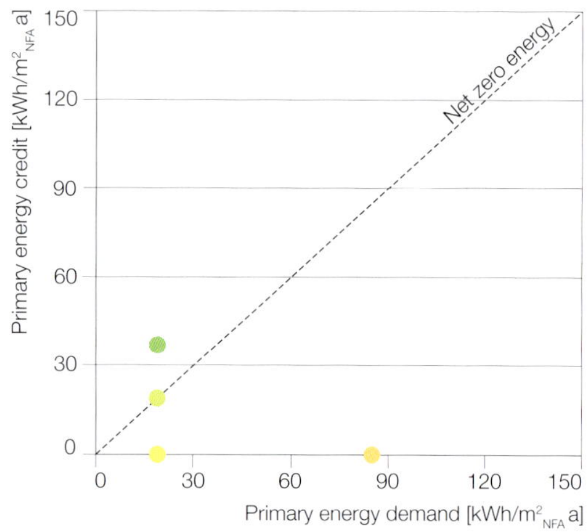

The Net-ZEB-18 Standard comprises the following aspects:

- Annual total primary energy consumption for all loads (80 kWh/m²a)
- Self-demand covered by monthly creditable energy yield (62 kWh/m²a)
- Residual consumption balance by monthly surplus
- Annual energy surplus, public grid feed-in (18 kWh/m²)

Low consumption results from the building's efficiency and predominant daytime use. Consumption can be more than offset in only eight months by yields from the large photo-voltaic system.

Primary energy factors as per DIN 18 599 (see Fig. A 2.07, p. 31)

ences and heat floors on which the children often play.
During planning, the Monheim day care centre project was awarded the Energy-Optimised Construction 2009 Architecture with Energy (Energieoptimiertes Bauen 2009 – Architektur mit Energie) prize by the Federal Ministry of Economics and Technology. After completion, it received the Design Excellence Award 2010 (Auszeichnung guter Bauten 2010) from the German Association of Architects (BDA) and the 2011 Green Building Award sponsored by the EU.

B 21.10 Energetic characterisation
B 21.11 Diagram of monthly balance, primary energy
B 21.12 Building and energy parameters (values refer to net floor area, NFA)

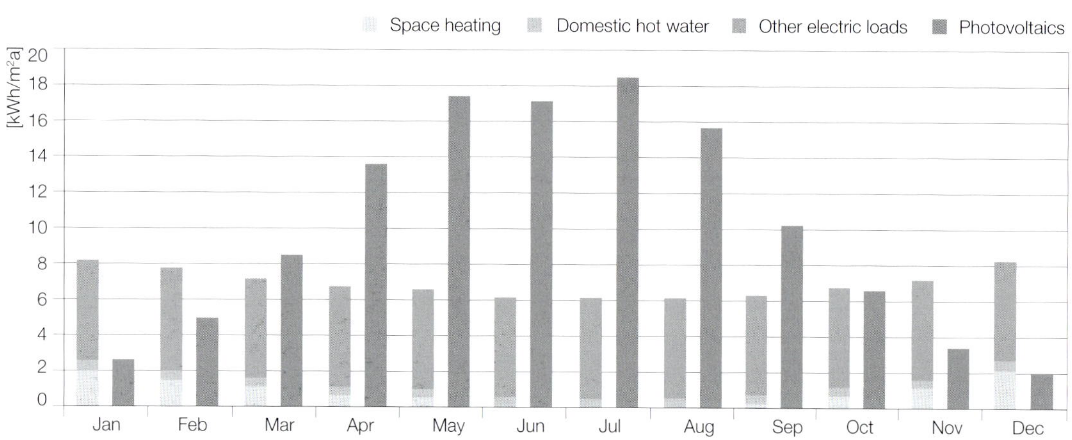

B 21.10
B 21.11
B 21.12

SITE	Monheim (D)
Annual global radiation at site	900 kWh/m²a
Annual mean temperature at site	10.0 °C
Urban ontext	suburban

BUILDING ENVELOPE PARAMETERS	W/m²K
U-value, exterior walls	0.14
U-value, windows (including frames)	0.94
U-value, roof	0.10
U-value, skylights (including frames)0.94	
U-value, ground floor slab	0.09
Mean U-value, building envelope	0.15

BUILDING EQUIPMENT PARAMETERS	
Solar collector area	22 m²
Area per m²	0.02 m²/m²
Thermal storage tank volume	1000 L
Storage tank volume per m²	1,00 L/m²
Photovoltaic system area	344 m²
System area per m²	0.34 m²/m²
Photovoltaic capacity	49 kW_p
Capacity per m²	50.60 W_p/m^2

GRID INFRASTRUCTURE AND ENERGY SOURCES	
Supply infrastructure	electricity grid
Energy source supply	electricity
Feed-in infrastructure	electricity grid
Feed-in energy source	electricity

DESIGN STRATEGIES, CONCEPTUAL FOCUS
Passive house components, mechanical ventilation with heat recovery, photovoltaics, solar thermal energy, geothermal heat pump, daylight optimisation

BUILDING PARAMETERS	
Net floor area, NFA	969 m²
Gross floor area, GFA	1218 m²
Gross volume, V	5105 m³
Building envelope, A	2877 m²
Surface to volume ratio, A/V	0.56 m²/m³
Building costs (net, construction/ technical systems)	2,200 €/m² (2009)
Number of units	1
Total number of users	71 (60 children + 11 staff)

CONSUMPTION PARAMETERS (2010)	kWh/m²a
Space heating consumption	16
Water heating demand	8
Site energy for heat (incl. hot water)	6
Electricity demand	25
Total primary energy consumption	80
Total primary energy generation (simulation)	116

This elementary school in Hohen Neuendorf features three wings with classrooms and a multi-use gym and was completed in 2011. It was built as part of the town's response to the demand for elementary school admission produced by the continuous growth of this region on the northern periphery of Berlin.
The energy concept of the 7400 m² building is based on the passive house concept and the use of locally available regenerative energy sources to cover demands. By means of biomass combined heat and power as well as photovoltaic arrays, both the calculated primary energy balance and the emissions balance are equalized, if consumption by the building services alone is taken into consideration. The school is, thus, one of the first in Germany to be built as a energy plus building.

DEVELOPMENT, DESIGN, AND STAKEHOLDERS
The stated aim of Hohen Neuendorf is to be the "green dot on top of the 'i' in Berlin". Consequently, as the client of the new elementary school in the district of Niederheide, a main concern of the municipal authorities was to implement top standards in terms of sustainability of construction and operation, while keeping in mind reasonable investment and building operation costs. The visible integration of the solar electricity plant as well as the sun protection and ventilation elements also turn this energy plus building into a visible landmark.

ELEMENTARY SCHOOL

Hohen Neuendorf, D 2011

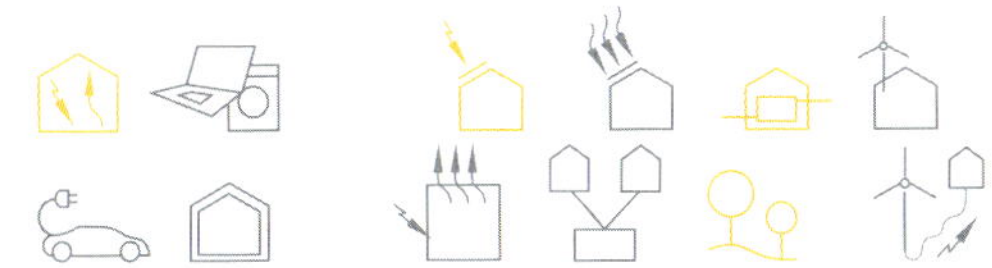

Client: Stadt Hohen Neuendorf
Architect: IBUS Architekten und Ingenieure
Prof. Ingo Lütkemeyer, Dr. Gustav Hillmann, Hans-Martin Schmid, Berlin/Bremen
Energy consultant: IBUS Architekten/BLS Energieplan, Berlin
Building services: BLS Energieplan, Berlin
Monitoring: sol·id·ar planungswerkstatt Berlin
HTW/IB, Prof. Dr. Friedrich Sick, Berlin
Main stakeholders: architect and energy consultant

B 22.01 View from south-west, before completion
B 22.02 Ground floor plan, scale 1:1500

1 Classroom with flex-room/group room
2 Special room
3 Teachers' room/administration
4 Restroom
5 Assembly hall
6 Kitchen
7 Sports hall
8 Equipment storage

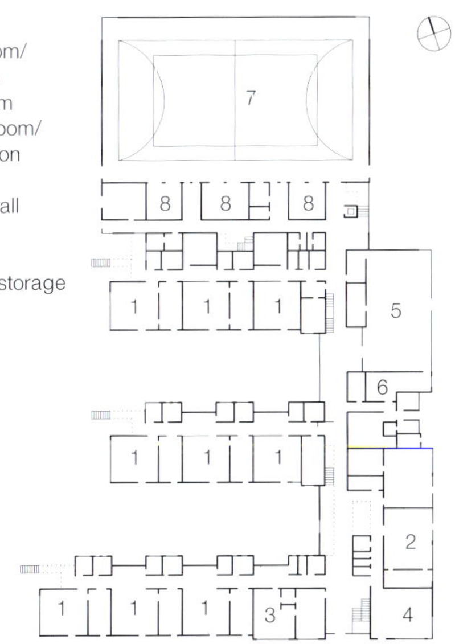

B 22.01
B 22.02

1 Classroom
2 Quiet zone or flex-/group room
3 Wardrobe/locker room
4 WC
5 Direct lighting
6 Indirect lighting
7 Daylight deflection/ glare protection
8 Supply air
9 Exhaust air
10 Natural ventilation

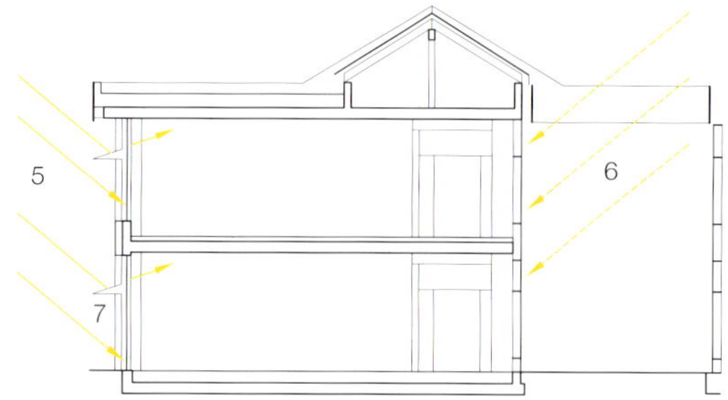

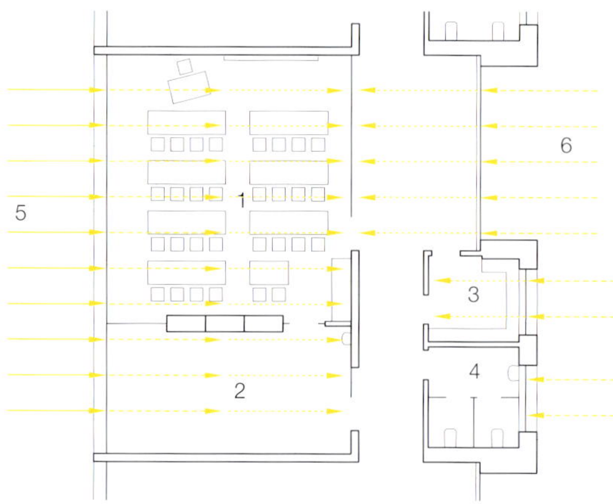

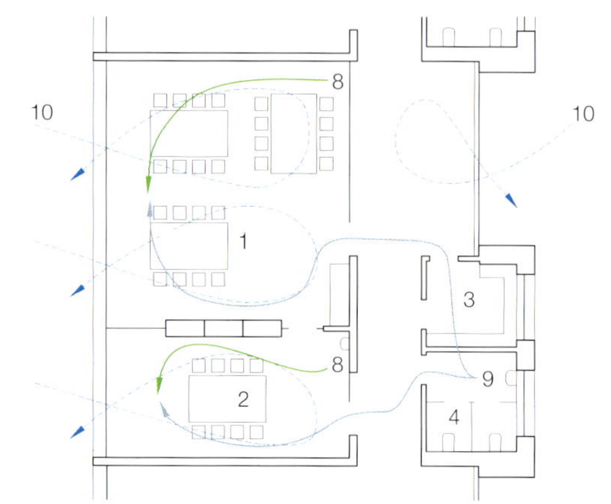

B 22.03
B 22.04
B 22.05

Thus, in the course of planning, the requirements for a project in the framework of the subsidy concept "energy optimised construction" (EnOB) of the German federal Ministry for Economics and Technology (BMWi) were achieved. Subsidies are granted for particularly innovative building measures and for monitoring over several years.

ARCHITECTURE The volume and form of the school complex are derived from its functional and technical requirements. The site forms part of an area of sports fields from which the school separates itself by a two-story wing. (Figs. B 22.01 and 02, p. 163). This wing with its long central, light-flooded hall forms the spine of the school. From this spine the school extends westwards, and the various functions are laid out like the teeth of a comb oriented towards the schoolyard. All the different functional zones of the school can be reached directly from the hall: the library, the special classrooms, and the assembly hall, which is a multi-purpose space (dining hall), as well as the relevant ancillary rooms are located on its eastern side. The classrooms are organised according to school years in three wings that project from the western side of the central hall. The administration is directly beside the main entrance at the southern end of the hall. The teachers' room is above it on the upper floor. The teaching areas in the classroom wings face south and are mostly organised along a single-loaded corridor system that allows daylight to enter from two sides, due to the partly glazed internal walls. The corridor zones and ancillary spaces face north.
For visitors to the school, the circulation hall provides access to the assembly hall and library beyond class hours. Until all the wings of the school are occupied, a neighbouring creche is permitted to use a number of rooms.
The sports hall with its ancillary spaces adjoins the third projecting classroom wing and closes off the volume towards the north. It has its own entrance and circulation area, allowing it also to be used beyond school hours, for instance by associations and clubs. The actual field, covering 1200 m² and divisible into three areas, and the equipment stores and service rooms are at ground floor level. The locker rooms and sanitary facilities are on the upper floor.

CONSTRUCTION AND INSULATION The elementary education concept that provides the basis for this building no longer defines teaching in its classic form, with the teacher in front of the class, but encourages self-reliance, alternative teaching scenarios, and groups of children of different ages. Consequently, the spaces must be adaptable and flexible to allow project work and the use of different media. "Home areas" developed for this purpose form a cluster consisting of a classroom, an attached "flex" or group room, and an individual corridor area with wardrobe and restrooms. These home areas allow children to identify with their own, reasonably sized space. All the teaching and ancillary rooms are

B 22.03 Section, home area, scale 1:250
B 22.04 Use of daylight, home area, scale 1:250
B 22.05 Ventilation concept, home area, scale 1:250

characterised by transparency and openness, can be individually designed, and encourage a sense of responsibility among the pupils for their area (Fig. B 22.04).
The solid load-bearing walls, like the roofs and floor slabs, consist of reinforced concrete. The double layer facade has an external brick skin and mineral wool insulation (20 cm) and achieves a U-value of 0.15 W/m²K. In order to minimise thermal bridges, vacuum insulation panels were used in the parapets below the windows in the south-facing classrooms and the east-facing special teaching rooms. The roof features 35 cm of mineral wool insulation under the extensive roof planting. In the ground floor slab 8 cm of insulation above the slab and additional rigid insulation beneath it achieve a U-value of 0.10 W/m²K. The highly insulated construction largely omits thermal bridges. The detailing is aimed at a high level of air tightness within the framework of the requirements for passive house subsidies. The windows comprise triple glazing in wood-aluminium frames and achieve a U-value of less than 0.8 W/m²K.

ENERGY EFFICIENCY The technical concept is directly tied to the spatial configuration. Good natural lighting in the classrooms for different teaching situations is made possible by the fact that light can enter from several sides. (Figs. B 22.03 and 04). Specific sun protection and daylight systems respond to the different orientations and the internal requirements of different teaching methods used. These systems include nanogel glazing that scatters light or electrochromic glazing that directs light and the transparency of which can manually altered.
The classrooms have a light control system that covers the entire space, with bands of light fittings running parallel to the windows and light sensors distributed throughout the space. This allows compensating the reduction in the amount of daylight, due to increasing distance from the windows, by minimal use of light and energy. As result, the lighting intensity in the entire room can be kept at a uniform level.
In the circulation areas, LED lights provide demand-driven basic and orientation lighting. If the necessary lighting level cannot be maintained by the LEDs alone, conventional and dimmable fluorescent strip lights are switched on by means of presence and light sensors. The light fixtures used for emergency lighting are all fitted with LEDs. Under current market conditions, this combination achieves an optimum cost-benefit ratio at the highest level of energy efficiency, while also minimising thermal loads in the summer months.

BUILDING SERVICES Each home area has a decentralised ventilation plant with heat recovery (82 %). This way, air transport routes and pressure losses can be kept to a minimum. Fresh air is provided in the classrooms, flows across the corridor areas, and is extracted in the wardrobes and restrooms. This means that the mechanical exhaust system and the amount of air that is also required in the restrooms can be used more efficiently. If necessary, the ventilation system can be switched to a second, higher setting. Ventilation appliances with double ventilators were planned so that the motors work with maximum efficiency at all settings.
In addition to the mechanical basic ventilation, free ventilation by means of automated vents extending up to ceiling height was also planned. The height of the vents has a positive influence on the flow profile, as warm stale air can pass through the upper area and, at the same time, cool fresh air can flow in through the lower area. These vents are suitable for short ventilation bursts during breaks from lessons, when they are automatically opened. Both forms of ventilation are combined in different ways according the demands of the users and the outdoor air conditions; they are centrally controlled and can also react to local demands. In addition, users can influence the indoor climate at any time by manually opening windows.
The automated openable windows (with protection against weather and intrusion), combined with the sizable storage mass of the building, enable natural night-time ventilation to counter overheating in summer without the use electrical energy. To maintain the storage effect of the floor and ceiling slabs, yet still enable soundproofing, broadband absorbers are used only along wall surfaces.

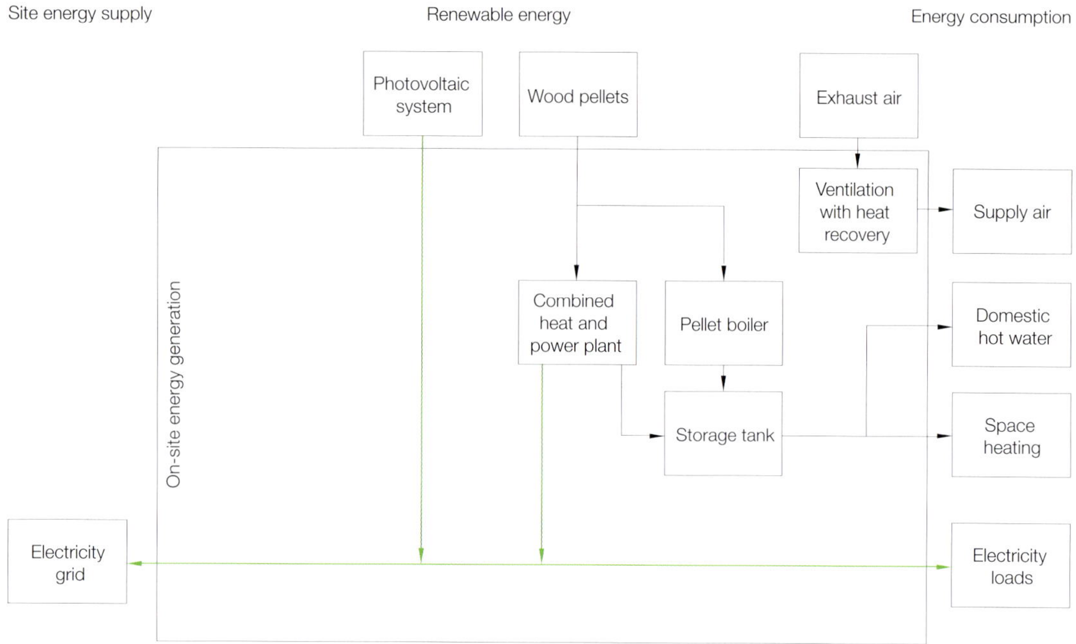

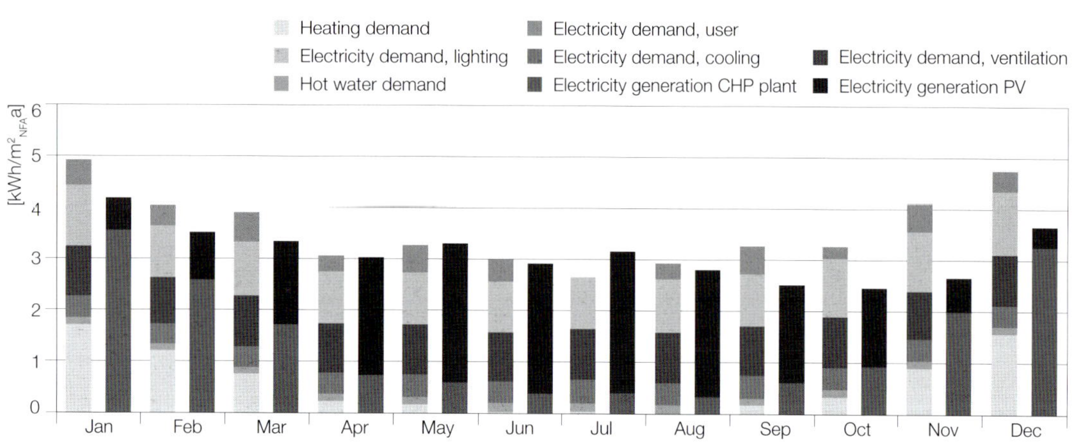

B 22.06
B 22.07

ENERGY SUPPLY In creating the energy concept the use of "slim building services" played a major role. The focused and economic use of building services technology reduces the life cycle costs of utilities and reduces the building's energy demands.
To achieve carbon neutrality and to meet the goals of the energy plus concept, the energy supply is based on sustainable generation of energy by renewable resources and the use of solar energy. A 220-kW_{th} wood pellet boiler covers the building's main heating load. A 10-kW_{th} pellet CHP (1.5 kW_{th}) was planned to assist the provision of hot water and to compensate for circulation losses. Like the 55-kW_p photovoltaic array on the pitched roofs, it generates power to cover self-demand and to be fed into the national grid (Fig. B 22.06).

ENERGY BALANCE The generation of electrical energy by the pellet CHP and the photovoltaic arrays is intended to offset the primary energy needs of the school and to produce more energy in the primary energy balance than is consumed. During planning, only the energy flows described in the Energy Saving Directive EnEV 2009 were used. Energy demands include heating and auxiliary energy for space heating and domestic hot water, auxiliary power for ventilation and cooling, as well as power for lighting. The building envelope is defined as the boundary for drawing up the balance. The balance includes the amounts of energy that flow across the balance boundary within a year. As a regenerative fuel wood, with its low primary energy factor of 0.2, lowers the total primary energy demands for space heating and hot water as well as ventilation, cooling and lighting (conversion factor 2.6) to just 39 $kWh/m^2_{NFA}a$ (without electricity feed-in from the CHP) and results in low CO_2-emissions of 9 $kg/m^2_{NFA}a$. The credits include the electricity feed from the photovoltaic array (21 $kWh_{prim}/m^2_{NFA}a$) and from the combined heat and power unit (18 $kWh_{prim}/m^2_{NFA}a$). The CO_2 equivalent credits amount to almost 10 kg/m^2a. Along with an assumed use-specific power requirement of 5 $kWh_{prim}/m^2_{NFA}a$ (not balanced in the design), a shortfall of about 6 $kWh_{prim}/m^2_{NFA}a$ or 2 kg $CO_2/m^2_{NFA}a$ remains. If energy use is not reduced, this requires the generation of additional solar power amounting to ca. 16 kW_p (Figs. B 22.07 and 08).

LESSONS LEARNED When the project was in its beginning stages, the intention was to carry out the calculation of energy requirements according to the German calculation method DIN V 18599 with the aid of a standard EnEV calculation software. However, tests of various software solutions produced strongly fluctuating results, particularly with regard to the wood pellet CHP unit. On this account, the energy requirements were calculated by using simulations for particular zones that were then applied to the entire building by extrapolation of calculated data.

As the building is scheduled for completion in 2011, data on experiences with the building in operation aren't yet available. A two-year monitoring phase is planned to enable optimisation of the building. The data for the balance will be recorded and evaluated by the Hochschule für Wirtschaft und Technik Berlin. The results will later be made available on the website of the federal ministry's EnOB subsidy concept.

B 22.06 Technical schematic of energy provision
B 22.07 Diagram with (monthly) demand and generation figures (synthesised)
B 22.08 Energy evaluation
B 22.09 Building and energy parameters (values refer to net floor area, NFA)

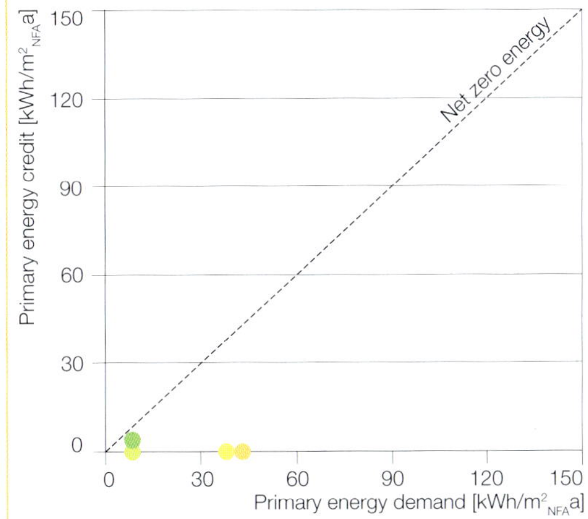

It was not possible to completely equalise all consumptions. In conceptual terms, user-specific consumers aren't intended for inclusion in the energy balance.

- Calculated annual total primary energy consumption (44 kWh/m^2a)
- Building-specific primary energy consumption (39 kWh/m^2a)
- Coverage of own needs by monthly chargeable yields (38 kWh/m^2a)
- Additional generation of energy (1 kWh/m^2)

The large number of monthly chargeable yields is due to the combination of solar electricity array and CHP as well as the conceptual dimensioning of the energy generators solely for balancing the building-specific consumption figures. Primary energy values according to DIN 18599 (Fig. A 2.07, p. 31)

SITE	Hohen Neuendorf (D)
Annual global radiation at site	1000 kWh/m^2a
Annual mean temperature at site	9.3 °C
Context	suburban
BUILDING ENVELOPE QUALITIES	W/m^2K
U-value, exterior walls	0.14
U-value, windows (including frames)	0.80
U-value, roof surface	0.11
U-value, skylights (including frames)	0.80
U-value, floor slab	0.10
Mean U-value, building envelope	0.20
BUILDING EQUIPMENT PARAMETERS	
Photovoltaic system area	412 m^2
System area per m^2	0.06 m^2/m^2
Photovoltaic capacity	55 kW_p
Capacity per m^2	8,40 W_p/m^2
Thermal storage volume	4300 l
Storage volume per m^2	0,66 l/m^2
CHP capacity	10 kW_{th}
Capacity per m^2	1,50 W_{th}/m^2
GRID INFRASTRUCTURE AND ENERGY SOURCES	
Supply infrastructure	electricity grid, delivery of wood pellets
Energy source supply	wood pellets, electricity
Feed-in infrastructure	electricity grid
Feed-in energy source	electricity

DESIGN STRATEGIES, CONCEPTUAL FOCUS
Passive house components, micro CHP with use of biomass, photovoltaics, hybrid ventilation concept, optimisation of daylight intake

BUILDING PARAMETERS	
Net floor area, NFA	6563 m^2
Gross floor area, GFA	7414 m^2
Gross volume V	38 184 m^3
Building envelope, A	15 021 m^2
Surface to volume ratio	0.39 m^2/m^3
Building costs (net, construction/technical systems)	1252 €/m^2 (2011)
Number of units	2
Total number of users	540
DEMAND PARAMETERS (Simulations 2010)	kWh/m^2a
Heating energy demand	19
Hot water heating	6
Site energy heat (incl. hot water)	35
Electricity demand	13
Total primary energy demand	44
Total primary energy generation	39

B 22.08
B 22.09

SOLAR DECATHLON EUROPE

Madrid, E 2010

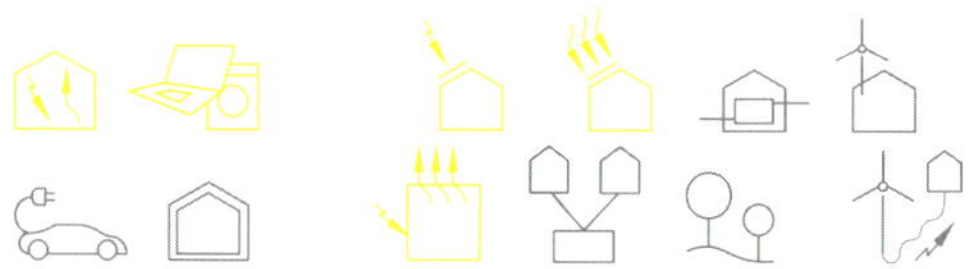

Team Virginia: Virginia Polytechnic Institute and State University, Blacksburg
Team Rosenheim: Hochschule für angewandte Wissenschaften Rosenheim
Team Stuttgart: Hochschule für Technik Stuttgart
Team Wuppertal: Bergische Universität Wuppertal
Team Berlin: Hochschule für Technik und Wirtschaft Berlin, Universität der Künste Berlin and Beuth Hochschule für Technik Berlin

To encourage the development of innovate energy concepts, in October 2002 the U.S. Department of Energy (DOE) set up a competition – the Solar Decathlon. A selection university teams from all across the globe and their solar houses compete against each other under the same climatic conditions within a practical test.
Since 2005 the Solar Decathlon is held bi-annually on the National Mall in Washington. In 2008 the starting signal was given for the Solar Decathlon Europe 2010 in Madrid. In 2012 the competition will be held again in the Spanish capital.
In integrated project teams, the students link their know-how, originating in various different disciplines ranging from architecture to civil and mechanical engineering to economics and product and media design.

SOLAR DECATHLON – THE COMPETITION

Among more than 100 applicants, 21 university teams qualified in autumn 2008 for the Solar Decathlon Europe. They planned, developed, and carried out their buildings over a project period of 18 months, firstly at their respective universities. They were then transported to Madrid, where they were erected in the space of ten days and competed against one another for a further ten days (Fig. B 23.01). As, for various reasons, four participants dropped out of the competition at an early stage, a total of 17 teams from seven different countries (China, Finland, France, UK, Germany, Spain, USA) participated with their buildings. From Germany, four university teams from Rosenheim, Stuttgart, Wuppertal, and Berlin qualified. The competition winner was the team from Virginia Polytechnic Institute & State University, USA.

Strict competition rules that precisely defined the general conditions and all other requirements guaranteed the same preliminary conditions for all participants. The solar houses were to have a maximum footprint of 74 m² and should not exceed a height of 5.50 m. Other requirements related to comfort in the interior. Under the climatic conditions that prevail in Madrid in June, the month the competition was held, an interior temperature of 23 – 25 °C with 40 – 55 % humidity was to be achieved. Designed for a two-person household with completely fitted kitchen, bathroom, and work area, the buildings weren't only supposed to meet the technical requirements of a residential house, but also high expectations regarding innovative concepts for domestic comfort.
The name Solar Decathlon refers to the ten competition disciplines in which a sum of 1000 points can be achieved (Fig. B 23.02).
The individual teams could attain these points during the competition phase firstly by presenting

B 23.01

B 23.01 Villa Solar 2010: the first European Solar Decathlon was held in Madrid.
B 23.02 Competition structure and distribution of points in the Solar Decathlon Europe 2010. A total of 1000 points are awarded in 10 different disciplines.

Category	Contest	Points	Evaluation
Architecture	Architecture	120	Jury 120
	Engineering & Construction	80	Jury 80
Solar	Solar Systems & Hot Water	80	Jury 60; Task hot water consumption 20
	Electrical Energy Balance	120	Measurement positive energy balance 75; Measurement highest solar electricity yield 10; Measurement generation/consumption 35
Comfort	Comfort Conditions	120	Measurement temperature 70; Measurement air humidity 10; Measurement air quality 5; Task lighting intensity 20; Measurement acoustics 15
	Appliances & Functioning	120	Measurement fridge/deep freezer 10; Task doing laundry/ washing dishes 55; Task entertainment technology 10; Task cooking 35; Task dinner 10
Social and Economy	Communication & Social Awareness	80	Jury 55; Task presentation of the project 25
	Industrialisation & Market Viability	80	Jury 80
Strategy	Innovation	80	Jury 80
	Sustainability	120	Jury 120

B 23.02

their concept to an interdisciplinary jury of experts, and then by meeting energy and interior climate parameters defined in the regulations and measured on site. The demands by far exceed essential requirements for architectural and technical solutions, since significant emphasis was given to questions on logistics, economics, and sociocultural aspects. During the ten day competition phase metering devices recorded all relevant data of a simulated average two-person household. All energy consumption took place within a fixed scope to generate comparable consumption data. Meals were cooked daily, laundry washed, hot water provided, and home entertainment systems operated. The goal was to cover consumption to the maximum possible extent by solar electricity generated at the building itself and to feed a surplus of at least 40 kWh into the public grid.

ENERGY CONCEPT In contrast to the first competitions in which, until 2007, energy-autonomous buildings with large battery buffers were required, from 2009 onwards – and therefore also in Madrid in 2010 – the buildings were provided with a connection to the electricity grid. The requirements were changed to meet those of "net zero energy buildings". The measurements made during the competition were decisive to achieve a net energy plus balance. An equalised annual balance was confirmed via calculations.

Based on the competition location (Madrid) and the time of year (June), sun protection and efficient cooling of the buildings were decisive for formulating energy concepts. Without active cooling, the internal climate goals could not be achieved. Hot water was supplied by heat pumps, partially supported by solar collectors. Large photovoltaic arrays covered the electricity demands of the houses and generated the surplus electricity that had to be fed into the public grid.

TEAM VIRGINIA – LUMENHAUS The Lumenhaus is a residential structure capable of responding permanently to different environmental conditions and related user requirements. The optimised utilization

B 23.03 Team Virginia – Lumenhaus. The south facade with the layered, sliding facade elements and the projecting beams for "parking" the elements.
B 23.04 Team Rosenheim – IKAROS Bavaria. View form the north-west with winter garden and closed sun shades

of light is one of the main ideas in this project. Back in 2009 the team from Virginia Tech submitted this building to the US Solar Decathlon in Washington and achieved the 13th place. A year later, as the only American participant in the European competition, it emerged as the outright winner. In the architectural assessment, it shared 1st place with Aalto University Helsinki, Finland and the Universidad Politécnica de Cataluña, Spain. In the category "Communication & Social Awareness" it reached 2nd place and 3rd place in both the "Industrialisation & Market Viability" and "Comfort Conditions" disciplines.

ARCHITECTURE A steel frame comprises the basic structure for the house with its floor area of just 53 m². Its external dimensions are 13.8 by 4.7 m and it consists of a single prefabricated and mobile module, including the interior fittings (Fig. B 23.03). To make transport easier, the chassis of a flat-bed truck that was customised according to the dimensions of the module, and was physically integrated into the building. Thus, the house reflects typical American ideas on the flexibility of single family houses and comprises an approach towards a modern "mobile home".

Special features of this house include the southern and northern facades, consisting of layers of sliding elements. The internal layers are textile curtains as protection against glare and slender sliding glass doors, through which the interior can be opened up and extended into the exterior. Where improved thermal insulation is required, the next layer of translucent aerogel insulating panels (a highly porous rigid insulation) can be shifted in front of glazed elements. The sun protection system is particularly striking: it consists of perforated sheet metal panels as the outermost layer – the "Eclipsis System". These shading panels feature circular cut-outs, some of which are equipped with matching circular shapes that are fixed at different angles, creating a dynamic pattern. On the shading panel in front of the southern facade, the circular shapes are arranged to reflect light as much as possible. In contrast, the shading panel on the northern facade is more porous, so that as much diffuse light as possible can enter the house. During the day a play of light and shadow is created in the interior. Conversely, LEDs in various colours in the cavity of the insulation panels determine the external appearance of the house at night. Two projecting beams on the western side indicate the "parking position" of the sliding elements when opening the facade.
A striking symbol for the energy plus goal of this building is provided by the photovoltaic array on the flat roof, which can be tilted to the optimum angle.

ENERGY EFFICIENCY AND SUPPLY The thermal envelope in the opaque, non-sliding areas consists of polycarbonate sandwich panels (20 mm PC + 40 mm air cavity + 40 mm PC). The closed areas of wall as well as the floor and ceiling achieve U-values of 0.31 W/m²K. The U-value of the double glazed windows (north side U = 2.21; south side U = 3.00 W/m²K) can be improved by the sliding insulating panels. If the double skin panels filled with aerogel are closed, the glass facade achieves a U-value of up to 0.12 W/m²K. All movable facade elements are integrated in the building automation system, permitting automatic adaptation to changing weather conditions. If temperatures below 21 °C are measured, the insulating panels automatically slide in front of the facade. The sunshade system (fc-figure 0.4) closes in the case of direct solar intake in summer and opens in winter in order to utilise passive solar gains. If required, the sliding elements can be operated individually by smartphone and building automation. The windows of the western facade in the area of the kitchen also have electrochrome glazing that can be switched from transparent to dark blue, thus serving as sun protection. In addition to optimising daylight utilisation, only fluorescent or LED lamps are used.
A solid concrete floor with integrated heating system serves as thermal storage mass. Within the simple energy supply system, two heat pumps provide the house with heat and cooling. To pro-

B 23.04

duce hot water and to operate the underfloor heating, a reversible water-water heat pump (7.19 kW_{th}) along with a 160 l hot water tank is used. Originally, the use of geothermal probes was planned; during the competition, this was only simulated. An air-water heat pump (5.43 kW_{th}) in combination with a mechanical ventilation system with heat recovery provides heating and cooling by means of supply air.
The entire generation of electricity is by means of a photovoltaic array measuring 47 m^2 on the flat roof. 42 photovoltaic modules are mounted on adjustable frames, the angles of which can be changed for ideal orientation according to season and location. The energy yield is improved by the use of modules with so-called double HIT cells. This means that electrical energy can be generated on both the front and back of the module. The light-coloured roof surface reflects solar radiation to the underside or back of the modules. This way the energy yield on the same area can be increased by around 30%. Overall, by using the reverse side of the modules, the capacity of the system is Increased from 8 to 10.4 kW_p.

TEAM ROSENHEIM – IKAROS BAVARIA This solar house, designed according to the principles of a passive and solar energy house, consists of four modules. In the competition, this team achieved 2nd place in the overall assessment as well as three 1st places in the disciplines “Electrical Energy Balance”, “Comfort Conditions” and “Appliances and Functioning”. The IKAROS Bavaria team also won first place in the special category “Lighting Design”.

ARCHITECTURE The building features about 65 m^2 of living space and consists of four modules in highly insulated timber frame construction. A system of modular spaces with a high level of pre-fabrication was adapted to enable transportation and the limited amount of time for erecting and dismantling the building. All dimensions were based on standard truck transport sizes. The modules with integrated interior equipment and technical services achieved a prefabrication level of more than 90%. On the building site, the individual elements of the building were simply assembled. The small interior can be used efficiently due to flexible furniture systems and the use of the same floor area for a number of different domestic functions. Full-height areas of glazing open the living area to the outdoors and ensure sufficient daylight intake (Fig. B 23.04). By means of a specially developed folding system for privacy and shading the interior, residents can include or exclude the surrounding environment. The folding sun protection system is made of projecting panels that can be positioned at whatever height is required or can be lowered into the ground. According to the sun's position, the building continuously assumes a different visual appearance.

ENERGY EFFICIENCY AND SUPPLY High cooling loads are avoided through efficient insulation, the highest possible degree of air tightness, the efficient sun protection system and active ventilation of the living areas with cold recovery. By the use of vacuum insulation panels (46 mm) and an additional flax insulation (80–120 mm) infilled between the timber frame construction, the entire closed building envelope achieves U-values below 0.1 W/m^2K. To meet all requirements in terms of noise, heat, and sun protection, special triple glazing was used for the large windows on the southern and northern facades.
The remaining cooling loads are covered via passive systems: firstly through radiative cooling that employs the surface of the roof-mount photovoltaic modules, secondly by the use of phase change materials (see phase change materials, p. 19). In cool night-time hours with clear skies, the roof surface is continuously moisturized with water. Through convection, radiation exchange with the night-time sky, and evaporation the water cools and is then collected in an insulated storage tank. During the day it feeds the heating or cooling ceiling in the interior. This radiative cooling is aided by a ventilation duct beneath the building with integrated phase change material (PCM). During the day warm air from the rooms is led through the duct, transfers its heat to the PCM and then enters the interior again at a lower temperature. The PCM is re-cooled at night by means of cool outside air. Peak loads are dealt

B 23.05 Team Stuttgart – home+. View from the south-east looking towards the loggia: glass recesses connect the closed modules.
B 23.06 Exemplary technical schematic of energy provision of the Wuppertal house
B 23.07 Team Wuppertal – A House for Europe. The southern facade with solar wall – PV modules with black, monocrystalline and blue polycrystalline cells produce a dynamic pixel graphics effect.

B 23.05
B 23.06

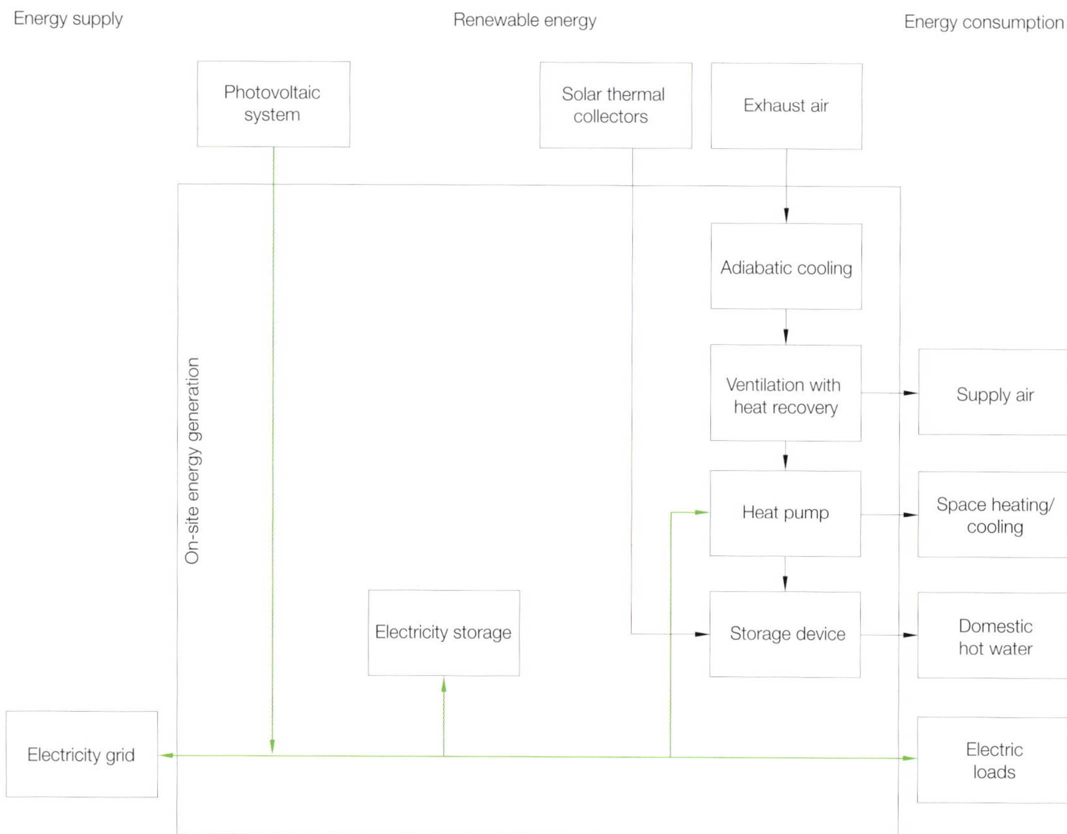

with by a reversible brine-water heat pump. The heating demand for space heating and hot water is covered by this reversible heat pump. In the summer months the waste heat produced by the cooling is used to heat hot water.
In addition to economical household appliances, particular attention was paid to energy-saving LED lighting. The entire lighting system has a connected load of only 200 W.
The solar electricity system that covers almost 70 m^2 (12.6 kW$_p$) is discretely integrated in the entire flat roof. Even at the Rosenheim location, the system produced more solar electricity in the annual balance that was consumed. As the hot water demand was covered by the heat pump, a thermal solar system was not employed.

TEAM STUTTGART – HOME+ home+ is a residential structure that employs important energy components as major design elements. In the overall assessment, the building achieved 3rd place. In the disciplines "Engineering & Construction", "Solar Systems & Hot Water", as well as "Innovation" the solar house was awarded 1st place. In the category "Appliances & Functioning" it was awarded the 2nd prize and in the category "Sustainability" the 3rd prize.

ARCHITECTURE The compact and well-insulated volume of the building with a living area of 48 m^2 is structured into four timber frame modules prefabri-

B 23.07

cated in the factory and connected by steel-glass recesses.The external appearance of the building is determined by an electricity-generating envelope of partly polycrystalline bronze and gold coloured solar cells on the eastern and western facade and on the roof (Fig. B 23.05). Towards the centre of the roof, the coloured solar cells are replaced by large and more efficient mono-crystalline photovoltaic modules. In the area of the building recesses, vacuum tube collectors are placed above the horizontal glazing. In addition to heating water, these also serve as sun protection; their angled absorber surfaces resemble a shed roof.
One of the recesses continues upwards and forms the so-called energy tower which, together with the wind and evaporative cooling, contributes to a pleasant internal climate in hot, dry regions. It applies traditional principles and building forms derived from historic models from respective regions, such as the wind towers in Arab countries (see Masdar Urban Planning Project, p. 108ff.). Its adaptation to present-day materials and technologies produces a central element that makes it possible to achieve high comfort levels with low energy consumption.

ENERGY EFFICIENCY AND SUPPLY The modules are thermally insulated by means of vacuum insulation panels (40 mm) with additional sheep wool in the area of the ceiling (total 280 mm) and floor (240 mm). Through the use of recycled polyurethane in the substructure of the insulating panels, the wall construction achieves a U-value of 0.13 W/m^2K, and the ceiling and floor construction a U-value of 0.10 W/m^2K. All windows are glazed with triple-pane, low-e glass.
On the west and east-facing areas of vertical glazing, exterior fabric sun protection finds use, while an external curtain of glass fibre material fulfils this function on the southern facade. The overhang of the terrace module in the south also provides seasonal shading.
In the interior of the energy tower, suspended fabric is evenly sprinkled with water from perforated pipes. Air that flows from top to bottom is cooled by means of evaporative cooling and then enters the interior. Warm air leaves the building through openings in the roof beside the energy tower, the so-called solar chimneys (see Masdar Urban Development Project, p. 108ff.). Only if temperatures and air humidity levels are very high, active cooling is employed by means of an adiabatic ventilation appliance with heat recovery and a reversible heat pump.

To give this lightweight building additional thermal inertia, PCM is integrated in the suspended ceiling. The heat absorbed in the interior during the day is extracted again at night by flowing cold water. To provide the cooling water required at minimum energy use, the long-wave radiation of the roof surfaces towards the cold night-time sky is utilised by leading water along the back of the PV modules, and thus cooling it. These photovoltaic thermal (PV/T) collectors were specially developed for night-time radiation cooling and cool a storage tank with a volume of 1.2 m^3 that feeds the underfloor cooling during the day and serves as a heat sink for the heat pump.
The entire opaque building envelope generates electricity by means of coloured polycrystalline PV cells in the east and west and the PV/T roof modules with mono-crystalline cells that provide the major part of the annual electricity yield (6 kW_p at the façades, 6 kW_p on the roof). Solar thermal collectors are used to provide hot water and, in addition to supplemental water heating at the subsequent location in Stuttgart, to heat the return cold storage tank as a heat source for the heat pump.

TEAM WUPPERTAL – A HOUSE FOR EUROPE

From the very start, the guiding principle of the Wuppertal project was to develop a building that is suited to any location in Europe, from north to south – a House for Europe, in other words. It is aimed, in part, at the market segment of holiday parks where vacationers can experience "living in a energy plus house" and become motivated to use innovative energy-efficient technologies at home. Minimal use of resources by both the house and its residents allows an equalised energy balance to be achieved at locations from Madrid to Copenhagen, solely by means of solar energy.

B 23.08a
B 23.08b

In the competition, the Wuppertal house was awarded 2nd prize in the category "Architecture". The team also achieved 3rd place in the category "Household Appliances & Functionality" and a 2nd place in the special category "Lighting Design". In the overall assessment, this house achieved 6th place. It also obtained the "Award for Good Buildings 2010" from the BDA Wuppertal.

ARCHITECTURE Two active solar wall panels are the main elements of this building and strikingly symbolise the architectural and energy concept (Fig. B 23.07). The interior, with a floor area of about 50 m^2, extends between these two long solar walls; by means of sliding glass elements, it can be completely opened to the east and the west and to the outdoor space. The interior is horizontally delimited by a volume that rests on the solar walls and spans the space along its entire length of 12 m without internal columns. It defines the partly two-story interior as well as a roof patio. The entire building, including the terraces, is designed to permit assembly via 34 large elements. The upper area of the building envelope has a facade of textile mesh with interior ventilation gap. In front of the sliding windows on either side of the building, curtains made of the same material serve as sun protection.
The two solar walls illustrate the spatial connection between inside and outside and integrate photovoltaic and solar thermal systems as design elements. 115 individually designed PV modules with different cell types and transparent laminate on the back define the appearance of the southern wall. The northern wall features vacuum tube collectors facing south. The dominant element in the interior is a multifunctional, accessible furniture unit that incorporates all domestic functions as well as the central elements of the building services on two levels.

ENERGY EFFICIENCY AND SUPPLY The highly insulated, air-tight building envelope consists of a wood I-beam construction with mineral wood insulation (300 mm with WLG 032). Vacuum insulation panels (80 mm) are used in the area of the free-spanning beams made of glued laminated timber. Thus, the entire opaque building envelope achieves a U-value of 0.1 W/m^2K. The sliding windows with triple low-e glazing have a U-value of around 0.8 W/m^2K including frames.
The core of the building services is a compact ventilation appliance that, in addition to providing active ventilation, also performs all necessary functions such as heating, cooling, and the provision of hot water. A heat pump with solar thermal support delivers the energy needed for space heating and hot water. By means of hydraulic changeover, it can switch from heating to cooling. The heat exchanger of this appliance has a heat recovery rate of over 80 % in winter. In summer it is used to cool supply air by means of indirect evaporative cooling (adiabatic cooling).

In the interior of the building, hollow panel components with PCM increase the thermal mass and stabilise the indoor climate. The regeneration of the PCM is by means of night-time ventilation through automatic vents. A water-based cooling and heating surface integrated in the floor supports the active heating or cooling of the air.

To minimise electricity consumption, particular attention was paid to the use of high-efficiency appliances for building services, economic household appliances, and innovative LED lighting. As an experimental approach to reducing energy consumption, a user-active illuminated ceiling was included. By means of motion sensors, the light follows the inhabitants as they move through the space, so that electricity for lighting is used only where light is actually needed.
Electricity is generated by a solar electricity plant on the roof (6.3 kW_p) and in the southern wall (3.9 kW_p). The energy management aims at covering the maximum possible proportion of the electricity consumption through the solar electricity generated in or on the building itself. In addition, the house has a battery pack with a storage capacity of 7.2 kWh that can store approximately the amount of electricity required in one day (see The Problem of Long-term Storage, p. 21f.). The battery buffer increases the annual direct coverage rate with self-generated electricity from 50 to 90 % at the Madrid location.

B 23.08 Team Berlin – Living EQUIA. View from south – the dark external skin combines wood, photovoltaics, and solar thermal energy.
a sunshades, open
b sunshades, closed
B 23.09 Energy evaluation of the four German participants

TEAM BERLIN – LIVING EQUIA The particular goals of the team Living EQUIA were the use and integration of technical innovations in a traditional central European building form. In the competition, this project achieved 10th place in the overall assessment and 2nd place in the category "Solar Systems & Hot Water". For its convincing public relations work, the Team Berlin was awarded 3rd prize in the category "Communication".

ARCHITECTURE The Berlin solar house with its pitched roof recalls the form of a traditional house. By rotating the ridge line by 12 °C, a dynamic building form is created (Fig. B 23.08). The 29 °C pitch of the roof provides the greatest possible solar electricity yields for different locations. Two light axes serve as architectural design elements to define the various functional areas in the interior. On an area of 49 m², the open living room extends upwards towards the finished underside of the roof construction. The bathroom, kitchen, and utility room are combined in a functional volume. The construction is based on a timber panel system that permits dismantling the building into individual elements for transport. Walls, floor, and roof are divided into 31 panels. While the continuous gaps of the light axes visibly separate the building parts, they are still connected, by means of two load bearing three-hinge frames made of steel profiles. All panel elements consist of a timber frame construction with wood fibre insulation (240 mm). In the interior, adobe panels cover the walls and ceiling. Additional thermal storage mass is provided by adding PCM to the adobe panels along the walls. The ceiling panels are used to activate the building components by means of an integrated duct system. The building envelope is comprised of a curtain wall facade with burnt larch cladding with interior ventilation gap. In conjunction with the solar electricity modules integrated in the roof surface with black foil reverse, this creates a uniform dark envelope that contrasts with the light interior.

ENERGY EFFICIENCY AND SUPPLY With a U-value of 0.14 W/m²K for the opaque building parts and a Uw-value of 0,75 W/m²K for the triple low-e glazing, the building envelope achieves passive house standard. Folding shutters with integrated PV cells in front of the windows on the southern and western facades efficiently combine the provision of sun protection with the generation of solar electricity. On an area of 8 m² flat collectors on the southern facade produce heat, while radiating areas on the northern surface of the roof are used to produce cooling at night and, on winter days with low solar radiation, to generate low temperature heat for the heat pump. The flat collectors feed a 450 l combi-storage tank with a stratified loading system to heat process water and support the heating system. A further water storage tank with a capacity of 1.4 m³ below the terrace is available as a cooling and buffer storage for the heat pump. The radiating area consists of 26 square metal plates, each 1 m²

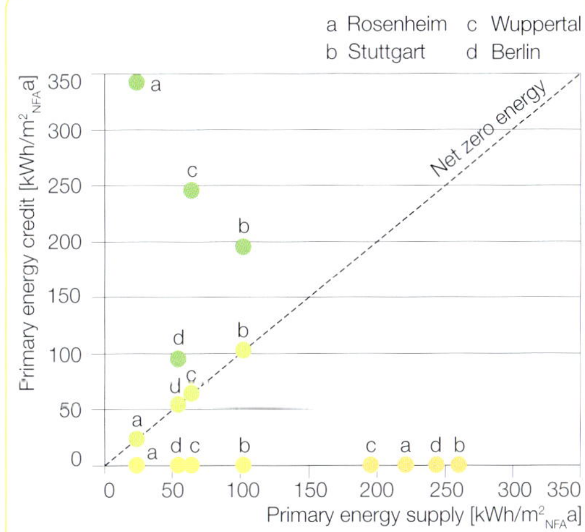

For the four German Solar Decathlon houses standards between Net-ZEB 97 and Net-ZEB 18 are achieved at their home locations:
- Calculated annual total primary energy consumption including household electricity
- Self-demand coverage by monthly chargeable yields
- Seasonal balance of remaining consumption
- Annual energy plus

The high coverage of self-consumption results from the comparatively large solar electricity plants.

B 23.09

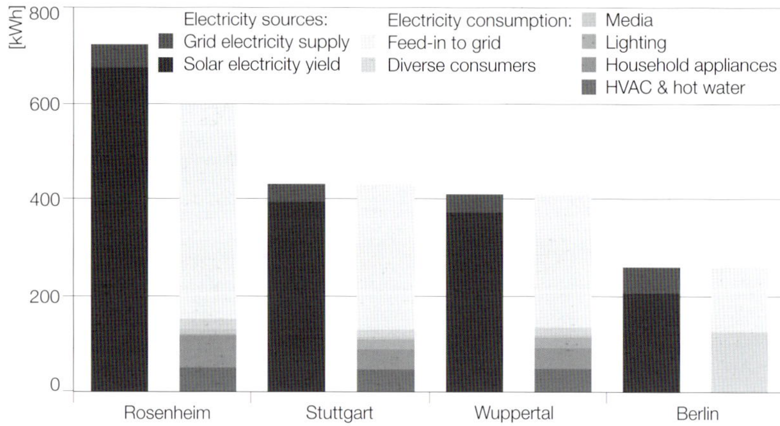

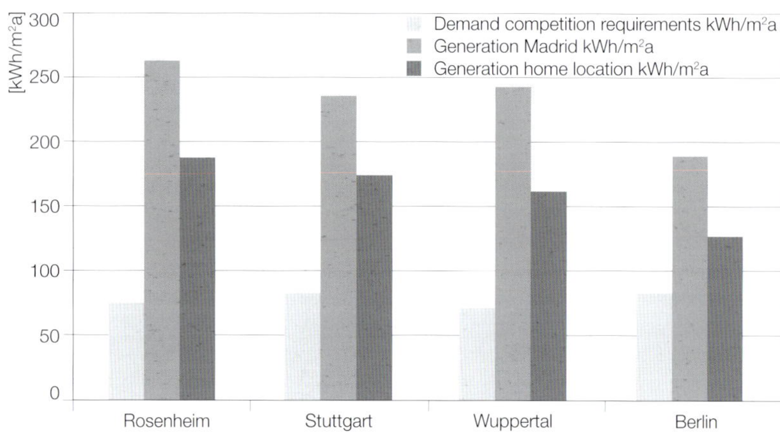

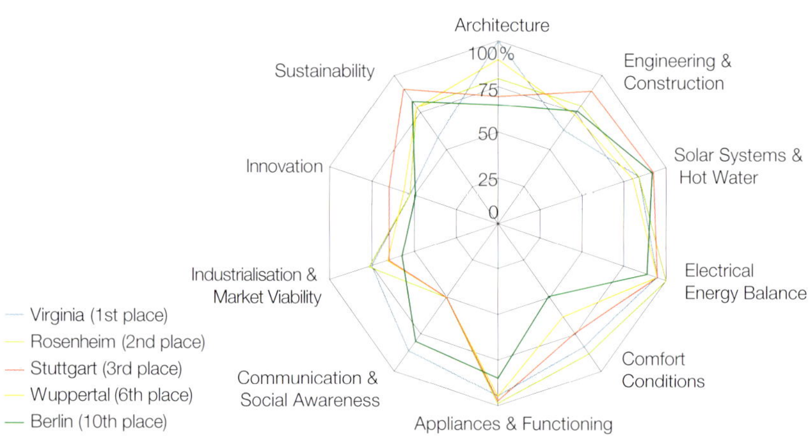

B 23.10
B 23.11
B 23.12

in area. The distribution of heat or cold is primarily by means of the adobe panels with pipework filled with water in the ceiling. For quick conditioning of the living room, there is a ventilation appliance with sorption wheel for combined heat and moisture recovery.
The solar electricity plant on the roof has an output of 4.6 kW_p. The 16 vertical PV modules of the sun-shade elements achieved a total output of 1.1 kW_p.

THE RESULTS AFTER 10 DAYS IN MADRID

Given the sunny weather and the very large solar electricity plants, the competition requirement to feed a surplus of 40 kWh into the public grid in eight days did not represent a problem for any of the teams. In relation to the living area the installed solar electricity systems were three to four times larger than is usual in single-family houses in Germany (see p. 54). With a surplus of 550 kWh after eight days the Rosenheim house achieved the highest yield, which was five times the figure achieved by the winning building at the Solar Decathlon 2009 in Washington, the house by the Technische Universität Darmstadt.
Most of the consumption of the houses occurs during times of the day when there are high solar energy yields. Only the consumption in the evening, above all for lighting, must be covered by grid electricity. The battery buffer installed by the Wuppertal and Berlin teams was not used, as the organisers decided on short notice that, as regards the temporal harmonisation of consumption and generation, only the period up to 8 p.m. was to be considered in the assessment. If the batteries had been used, these two houses could also operate without using any electricity from the grid (see Fig. A 2.19, p. 38).
In the calculated annual balance of operating energy, the four German projects achieve a clear plus at both locations, Madrid and Germany. In Madrid, an electrical energy demand of between 70 and 90 $kWh/m^2_{NFA}a$ was matched by yields of between 170 and 270 $kWh/m^2_{NFA}a$; in Germany, where the demand figures are similar, the yields are between 110 and 180 $kWh/m^2_{NFA}a$.
A profile diagram shows the strengths and weaknesses of the four German teams in comparison with the winning team from Virginia Tech, USA (Fig. B 23.12).

B 23.13

	Virginia	Rosenheim	Stuttgart	Wuppertal	Berlin
Floor area	52.7 m^2	65 m^2	48 m^2	48.5 m^2	48 m^2
Gross volume	218 m^3	250 m^3	200 m^3	290 m^3	220 m^3
Area of building envelope	280 m^2	260 m^2	206 m^2	280 m^2	296 m^2
Proportion of windows	21.3 %	13 %	23 %	25 %	13.9 %
Envelope to volume ratio	1.28	1.04	1.03	0.96	1.34
g-value windows	0.70	0.35	0.50; 0.40 (North)	0.55	0.41
τ_{vis}	no data	0.55	0.70; 0.60 (North)	0.72	–
U-value of opaque building envelope	0.31 W/m^2K	0.095–0.098 W/m^2K	0.1–0.13 W/m^2K	0.1 W/m^2K	0.1–0.14 W/m^2K
U-value of windows	2.21–3.00 W/m^2K *	0.71 W/m^2K	0.67 W/m^2K	0.8 W/m^2K	0.75 W/m^2K
Mean U-value (H_T')	0.8 W/m^2K	0.14 W/m^2K	0.2 W/m^2K	0.2 W/m^2K	0.14 W/m^2K
Electricity demand under competition conditions	no data	4500 kWh/a	3800 kWh/a	3500 kWh/a	4200 kWh/a
PV area	47 m^2	69 m^2	100.2 m^2 Roof 50.5 m^2 Facade 49.5 m^2	73 m^2 Roof 40 m^2 Facade 33 m^2	42 m^2 Roof 34 m^2 Facade 8 m^2
Installed capacity	10.4 kW_p	12.6 kW_p	12.0 kW_p Roof 6.0 kW_p Facade 6.0 kW_p	10.2 kW_p Roof 6.4 kW_p Facade 3.8 kW_p	5.69 kW_p Roof 4.59 kW_p Facade sunshade 1.1 kW_p
PV yield	Madrid 16,500 kWh/a Virginia 12,700 kWh/a	Madrid 16,000 kWh/a Rosenheim 11,000 kWh/a	Madrid 11,400 kWh/a Stuttgart 7500 kWh/a	Madrid 12,500 kWh/a Wuppertal 7000 kWh/a	Madrid 8300 kWh/a Berlin 5300 kWh/a
Battery	–	–	–	7.2 kWh	37 kWh
Solar thermal energy	–	–	6.6 m^2 Vacuum tube collectors	6 m^2 Vacuum tube collectors	8 m^2 flat-plate collectors

* With sliding aerogel-insulation panels closed U-value 0.12 W/m^2K

B 23.10 Energy balance of the four German buildings during the eight-day competition period. For the Berlin building, no breakdown of the consumption figures is available.

B 23.11 Comparison of the consumption and solar electricity generation at the competition location Madrid and the respective home location in Germany

B 23.12 Competition disciplines in comparison between the four German teams and the winning team from Virginia Tech. The relative results are depicted, whereby 100 % represents the full number of points in a discipline.

B 23.13 Comparison of the results of the four houses

PROJECT LIST

SMALL RESIDENTIAL BUILDINGS	Single-family house, Thening (A)
	Velux Sunlight House, Pressbaum (A)
	Renovation of dwelling house, Stossau (A)
	Riverdale NetZero Project, Edmonton (CDN)
	R128, Stuttgart (D)
	Haus der Zukunft, Regensburg (D)
	Single-family house, Leonberg (D)
	Zero emissions house Pola-Roid, Emmendingen (D)
	Single-family house, Pforzheim (D)
	Energy flex house, Taastrup (DK)
	SOLTAGhouse, Horsholm (DK)
	Little Greenie, Golden Bay (NZ)
LARGE RESIDENTIAL BUILDINGS	Abondance house, Montréal (CDN)
	House for two families, Zurich (CH)
	Sunny Woods, Zurich (CH)
	Berg am Laim, Munich (D)
	elementar, Tubingen (D)
	Bolig+, Aalborg (DK)
HOUSING DEVELOPMENTS	Sundays, Gleisdorf (A)
	SunnyWatt, Watt (CH)
	Eulachhof housing development, Winterthur (CH)
	Kronsberg housing development, Hannover (D)
	Messestadt Riem, Munich (D)
	Bad Aibling housing development, Bad Aibling (D)

	BALANCED ENERGY CONSUMPTION				ENERGY SUPPLY							
	Technical building services	User-specific consumption	Electro-mobility	Embodied energy	Solar electrical power	Solar-thermal power	CHP	Wind power	Heat pump	Heating grid	Biomass	Green power
www.plusenergiehaus.at	■	■			■	■			■			
www.velux.at/ueber_velux/sunlighthouse	■	■		■	■	■			■			
www.aee.or.at	■	■			■	■					■	
www.riverdalenetzero.ca	■	■			■	■			■			
www.wernersobek.com	■	■			■				■			
www.fabi-architekten.de/projekte/hausderzukunft_plus	■	■			■	■			■			
	■	■	■		■	■			■			
www.solares-bauen.de/	■	■			■	■						
	■	■			■				■			
www.energyflexhouse.dk	■	■	■		■	■			■			
www.soltag.net	■				■	■			■			
www.goldenbayhideaway.co.nz/abodes/little_greenie	■	■			■	■						
www.ecocite.ca/home.html	■	■			■	■			■			
www.kaempfen.com	■				■				■			
www.kaempfen.com	■				■	■			■			
www.nullenergieprojekt.de	■	■			■	■			■			
www.stahl-weiss.de/projekte	■	■			■	■					■	
www.boligplus.org	■	■			■	■	■					
www.aee-intec.at	■	■			■	■				■	■	
www.kaempfen.com	■				■	■			■			
www.eulachhof.ch	■				■				■	■	■	
www.cepheus.de/Kurzberichte/Kurzbericht-PI18.html	■	■					■			■		■
www.nullenergieprojekt.de	■	■			■	■				■		
www.eneff-stadt.info/de/pilotprojekte	■	■			■	■	■		■	■	■	

OFFICE BUILDINGS	Halle 8, Freiburg (D)
	Company headquarters, Wörrstadt (D)
	Solar Info Centre, Freiburg (D)
	IBA Dock, Hamburg (D)
	Company headquarters, Mainz (D)
	Elithis tower, Dijon (F)
	Green Office, Meudon (F)
	Naturalia Bau, Merano (I)
	Solar XXI, Lissabon (P)
	Acciona, Egües (E)
	National Renewable Energy Lab, Golden (USA)
	Aldo Leopold Legacy Center, Baraboo (USA)
	IdeasZ2, San Jose (USA)
PRODUCTION AND ADMINISTRATION	Solarfabrik, Freiburg (D)
	Inverter factory, Niestetal (D)
EDUCATIONAL BUILDINGS	School renovation, Schwanenstadt (A)
	Solar Academy, Niestetal (D)
	Green Lighthouse, Kopenhagen (DK)
	École Limeil, Paris (F)
	Elementary school, Lajen (I)
	Christiaan Huygens College, Eindhoven (NL)
	Oberlin College, Oberlin (USA)
RETAIL AND ACCOMMODATION	Hotel, Vienna (A)
	Hotel, Freiburg (D)
	Supermarket, Thening (A)

	BALANCED ENERGY CONSUMPTION				ENERGY SUPPLY							
	Technical building services	User-specific consumption	Electro-mobility	Embodied energy	Solar electrical power	Solar-thermal power	CHP	Wind power	Heat pump	Heating grid	Biomass	Green power
www.halle8-freiburg.de	■	■			■						■	
www.juwi.de/ueber_uns/standorte_weltweit/woerrstadt.html	■	■			■	■					■	
www.solares-bauen.de	■	■			■	■	■			■		■
www.iba-hamburg.de	■				■	■			■			
www.werner-mertz.de	■				■				■			
www.arte-charpentier.fr/en/index.htm	■	■			■							
www.green-office.fr	■	■			■							
www.naturalia-bau.it	■	■			■				■	■		
www.ineti.pt	■	■			■	■						
www.accionaenergia.es/	■				■	■						
www.nrel.gov	■	■			■	■					■	
www.aldoleopold.org/legacycenter/	■	■			■				■			
www.z2building.com	■	■			■				■			
www.solar-fabrik.de/unternehmen/firmenprofil/nullemissionsfabrik	■	■			■		■				■	
www.sma.de/de/unternehmen/co2-neutrale-produktion.html	■				■		■					■
www.hausderzukunft.at/results.html/id2761	■				■						■	
www.hhs-architekten.de/projektgalerie	■	■			■		■				■	
www.velux.com/sustainable_living/model_home_2020	■			■	■	■			■			
www.limeil-brevannes.fr	■				■	■					■	
	■	■			■	■			■			
www.rau.eu	■				■	■						
www.oberlin.edu/ajlc	■	■			■				■			
www.hotelstadthalle.at	■	■			■	■		■				■
www.solares-bauen.de	■	■			■	■		■			■	■
www.poppeprehal.at/	■				■				■			

GLOSSARY ABBREVIATIONS

ABSORPTION CHILLING MACHINE The production of cold by the use of heat energy when economical waste heat from other processes is available. Through the temperature-dependent solubility of two substances a coolant is absorbed at low temperature in a second substance in a solvent circuit and desorbed at higher temperatures.

ADIABATIC COOLING Removal of heat from the air by the evaporation of water. As a rule indirect evaporative cooling is used to avoid unnecessarily increasing the moisture content in the room air and to prevent hygiene problems. Only the exhaust air is moistened. The cooling of the intake air is carried out indirectly by means of heat exchange with the exhaust air.

AIR CHANGE Continuous replacement of room air. The air change rate n (1/h) is a unit for measuring air change. It indicates how often the air in space is replaced in one hour.

AMORTISATION The initial expenditure for a facility or plant (e.g. solar electricity plant) is offset by the yields produced over a certain time frame. This entire period is described as the amortisation period.

ANNUAL COEFFICIENT OF PERFORMANCE The annual coefficient of performance describes the relationship between the heat produced (heat pump) or absorbed (cooling machine) in relation to the energy used to operate the heat pump or cooling machine. Momentary values (performance) are presented as performance figures (see COP, p. 180).

BIPV/T Building integrated photovoltaic thermal array. Photovoltaic system that is coupled with an air duct beneath the photovoltaic modules to preheat fresh air for inside use (as a heat source, for passive heating or to preheat the domestic hot water).

BLOWER-DOOR TEST Measurement of the air-tightness of buildings as a means of ensuring quality. With windows and doors closed, a constant pressure difference between inside and outside is produced by using a fan. By measuring the volume flow produced by the blower the air change rate n_{50} (at a pressure difference of 50 Pa) is calculated. Can also be used as a diagnostic tool to identify defects by combining this process with thermography, fog, or air speed measurements (thermo-anemometer).

BUFFER STORAGE TANK Heat storage for integration in heating and cooling supply plants. It separates consumption from generation and therefore allows influence to be exerted on the installed output and the achievable coverage rates of individual generation.

BUILDING AUTOMATION Generic term for an entire range of instruments for centralised automatic control, regulation, and monitoring of the building technical services. Often equated with the software with which the building is monitored and controlled – on the management level. There are various building control systems that are manufacturer-specific and a few that are manufacturer-independent. These communicate with the automation technology in the building by means of manufacturer-specific or standardized interfaces.

BUS SYSTEM In data processing a bus is a subsystem used for the exchange of data between different hardware components. In buildings, lighting, sun shade, supervision of windows and doors, heating, air conditioning, and ventilation can be operated by such a system.

CO_2 EQUIVALENT The global warming potential – GWP or CO_2 equivalent – indicates how much a specific amount of a greenhouse gas contributes to the greenhouse effect. Conversion into an effective equivalent amount of CO_2.

COEFFICIENT OF PERFORMANCE Most important coefficient for describing the output of a heat pump or compression cooling machine. It describes the relationship between the heat given off and the electrical energy required. Depending on the system technology, different auxiliary energies are included in the balance (pumping, regulating etc.). In manufacturers' data the COP refers to a specific operating point (constant temperatures and operation at full swing), and thus represents the so-called performance figure. In measuring the consumption of energy by plants in operation in real buildings, as a result of the varying operation points the COP becomes a mean value, the work figure. While the same aggregates have the same performance figures, in use they can have very different work figures (see annual coefficient of performance, p. 181).

COMBINED HEAT AND POWER UNIT (CHP) Plants that produce both electrical current and heat. The waste heat from the production of electricity is used locally as heat. This achieves a high exploitation of the primary energy used. In the form of a micro CHP these units can also be designed for use in individual buildings. Without using the heat, CHPs cannot be run in an energy-efficient way.

COMFORT, THERMAL This term from the field of building physics describes the relationship between internal climate in a room and the well-being of individuals in this room. Air and surfaces temperature, air humidity, air movement and air quality, lighting and acoustics define the comfort area. Activity and clothing, but also the time of year, age, and gender are human-related, and therefore have a subjective effect on the sense of comfort.

COMPRESSION COOLING MACHINE Built form of a cooling machine. It uses the evaporation heat absorbed by a refrigerant in changing its aggregate state. By the removal of heat from the surroundings (cooling) the refrigerant is evaporated under low pressure. The vapour is compressed by an electrically operated condenser and condenses back into a liquid state, giving off heat in the process (re-cooling plant). The liquid refrigerant is led to a throttle member, which again reduces the pressure: a cycle is established.

CONCRETE CORE ACTIVATION Use of the mass of concrete building parts to condition a space, generally by integrating pipework filled with water. The large heat transferring surfaces make it possible to effectively use the relatively small temperature differences of natural heat sources and sinks.

DIN V 18599 This series of standards, published in early 2007, describes a calculation method for an energy-based evaluation of buildings that incorporates useful site and primary energy demands for space heating and the provision of hot water as well as for cooling and lighting. The new standard forms the basis for the Energy Pass and also serves as a calculation method for the revised Energy Saving Directive (EnEV). The project under examination is compared with a building with standard parameters.

EFFICIENCY Relationship between the useful output and the input energy (expenditure). The degree of efficiency is, thus, used to measure the efficiency of the transformation and transfer of energy. For energy generation plants it describes the relationship between useful energy output and expended energy. The difference between the energy provided and delivered is the loss.

EMBODIED ENERGY Energy required to produce, maintain, and dispose of a building over its entire period of use.

EN 13779 This European norm formulates general principles on the theme of ventilation in non-residential buildings and specifies the requirements for ventilation and cooling plants as well as space cooling systems.

EN 15251 This EU standard formulates parameters for indoor climate (air quality, temperature, light, and acoustics) to interpret and evaluate the energy efficiency of buildings. In particular, comfort classifications are defined for summer indoor climates in buildings without active cooling. These comfort classifications take into account the adaptation of the comfort area under higher outdoor temperatues. They are based on extensive user surveys in buildings.

EN 15603 European standard for the energy efficiency of buildings, the calculation of the total energy demand, and for determining specific energy values.

ENERGY CONSUMPTION [kWh/a] In contrast to energy demand energy consumption is a measured amount of energy that is consumed within a certain period of time. Therefore, the energy consumption figure takes the real climate and user behavior into account. With a standardized climate adjustment the influence of the changing outdoor climate can be neutralised. In terms of physics, however, energy cannot be consumed but only changed into another form of energy (for CH: [MJ/a]).

ENERGY DEMAND [kwh/a] Amount of energy which a building needs under defined conditions in a certain period. The energy demand is a calculated figure and is established according to a defined process. Due to influences such as weather, user behaviour, or the operation of the plant, the actual site energy consumption can differ from the calculated energy demand (for CH: [MJ/a]).

ENERGY VALUE [kWh/m², kWh/Person] Calculated or measured energy expenditure for an area or a person and a period of time, with specification of which area is meant. This book uses the net floor area throughout as a reference unit (for CH: [MJ/a]).

EPBD (ENERGY PERFORMANCE OF BUILDINGS DIRECTIVE) Guideline for member states of the European Union from 2002. It formed the basis for the introduction of a comparable method for calculating overall energy efficiency of buildings and gave rise to the creation of energy certificates for new and existing buildings. In the new version 2010 the standard "nearly zero energy buildings" was introduced as a target for 2020.

EXTENT OF COVERAGE Percentage figure that specifies how much of the necessary energy expenditure a technical plant can cover in a defined observation period.

FEED-IN REMUNERATION Monetary equivalent value for energy fed into a grid, generally electricity. The remuneration amounts are generally defined by political outline circumstances and are fixed for a certain period of time. They differ significantly between different countries and, to some extent, also regions.

GEMIS (GLOBAL EMISSION MODEL INTEGRATED SYSTEMS) Instrument for comparative analysis of environmental effects of the provision and use of energy. The GEMIS data bank contains information on the provision of fossil fuels, regenerative energies, nuclear energies, of heat, electricity, renewable raw materials, and hydrogen.

GEOTHERMAL ENERGY Heat energy stored in the earth in relation to extraction and use. It is one of the renewable energies and is generally used for heating, mostly in combination with a heat pump. Direct use is possible only with higher ground temperatures or greater depth.

GLARE PROTECTION Glare protection systems ensure that daylight is as diffuse as possible, so that sunshine, bright clouds, or reflections from light-coloured surfaces do not irritate users when in their field of vision. Because of its extreme brightness, if the sun is in the user's field of vision, it is advisable to screen it off by the use of opaque materials.

GLOBAL RADIATION [W/m²] Amount of solar energy that, related to a horizontal surface, impacts the surface of the earth. It consists of direct and diffuse radiation (scattering by clouds, water, and dust particles). Global radiation can also be given as an annual sum [kWh/m²a].

GROUND PROBE Tube with double or more sleeves which, in a process borrowed from well boring, is introduced into the ground. It serves to extract heat from the ground, generally used for heating purposes by means of a heat pump. Ground probes are suitable for accessing the ground as a summertime heat sink. For use over a considerable number of years, those systems in which the extraction of heat in summer is offset by the introduction of heat in summer are ideal.

HEAT EXCHANGER A piece of equipment with which heat energy from one medium (liquid or gaseous) can be transferred to another. In a central ventilation plant with heat recovery the heat exchanger transfers the heat that is extracted from the space along with the exhaust air to the intake air. The intake air is thus pre-warmed before it enters the space.

HEAT PUMP A device or machine that draws heat from the ground, outside air, or expelled air and transfers it to a heating system. Generally powered by electrical energy (compression heat pump).

HEAT RECOVERY Use of thermal energy that would otherwise leave the building unused through the ventilation system. By means of a heat exchanger the energy content of the waste air is used in order to condition the intake air. Heat can also be recovered from waste water (with a heat pump).

HEAT STORAGE ABILITY [Wh/m²K] Heat storage ability is calculated as the product of the specific heat storage capacity, the gross density, and the layer thickness of the building element under consideration. In short intervals of time (hours or a few days) only a fraction of the total heat storage ability of a building component becomes effective, as the heat exchange processes with the surrounding air are weak.

HEAT TRANSFER COEFFICIENT see U-value, p. 182

INTEGRATED OVERFLOW OPENING Used in the controlled ventilation of buildings. These are openings in walls or doors that connect intake air areas with exhaust air zones

INVERTER Electrical appliance used for transforming direct current (e.g. from solar electricity plants) into the alternating current generally used in buildings with minimum losses.

INSULAR SYSTEM autarkic electricity supply without any connection to a grid infrastructure.

LIFE CYCLE ANALYSIS (ECOLOGICAL BALANCE) Analysis of the environmental impact of a building during ist entire life. Subdivided into the life cycle phases new building, use, renovation, and demolition. The environmental impact includes everything of environmental relevance, ranging from the environment to emissions.

LOAD MANAGEMENT By switching energy consumers in a building on and off in a planned way the maximum output and time of use can be influenced. This has an influence on the extent of self-demand coverage that can be achieved.

MONITORING Systematic time-resolved recording, analysis and evaluation of a building's operating data with the help of a data recording system, as a rule part of the central building control technology.

NET ZERO ENERGY BUILDING A building the calculated or measured annual energy balance of which amounts to zero. Energy is produced by the building's own plants (generally in the form of solar energy plants). In contrast to energy-autarkic buildings, zero energy houses are connected to energy grids that they use to balance energy supply and demand in terms of time and output.

PASSIVE HOUSE Building concept the requirements of which were formulated by an Institute (Passivhaus – Institut Darmstadt) rather than by a public authority. Buildings in which a comfortable temperature is achieved in winter and summer without the use of a separate heating or air-conditioning system. The maximum figure for the heating power requirement is 10 W/m². Under the prevailing climatic conditions in central Europe, this produces a space heating demand of maximum 15 kWh/m²a; the upper limit for the primary energy demand including hot water and household electricity is 120 kWh/m²a. Calculated using the passive house projection package – PHPP.

PCM (PHASE CHANGE MATERIALS) store heat energy, in that a phase transition (generally from solid to liquid) is stimulated by introducing heat in a defined temperature range. Paraffins or salt hydrates are most often used. With PCMs a significant amount of heat can be stored in a small space (storage containers) or deposited in building materials (e.g. through micro-capsulation) in order to increase the heat capacity that can be activated in these spaces. If PCMs are fitted in dry-wall panels or internal plaster the thermal inertia of a space can be increased without using up much room and even if light-weight construction is employed.

PELLET HEATING The burning of wood pellets (small beads of compressed sawdust and wood chips) in domestic stoves or central heating boilers. If fed by machine, central heating systems with pellets can achieve a high degree of automation. In terms of operation and maintenance, they are comparable to gas or oil heating systems with the same capacity. Compared to other forms of heating with wood the amount of ash produced is low.

PHOTOVOLTAIC The direct transformation of solar radiation energy into electrical energy. The energy transformation takes place in solar cells that are connected in what are known as solar modules (PV modules).

PHOTOVOLTAIC-THERMAL COLLECTORS (PV/T) Combination of electrical and thermal use of solar energy in a single building element. Generally PV modules are cooled by a water circuit. The homogeneous appearance (resembles a single building element) is an advantage. The disadvantage is that only moderate temperature levels can be achieved without negatively affecting the efficiency of solar electricity generation.

PHPP (PASSIVE HOUSE PROJECT PACKAGE) see Passive House p. 184

POWER-HEAT-CHILLING COUPLING Further development of a CHP by using waste heat to operate a thermally driven cooling machine (see absorption cooling machine p. 180).

POWER-HEAT-COUPLING See CHP p. 180

PRIMARY ENERGY [J, Wh] Energy in the forms in which it is found in nature. These include primary energy sources such as coal and lignite, mineral oil and natural gas, as well as renewable energy sources. In most cases this primary energy has to be transformed, for instance by coal mining, (heating) power stations etc. into secondary energy (coke, briquettes, electricity, district heating, heating oil or petrol). The energy at the place where it is consumed is the site energy, which in turn is transformed into useful energy, heating and process heat, light or mechanical energy.

PUBLIC PRIVATE PARTNERSHIP (PPP) Collaboration, regulated by contracts, between public (e.g. authorities) and private (e.g. corporations) stakeholders or constellations of actors, in which the necessary resources (e.g. capital, production facilities, staff etc.) are provided by the partners for joint use in a common organizational context.

SANKEY DIAGRAM A specific type of energy flow diagram. The width of the arrows used is in proportion to the flow quantity. Used to depict energy supply and energy consumption.

SIA 380/4 Guideline on the use of electrical energy from the Association of Swiss Engineers and Architects.

SIA 2031 Guideline for an energy pass for buildings in Switzerland on the basis of the European energy efficiency guidelines. The SIA 3031 provides a basis for a uniform declaration of energy consumption and the associated emission of greenhouse gases by buildings.

SITE ENERGY The part of the primary energy, e.g. electricity, wood pellets, heating oil, district heating that is locally available to the end user. Transport, distribution and conversion losses are deducted.

SMART GRID Intelligent electricity grid that enables communication between consumers (in buildings), storage (batteries, pump storage facilities), and generators (power stations, but also buildings). The goals include a high degree of operational reliability and a supply with renewable energy.

SOLAR GAINS Also known as solar heat gains or passive use of solar energy. Describes the amounts of heat that, as a result of solar radiation on the transparent areas of a building, contribute to warming the interior, and thus reduce the heating demand.

SOLAR RADIATION see global radiation p. 181.

SOLAR TRACKER Single or dual axis solar energy plants. They are automatically operated and can follow the direction of solar radiation, thus increasing the energy yield per surface area.

SPACE HEATING DEMAND [kWh/a, kWh/m²a] An amount of energy established by calculations. It is the minimum amount that must be introduced to the building during the heating period in order to cover heat losses at the required indoor temperature. It is the sum of the transmission and ventilation heat losses minus solar and internal gains under physically ideal, loss-free introduction of heat to the room air.

STORAGE MASS The mass of a building that is thermally effective and that influences the thermal behaviour of the building through its thermal inertia. It can store heat and discharge it later.

SUN PROTECTION Sun-shade or protection systems that reduce the amount of solar radiation in a space, and thus prevent overheating. Their effect can be expressed in figures by means of the total energy transmission (g-value) of the facade. This g-value of the entire construction includes radiation transmission as well as the secondary release of heat to the inside.

SURFACE TO VOLUME RATIO [m⁻¹] Building physics parameter that describes the compactness of a building. The relationship is calculated as a quotient of all heat-transferring parts of the building envelope, including those in contact with the earth, and the external dimensions of the heated building volume.

TABS Thermally active building systems (see concrete core activation) utilise building mass to regulate temperature. They can be used alone or to supplement space heating or cooling.

TRANSMISSION HEAT LOSSES [W/K] Transmission heat losses arise as a result of the transport of heat through the surfaces that enclose heated spaces (roof, external walls, floor slab, basement ceiling slab), as well as through heat bridges.

TOTAL ENERGY TRANSMISSION (G-VALUE) Describes the energy transmission of windows from outside to inside. A high g-value means higher solar heat gains in winter, but in summer can lead to the overheating of a room.

U-VALUE [W/m²K] Overall heat transfer coefficient and, therefore, the unit used to measure the transfer of heat through a building element. It indicates how much energy per 1 Kelvin is transferred over one square metre of the building element.

VACUUM INSULATION The transport of heat through the gas molecules of the air can be hindered by the creation of a vacuum. As convection accounts for the major share of heat transfer, whereas heat conduction and radiation play only minor roles, systems employing this principle provide extremely good insulation performance. Standard vacuum insulation panels (VIP) insulate about ten times as efficiently as conventional mineral fibre or hard foam insulation.

VACUUM TUBE COLLECTORS An arrangement of numbers of evacuated glass tubes to form a collector module for the thermal exploitation of solar energy. Due to the extremely good insulating effect of the vacuum around the absorbers, higher temperatures and degrees of efficiency are achieved than in standard flat plate collectors. On account of the additional cost of vacuum tube collectors, flat plate collectors are often more economical to use, but require a larger area.

VENTILATION HEAT LOSSES Ventilation heat losses occur when warm room air is exchanged for cold outdoor air. The air exchange is required for reasons of hygiene so as to change the vitiated room air. As the result of leaky junctions and joints in a building, uncontrolled ventilation heat losses can considerably increase the heating demand. Controlled ventilation with heat recovery reduces ventilation heat losses.

VENTILATION PLANTS Systems for an adjustable or automatic air change that is independent of the actual wind and temperature conditions, in contrast to natural ventilation through windows. Uses at least one fan (exhaust air plant). A central ventilation plant serves a number of different rooms, an apartment, or an entire house, is located in the building services room, and the exchange of air is by means of a duct network. De-centrally located ventilation appliances ventilate only individual rooms. As a rule intake and exhaust plants are used with two fans and heat recovery. This is efficient only with low electricity consumption.

WATT-PEAK [MW_p] Maximum output of a PV module or PV array under what are called standard test conditions. In practice, these conditions typically never occur, but are laboratory figures. The maximum output levels recorded in practice are lower.

ABBREVIATIONS

A	Area of building envelope
BAFA	Federal Office of Economics and Export Control
BauO	State/Regional building regulations
CFA	Conditioned floor area
CHP	Combined heat and power (plant)
BMU	Federal Ministry for the Environment, Nature Conservation and Nuclear Safety
DHW	Domestic hot water
EEG	Renewable Energies directive
EEX	European Energy Exchange, market place for energy and energy-associated products
EnEV	Energy Saving Directive
EnWG	Energy Industry Act
EVU	Energy distribution company
GFA	Gross floor area
g-value	Overall energy transmittance
GEMIS	Global Emission Model for Integrated Systems
HVAC	Heating, ventilating and air conditioning
HW	Hot water
ISE	Fraunhofer Institute for Solar Energy Systems
IWU	Institut Wohnen und Umwelt
KfW	Kreditanstalt fur Wiederaufbau
KWKG	Cogeneration legislation
kW_p	Kilowatt peak
kW_{th}	Kilowatt thermal
MFH	Multi-family house
MW	Megawatt
MW_{el}	Megawatt electrical
MWh	Megawatt hours
MWh_{el}	Megawatt hours electrical
MWh_{th}	Megawatt hours thermal
NFA	Net floor area
PCM	Phase Change Material
PE	Primary energy
PH	Passive house
PHPP	Passive house project package
PV	Photovoltaic
R & D	Research and development
SFH	Single-family house
T_{innen}	Indoor temperature (in °C)
T_{aussen}	Outdoor temperature (in °C)
U-value	Heat transfer coefficient
V	Gross volume
VDE	Association for Electrical, Electronic & Information Technologies
ZEB	Zero energy building

INDICES

BFA	Gross floor area
el	electrical
End	End energy
f	frame
g	Glass
NFA	Net floor area
p	peak
prim	primary energy
th	thermal
w	window

SYMBOLS

A/V	[m^{-1}]	Surface area to volume ratio
A_N	[m^2]	Useful floor area in the energy saving directive
H_T	[W/m^2K]	Transmission heat loss coeff.
H_V	[W/m^2K]	Ventilation heat loss coeff.
Q	[kWh]	Heat
U	[W/m^2K]	Heat transfer coeff., U-value
U_f	[W/m^2K]	Heat transfer coeff., U-value window frames
U_g	[W/m^2K]	Heat transfer coeff., U-value glass
U_w	[W/m^2K]	Heat transfer coeff., U-value windows

LITERATURE

OVERVIEW OF PROJECTS AND THEIR CHARACTERISTICS (1)

- International Energy Agency – IEA: Solar Heating & Cooling Programme + Energy Conservation in Buildings and Community Systems Programme. www.iea-shc.org/task40/subtask/index.html. Accessed May 2, 2011
- Musall, Eike; et al.: Net Zero energy solar buildings: an overview and analysis on worldwide building projects. Tagungsband EuroSun conference Graz 2010
- Musall, Eike: Emissionslos Wohnen – Quartier "Am Müggenberg", Arnsberg-Neheim. Hamburg 2009
- www.enob.info/nullenergie. Accessed September 5, 2011
- www.buildingdata.energy.gov/ZEB

PROJECT 01 RESIDENTIAL HOUSE, RIEHEN (CH)

- Schweizer Solarpreis 2008. www.solaragentur.ch. Accessed May 5, 2011
- Förderprogramm P+D BAUTEN. www.pd-bauten.bs.ch. Accessed May 5, 2011
- Niederhäusern, Anita: Das Haus als Kraftwerk. In: Nachhaltig Bauen 01/2009, p. 6–7
- Setz-Architektur, www.setz-architektur.ch. Accessed April 29, 2011
- Familie Wenk-Furter: Wohnen im Kraftwerk. In: Riehener Zeitung. Accessed February 5, 2010
- www.renggli-holz.ch. Accessed May 2, 2011

PROJECT 02 ÉCOTERRA HOME, EASTMAN (AUS)

- Chen, YuXiang; Athienitis, Andreas; Galal, Khaled: Modeling, design and thermal performance of a BIPV/T system thermally coupled with a ventilated concrete slab in a low energy solar house: Part 1. In: Solar Energy 84/2011, p. 1892–1907
- Chen, YuXiang; Athienitis, Andreas; Galal, Khaled: Modeling, design and thermal performance of a BIPV/T system thermally coupled with a ventilated concrete slab in a low energy solar house: Part 2. In: Solar Energy 84/2011, p. 1908–1919
- Doiron, Matthew Anthony: Whole-Building Energy Analysis and Lessons Learned for a Near Net-Zero Energy Solar House. Master's Thesis. Concordia University, Montréal 2010
- Doiron, Matthew Anthony; O'Brien, William; Athienitis, Andreas: Energy Performance, Comfort and Lessons Learned From a Near Net-Zero Energy Solar House. ASHRAE Transactions, vol. 117, 02/2011(under review)

PROJECT 03 LIGHTHOUSE, WATFORD (GB)

- Building Research Establishment (BRE): The Government's Standard Assessment Procedure for Energy Rating of Dwellings. Watford 2008
- Committee on Climate Change: Climate Change Act. www.theccc.org.uk/about-the-ccc/climate-change-act. Accessed November 17, 2010.
- Communities of Local Government: Code for Sustainable Homes. Technical Guide. London 2008
- Kingspan Group: Climate for Challenge. Cavan 2008
- Kingspan Off-Site: Energy. www.kingspanlighthouse.com/energy.htm. Accessed November 17, 2010

PROJECT 04 HOME FOR LIFE, LYSTRUP (DK)

- aart architects: www.aart.dk/projects/?id=101&cat=4&view=images. Accessed May 5, 2011
- www.velux.com/sustainable_living/model_home_2020/home_for_life/default.aspx. Accessed May 5, 2011
- www.velux.com/Sustainable_living/Model_Home_2020/Press/Images_and_literature/default.aspx. Accessed May 5, 2011
- www.velux.com/modelhome2020. Accessed May 5, 2011
- Broschüre Velux Model Home 2020. 6 experiments from 2009 to 2011
- Broschüre Velux Model Home 2020. home for life
- Broschüre Velux Model Home 2020. Update No. 4, December 2009
- Broschüre Velux Model Home 2020. Update No. 5, May 2010
- Broschüre Velux Model Home 2020. Update No. 6, October 2010
- Broschüre Velux Model Home 2020. Update No. 7, December 2010
- Schoof, Jakob: Das Haus, das mehr gibt, als es verbraucht. In: Daylight and Architecture 13, 2010, p. 64–73
- IEA Task 39: Solar heating and cooling programme. Polymeric Materials for Solar Thermal Applications, »Home for life« concept house. Aarhus 2009
- www.activehouse.info/cases/home-life. Accessed May 5, 2011
- www.velfac.co.uk/Global/Home_for_life . Accessed May 5, 2011
- www.velfac.co.uk/velfac-uk/data.nsf/q/Diary?OpenDocument&Click=. Accessed May 5, 2011
- www.detail.de/artikel_bolig-for-livet_23967_De.htm. Accessed May 5, 2011
- www.alexandra.dk. Accessed May 5, 2011
- www.iha.dk. Accessed May 5, 2011
- www.sbi.dk/en/publications/programs_models/beat-2002 Accessed May 5, 2011
- Solar complete Komplettheizungssystem, Technische Daten der Wärmepumpe: www.sonnenkraft.com. Accessed May 5, 2011

PROJECT 05 KRAFTWERK B, BENNAU (CH)

- Schweizer Solarpreis 2009. www.solaragentur.ch. Accessed May 5, 2011
- Norman Forster Award 2010
- Mehrfamilienhaus in Bennau. In: Detail Green 1/2010. p. 38–41
- Piccolotto, Moreno: Synthese aus Energietechnik und Architektur. In: Detail Green 1/2010, p. 42–45
- Niederhäusern, Anita: Prunkstück an Effizienz. In: HK-Gebäudetechnik 4/2010, p. 40–43
- Piccolotto, Moreno; Vogelsang, Janine: Wohnen im Kraftwerk. In: DBZ 09/2010, p. 38–43
- grab-architekten: Ein modernes Wohnhaus mit Energie-Überschuss. In: Schweizer Holz-Zeitung Nr. 22/30, 2008, p. 16–18
- Gütermann, Andreas: Plus-Energie-Mehrfamilienhaus in Bennau. Erste Erfahrungen. Brenet-Status-Seminar 2010
- www.detail.de/artikel_kraftwerk-b-bennau-grab-architekten_24281_De.htm. Accessed May 5, 2011, Text: Cornelia Bauer/Jakob Schoof, Interview: Cornelia Bauer, Intep – Integrale Planung GmbH

- www.minergie.ch/buildings/de/details.php?gid=SZ-001-P-ECO. Accessed May 5, 2011
- issuu.com/hk-gt/docs/hkgt_201004/1?mode=a_p. Accessed May 5, 2011
- www.dbz.de/artikel/dbz_Wohnen_im_Kraftwerk_Mehrfamilienhaus_Bennau_CH_985379.html. Accessed May 5, 2011

PROJECT 07 KLEEHÄUSER, FREIBURG/VAUBAN (D)
- www.zero-haus.de. Accessed May 5, 2011
- www.2000watt-gesellschaft.ch. Accessed May 5, 2011
- www.kleehaeuser.de. Accessed May 5, 2011
- Lange, Jörg: Exposee der Kleehäuser. Freiburg 2010

PROJECT 09 SOLAR COMMUNITY, FREIBURG (D)
- Heinze, Mira; Voss, Karsten: GOAL. ZERO ENERGY BUILDING. Exemplary Experience Based on the Solar Estate Solarsiedlung Freiburg am Schlierberg. In: Journal of Green Building 04/2009
- Heinze, Mira; Voss, Karsten: Ziel Null Energie. Erfahrungen am Beispiel der Solarsiedlung Freiburg am Schlierberg. In: DBZ 01/2009, p. 72–74
- Bernard, Soara; Halbtuch, Andreas: Research Solarsiedlung am Schlierberg in Freiburg. Bergische Universität Wuppertal 2008
- Angeli, Thomas: Sonnen- und Schattenseiten, Bauen und Wohnen Beobachter Kompakt 22/2007, p. 51–56

PROJECT 10 ENERGY PLUS COMMUNITY, WEIZ (A)
- Ökostromverordnung 2010 (ÖSVO 2010). www.e-control.at. Accessed May 5, 2011

PROJECT 11 BEDZED, SUTTON NEAR LONDON (GB)
- BioRegional Development Group, Beddington Zero Energy Development: Total Energy Strategy Including Green Transport Plan. Wallington 1999
- BioRegional Development Group, Beddington Zero Energy Development: Case Study Report. Wallington 2002
- BioRegional Development Group, BedZED: Toolkit Part II. A Practical Guide to Producing Affordable Carbon Neutral Developments. Wallington 2003
- BioRegional Development Group, BedZED Seven Years On: The Impact of the UK's Best Known Eco-village and its Residents. Wallington 2009
- Building Research Establishment, General Information Report 89: BedZED. Beddington Zero Energy Development Sutton. Watford 2002
- Twinn, Chris: BedZED. In: The Arup Journal 01/2003, p.10–16
- Hegger, Manfred; et al..: Energie Atlas. Munich 2007, p. 252–253

OVERVIEW OF PROJECTS AND THEIR CHARACTERISTICS (2)
- Bundesministerium für Verkehr, Bau und Stadtentwicklung: Bekanntmachung der Regeln für Energieverbrauchskennwerte und der Vergleichswerte im Nichtwohngebäudebestand. Berlin 2009
- International Energy Agency – IEA: Solar Heating & Cooling Programme + Energy Conservation in Buildings and Community Systems Programme. www.iea-shc.org/task40/subtask/index.html. Accessed May 5, 2011
- Musall, Eike; et al.: Net Zero energy solar buildings: an overview and analysis on worldwide building projects. Tagungsband EuroSun conference Graz 2010
- Voss, Karsten; et al.: Bürogebäude mit Zukunft. Berlin 2006
- www.enob.info/nullenergie. Accessed May 5, 2011
- www.buildingdata.energy.gov/ZEB

PROJECT 13 CORPORATE HEADQUARTERS, KEMPTTHAL (CH)
- Factsheet 2010. www.marche-international.ch. Accessed May 5, 2011
- Factsheet 2007. www.marche-international.com. Accessed May 5, 2011
- Schweizer Solarpreis 2007. www.solaragentur.ch. Accessed May 5, 2011
- EKZ Energiekontrakting. www.ekz.ch. Accessed May 5, 2011
- www.glassx.ch. Accessed May 5, 2011

PROJECT 14 WWF-HEADQUARTERS, ZEIST (NL)
- Moreno, Concha; et al.: Broschüre zum EUROPEAN ENVIRONMENTAL DESIGN AWARD. Madrid 2009
- Sokol, David: Where the Wild Things Are. In: GreenSource Magazine, Mai/Juni 2009, p. 73ff.
- Nendzig, Ursel; Tesselaar, Milo: Der Aktivist. In: BIORAMA – Magazin für nachhaltigen Lebensstil, 11/2010, p. 24–26
- Hannemann, Matthias: Was Menschen bewegt – Architektur. In: BRAND EINS 06/10, p. 130ff.
- Hillen, Marieke: Oneplanetarchitecture – design with awareness. In: HOLLANDQUALITIME 2010, p. 104ff.
- Ottosson, David: Rhythm is an Architect. In: Loft magazine 13/2010, p. 56ff.

PROJECT 16 PIXEL BUILDING, MELBOURNE (AUS)
- www.pixelbuilding.com.au. Accessed May 5, 2011
- www.gbca.org.au/green-star/pixel/13600.htm. Accessed May 5, 2011
- www.airah.org.au/Ecolibrium/AM/ContentManagerNet/ContentDisplay.aspx?Section=Ecolibrium&ContentID=5654. Accessed May 5, 2011
- Torcellini, Paul; et al.: Zero Energy Buildings. A Critical Look at the Definition. Golden 2006.

PROJECT 18 ZERO EMISSIONS FACTORY, BRAUNSCHWEIG (D)
- Herkel, Sebastian: Nullemissionsfabrik Solvis. In: EB-Zeitschrift für EnergieEffizientes Bauen 03/2002, p. 34ff.
- Jäger, Helmut: Nullemissionsfabrik Solvis. Förderkonzept solar optimiertes Bauen des Bundesministeriums für Wirtschaft und Technologie. Braunschweig 2003
- Herkel, Sebastian; Riecks, Dietmar; Ufheil, Martin: Verwaltungs- und Produktionsgebäude Solvis. Veröffentlichung zur Otti-Tagung. Freiburg 2002
- www.solares-bauen.de/projekte. Accessed May 5, 2011
- www.enob.info/de/forschungsfelder/enbau. Accessed May 5, 2011

PROJECT 20 UNIVERSITY BUILDING, ST.-PIERRE (F)
- Garde, François; Lenoir Aurélie; David, Mathieu: Building design and energy performance of buildings in the French island of La Réunion. Clima2010-Kongress, Antalya 2010
- Garde, François; et al.: ENERPOS. A French research program for developing new methods for the design of zero energy buildings in hot climates. CESB, Prag 2007
- Garde, François; David, Mathieu; Lenoir Aurélie: Towards Net Zero Energy buildings in hot climates. Part I. ASHRAE-Winterkonferenz, Las Vegas 2011
- Lenoir, Aurélie; et al.: Presentation of the experimental feedback of a zero energy building under tropical climate. ISES Solar World-Kongress, Johannesburg 2009
- Lenoir, Aurélie; Thellier, Françoise; Garde, François: Towards Net Zero Energy buildings in Hot climates. Part II. ASHRAE Winterkonferenz, Las Vegas 2011

AUTHORS

KARSTEN VOSS, Prof. Dr.-Ing.
Bergische Universität Wuppertal
▷ research and teaching in the fields of building physics and technical building services
www.btga.uni-wuppertal.de

EIKE MUSALL, M.Sc.arch.
Bergische Universität Wuppertal I
▷ research and teaching in the fields of building physics and technical building services
www.btga.uni-wuppertal.de

ANDREAS K. ATHIENITIS, Prof. Dr.
Concordia University, Montréal / Department of Building, Civil and Environmental Engineering
▷ integration of solar energy systems; daylighting; modelling, simulation and design of sunspaces and demand management
www.bcee.concordia.ca
www.solarbuildings.ca/

SOARA BERNARD, M.Sc.arch., Dipl.-Ing.
Bergische Universität Wuppertal / University of Wuppertal
▷ research and teaching in the fields of building physics and technical building services
www.sdeurope.uni-wuppertal.de

ARMIN BINZ, Prof.
Fachhochschule Nordwestschweiz – Institute of Energy in Building
▷ energy-efficient building, thermal insulation, passive use of solar energy
www.fhnw.ch/iebau

DYLAN BRADY
studio505
▷ architecture and design
www.studio505.com.au

MARKO BRANDES, Dipl.-Ing.
BLS Energieplan GmbH
▷ energy-efficient and sustainable building technology
www.bls-energieplan.de

JOSÉ A. CANDANEDO, Ph.D. Student / Research Assistant
Concordia University, Montréal / Department of Building, Civil and Environmental Engineering
▷ Integration of renewable energy technologies, system identification, modelling and simulation of thermal energy storage and load management
www.bcee.concordia.ca
www.solarbuildings.ca

ARNULF DINKEL, Dipl.-Ing. Architekt
Fraunhofer-Institut für Solare Energiesysteme ISE, Freiburg
▷ international cooperation, energy-efficient building design and urban planning
www.ise.fraunhofer.de

SABINE ERBER, DI
Energieinstitut Voralberg
▷ solar architecture, passive house
www.energieinstitut.at
www.klimaaktivhaus.at

SHANE ESMORE, Director & Principal Sustainability Consultant
Umow Lai
▷ multi-disciplinary building services, engineering and sustainability
www.umowlai.com.au

NORBERT FISCH, Univ.-Prof. Dr.-Ing.
Technische Universität Braunschweig
▷ IGS – Institute of Building Services and Energy Design
www.igs.bau.tu-bs.de

FRANÇOIS GARDE, Prof.
University of La Réunion
▷ Laboratoire de Physique du Bâtiment et des Systèmes
www.lpbs.univ-reunion.fr/enerpos

SONJA GEIER, DI
AEE – Institute for Sustainable Technologies (AEE INTEC), Gleisdorf
▷ sustainable building
www.aee-intec.at

MICHAEL GIES, Dipl.-Ing.
Gies Architekten BDA, Freiburg
▷ architecture, town planning, transfer of knowledge
www.giesarchitekten.de

MONIKA HALL, Dr.-Ing.
Fachhochschule Nordwestschweiz – Institute of Energy in Building
▷ energy-efficient and sustainable building
www.fhnw.ch

FERDINAND HAMMERER
Marktgemeinde Wolfurt Immobilien GesmbH
▷ new buildings and refurbishment of public buildings
www.wolfurt.at

SEBASTIAN HERKEL, Dipl.-Ing. Maschinenbau
Fraunhofer Institute for Solar Energy Systems ISE, Freiburg
▷ solar building, simulation, monitoring
www.ise.fraunhofer.de

PATRICK JUNG, Dipl.-Ing., Visiting Prof.
Donau-Universität Krems,
IPJ Ingenieurbüro P. Jung GmbH
▷ concepts for innovative buildings
www.jung-ingenieure.de

FLORIAN KAGERER, Dipl.-Ing.
Fraunhofer Institute for Solar Energy Systems ISE, Freiburg
▷ solar building, simulation, monitoring
www.ise.fraunhofer.de

BEAT KÄMPFEN, Dipl. Architekt eth/sia, M.A. UC-Berkeley, CA
Architekturbüro Kämpfen für Architektur, Zürich
▷ energy-efficient and ecological design and building
www.kaempfen.com

JENS KRAUSE, Dipl.-Ing.
BLS Energieplan GmbH
▷ energy-efficient and sustainable building services
www.bls-energieplan.de

JÖRG LANGE, Dr.
Wohnungseigentümergemeinschaft (WEG)
▷ sustainable building
www.forum-vauban.de
www.vauban.de

HENRIK LANGEHEIN, Dipl.-Ing. (FH)
IGS – Institute of Building Services and Energy Design, Technische Universität Braunschweig
▷ work group plant optimisation
www.igs.bau.tu-bs.de

BEATRIX LEHNERT
Grocon
▷ property development, construction and funds management
www.Grocon.com.au
www.pixelbuilding.com.au

AURÉLIE LENOIR
University of La Réunion, Laboratory Piment
▷ monitoring and comfort in zero energy buildings
www.piment.univ-reunion.fr

INGO LÜTKEMEYER, Prof. Dipl.-Ing. Architekt
IBUS Architekten und Ingenieure, Berlin/Bremen
▷ sustainable design and building
www.IBUS-Architekten.de

Dr. MASA NOGUCHI, Senior Lecturer
MEARU Mackintosh School of Architecture
▷ Zero-energy Mass Custom Home (ZEMCH)
research and teaching
www.gsa.ac.uk

STEFAN PLESSER, Dipl.-Ing. Architekt
IGS – Institut für Gebäude- und Solartechnik, Technische Universität Braunschweig
▷ plant optimisation
www.igs.bau.tu-bs.de
www.energydesign-bs.de
www.synavision.de

DIETMAR RIECKS, Dipl.-Ing. Architekt BDA
Banz + Riecks Architekten BDA, Bochum
▷ energy-optimised industrial and commercial building
www.banz-riecks.de

AUDRIUS RINGAILA, Designer
MEARU Mackintosh School of Architecture
▷ Zero-energy Mass Custom Home (ZEMCH)
networking
www.gsa.ac.uk

LARS RÖSSING, Dipl.-Ing. Architekt
tr.architekten
▷ holistic architecture concepts and energy consulting
www.tr-architekten.com

TANJA SIEMS, Prof. Dr.-Ing.
Bergische Universität Wuppertal und T2 spatialwork ltd architecture and town planning, London
▷ town and infrastructure plannning in theory, research and practice
www.arch.uni-wuppertal.de/Forschungs_und_Lehrbereich/Staedtebau, www.t-2.org

KATHARINA SIMON, M.Sc.arch.
Bergische Universität Wuppertal
▷ urban planning and urban-scape
www.arch.uni-wuppertal.de/Forschungs_und_Lehrbereich/Staedtebau/

ARMIN THEMESSL, Ing.
AEE Arbeitsgemeinschaft ERNEUERBARE ENERGIE Kärnten
▷ energy consulting
www.aee.or.at

BERT TILICKE, Dipl.-Ing. Architekt/Landschaftsarchitekt
tr.architekten
▷ holistic architecture concepts and energy consulting
www.tr-architekten.com

MARTIN UFHEIL, Dipl.-Ing.
solares bauen GmbH, Freiburg
▷ energy concepts, simulation
www.solares-bauen.de

DAVID WALDREN
Grocon
▷ property development, construction and funds management
www.Grocon.com.au
www.pixelbuilding.com.au

GERHARD ZWEIER, Dipl.Ing. Architekt
Architekturbüro Gerhard Zweier
▷ energy-efficient architecture and refurbishment
www.zweier.at

PICTURE CREDITS

The authors and the publisher would like to thank everybody who has helped to create this book by providing their pictures, granting permission to reproduce them and supplying information. All diagrams and graphics in this book not below listed were specifically developed for this purpose by the authors.

Cover: Myrzik & Jarisch, Munich / Solon SE

A BACKGROUND INFORMATION

A 1.01–02 McKinsey & Company: Kosten und Potenziale zur Vermeidung von Treibhausgasemissionen in Deutschland. Sektorperspektive Gebäude. Düsseldorf 2007
A 1.03 Nigel Young, Foster + Partners, London
A 1.04 Tomas Riehle, Bergisch-Gladbach
A 1.05 Engelmann, Peter: Studentisches Wohnen im Passivhaus. Evaluierung energieeffizienter Studierendenwohnheime. Dissertation, Universität Wuppertal 2010
A 1.06 Arup/Gem Advertising & Publications
A 1.07–08 www.novatlantis.ch
A 1.09 S.O.L.I.D. Gesellschaft für Solarinstallation und Design GmbH, Graz
A 1.12 Griffith, B.; et al.: Assessment of the Technical Potential for Achieving Net Zero-Energy Buildings in the Commercial Sector. In: Technical Report NREL / TP-550-41957, 2007, p. 55
A 1.13 Dennis Schroeder, NREL National Renewable Energy Laboratory, Golden / Washington, D.C.
A 1.14 Güttinger, Herbert; et al.: Energie-Detailbilanz des Eawag Forum Chriesbach. Schweizer Bundesamt für Energie. Dübendorf 2009
A 1.15 Eawag, Dübendorf
A 1.16 http://www.enob.info/de/analysen/
A 1.18 Wagner, Waldemar; et al.: Energietechnische und baubiologische Begleituntersuchung Lehm-Bürogebäude Tattendorf. In: Berichte aus Energie- und Umweltforschung 65/2009. Edited by. Bundesministerium für Verkehr, Innovation und Technologie (BMVIT). Vienna 2010, p. 43
A 1.19–20 Voss, Karsten: Experimentelle und theoretische Analyse des thermischen Gebäudeverhaltens für das Energieautarke Solarhaus Freiburg. Dissertation, Stuttgart 1997
A 1.26 ETH-Studio Monte Rosa / Tonatiuh Ambrosetti, Zurich/Lausanne
A 1.27 Voss, Karsten; et al.: The self-sufficient solar house in Freiburg. Results of 3 years of operation. In: Solar Energy 58 (1–3) 1996, p. 17–23
A 1.28 Triolog, Freiburg
A 1.29 Goetzberger, Adolf; et al.: Mobile Blockheizkraftwerke. In: Erneuerbare Energie. 04 /1996, p. 12
A 1.30 Solar Impulse SA / Revillard, Lausanne

A 2.01 Doug Snower Photography, Chicago
A 2.04 SenerTec GmbH, Schweinfurt
A 2.06 Institut Wohnen und Umwelt: Kumulierter Energieaufwand und CO_2-Emissionsfaktoren verschiedener Energieträger und -versorgungen. Darmstadt 2009
A 2.10 Schweizer Ingenieur- und Architektenverein: SIA 2031. Energieausweis für Gebäude; SIA 2040: Effizienzpfad Energie. Zurich 2009
Bundesministerium für Verkehr, Bau und Stadtentwicklung (BMVBS): Bekanntmachung der Regeln für Energieverbrauchskennwerte und der Vergleichswerte im Nichtwohngebäudebestand. Berlin 2009
A 2.11 ILEK, Universität Stuttgart
A 2.13 Peter Engelmann, Wuppertal / Boston
A 2.14 Güttinger, Herbert; et al.: Energie-Detailbilanz des EWAG Forum Chriesbach. Schweizer Bundesamt für Energie. Dübendorf 2009;
Primas, Alex: Berechnung der Grauen Energie. Zurich 2010;
Hall, Monika: Fachhochschule Nordwestschweiz, information from client, 2010
A 2.15 Pedro Cabrito, Lisbon
A 2.16 John Petersen, Oberlin College
A 2.17 Calculated with EnerCalc, Markus Lichtmeß

A 3.01 Oliver Kern, Freiburg
A 3.05 Monika Hall, Fachhochschule Nordwestschweiz
A 3.06–07 Allreal Generalunternehmung AG (www.eulachhof.ch), Zurich
A 3.09–10 Armin Binz, Fachhochschule Nordwestschweiz
A 3.11 Adam Mørk, Copenhagen
A 3.12 Sonja Geier, Arbeitsgemeinschaft Erneuerbare Energie – Institut für Nachhaltige Technologien (AEE INTEC), Gleisdorf

B PROJECTS AND LESSONS LEARNED

B 1.03 Uwe Kroiss, Kirchberg-Thening
B 1.09 SCHANKULA Architekten, Munich
B 1.10, 12 Büro Kämpfen für Architektur, Zurich
B 1.11 Roland Halbe, Stuttgart
B 1.13 Energy and Climate, Danish Technological Institute

B 02.01 Agnieszka Koziol, Concordia University, Montreal

B 03.05 Kingspan Off Site, Kingscourt

B 04.02 Adam Mørk, Copenhagen
B 04.06 Martin Dyrløv, Copenhagen

B 05.09 Jakob Schoof, Munich

B 06.02, 08 Gesellschaft für Grund- und Hausbesitz, Heidelberg

B 07.10 Deutscher Mieterbund e.V., Berlin

B 08.01, 03, 08	Büro Kämpfen für Architektur, Zurich
B 09.06, 07	Soara Bernard, Andreas Halbtuch, Bergische Universität Wuppertal
B 10.02	Harald Eisenberger, Vienna
B 11.05	Dennis Gilbert, London/
B 11.08	Graham Gaunt, St. Ives
B 12.02	Foster + Partners, London
B 12.09	Roland Halbe, Stuttgart
B 12.11	Nigel Young, Foster + Partners, London
B 2.01	Walter Luttenberger, Gratkorn
B 2.02	Adam Mørk, Copenhagen
B 2.04	Solar-Fabrik AG, Freiburg
B 2.05	Dietmar Tollerian, Linz
B 2.07	SMA Solar Technology AG, Niestetal/ Constantin Meyer
B 2.08	SMA Solar Technology AG, Niestetal
B 2.10	Norbert van Onna, Veldhoven
B 2.11	Antonio Corchera, Santiago de Chile
B 13.02, 05, 10	Willi Kracher, Zurich
B 14.01, 07	Christian Richters, Münster
B 15.01	Armin Themeßl, Arbeitsgemeinschaft Erneuerbare Energie Kärnten (AEE), Villach
B 16.02, 06	John Gollings, St. Kilda
B 17.01, 04–07	Myrzik & Jarisch, Munich/Solon SE
B 18.02, 06–07 und 09	Christian Richters, Münster
B 19.02, 08	Gemeinde Wolfurt/Raimund Doerler, Wolfurt
B 19.08	Arbeitsgemeinschaft Erneuerbare Energie – Institut für Nachhaltige Technologien (AEE INTEC), Gleisdorf
B 20.02, 07	Jérôme Balleydier, St. Leu, La Réunion
B 21.02, 07–08	Antje Schröder, Stuttgart
B 22.01	Ingo Lütkemeyer, Bremen
B 23.01	Javier Alonso Huerta, Madrid
B 23.03	Team IKAROS Bavaria, Rosenheim
B 23.04	Jan Cremers, Munich
B 23.06	Peter Keil photography, Düsseldorf
B 23.07	Team Berlin living EQUIA/Markus Bachmann